FIBER OPTIC REFERENCE GUIDE

A Practical Guide to the Technology

by David R. Goff,
V.P. Engineering
Force, Incorporated

Editor:
Kimberly S. Hansen

Assistant Editor:
James G. Stewart

Focal Press
Boston Oxford Melbourne Singapore Toronto Munich New Delhi Tokyo

Focal Press is an imprint of Butterworth-Heinemann.

⚡ A member of the Reed Elsevier group.

Copyright © 1996 by Force, Incorporated.

Library of Congress Cataloging-in-Publication Data

Goff, David R.
    Fiber optic reference guide / by David R. Goff.
    p.   cm.
    Includes index.
    ISBN 0-240-80263-2 (paperback)
    1. Fiber optics.   2. Optical communications.   I. Title.
  TA1800.G64 1996
  621.36'92—dc20   95-51669
                                                                CIP

British Library Cataloguing-in-Publication Data
A catalogue record for this book is available from the British Library.

Cover design by Kimberly S. Hansen.

The publisher offers special discounts on bulk orders of this book.
For information, please contact:

Manager of Special Sales
Butterworth-Heinemann
313 Washington Street
Newton, MA 02158–1626
Tel: 617-928-2500
Fax: 617-928-2620

For information on all Focal Press publications available, contact our World Wide Web home page at:
http://www.bh.com/bh/

10 9 8 7 6 5 4

Printed in the United States of America

TABLE OF CONTENTS

ACKNOWLEDGEMENTS

This *Fiber Optic Reference Guide* is a comprehensive distillation of information obtained from a number of sources. In addition to the knowledge of Force, Incorporated engineers and technical writing staff, input and information was received from a number of companies. Force, Incorporated would like to gratefully acknowledge the time and information contributed by the following companies:

ABB Hafo, Incorporated

AMP, Incorporated

Amphenol, Incorporated

AT&T

Bellcore

BT&D Technologies

Corning, Incorporated

DiCon Fiberoptics, Incorporated

EG&G Judson

Epitaxx

Fermionics

Gould, Incorporated

Hewlett-Packard

Kaptron, Incorporated

Le Verre Fluoré

Melcor Corporation

Northern Lights Cable, Incorporated

Optical Cable Corporation

OZ Optics, Limited

Siecor Corporation

FOREWORD

Fiber optics is a fascinating and somewhat mysterious industry. A largely self-contained group of companies keeps fiber optic technology more or less shielded from outsiders. One of the goals of this book is to raise the curtain and let you see backstage. Fiber optics has been a major growth area for years, but, for three reasons, the growth has rarely matched the bold predictions made for it. First of all fiber optics has always suffered from the "chicken or egg first" syndrome. In other words, the technology would be available at a reasonable price once the demand for the technology was demonstrated. However, the demand wouldn't arise until affordable technology was in place. The second reason for the slow growth of fiber optics has been standards. In the early years, the industry suffered from too few standards. Lately, the industry suffers from too many standards that are nearly obsolete by the time they are introduced. Last, competition from other technologies has been intense. No one would have guessed ten years ago that video would be sent over twisted pair copper cable for substantial distances, yet today it can be done. The fiber optics community has overcome these obstacles which plague nearly all new technologies to make fiber optics a viable, keystone technology for the communications age.

Today, the technology and the demand for fiber optics are firmly in place. The rise of this industry's influence is illustrated in the time-line below. Where the telephone has taken roughly 80 years to become ubiquitous, fiber optic technology has claimed its place in our lives in a mere 18 years. Even the automobile experienced a more reluctant acceptance.

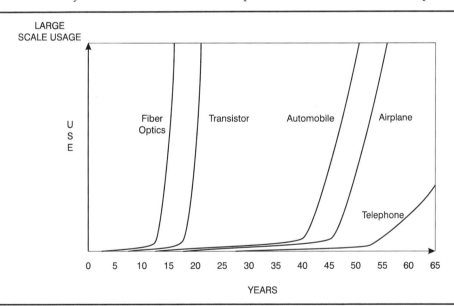

Time-line of the Acceptance of Various Technologies

Fiber optics requires a knowledge of optics, electronics, and mechanics, making this technology and industry very diverse and challenging. It is a breakthrough technology that has made a difference and will continue to influence and enrich our lives for years to come.

ABOUT THIS BOOK

This reference guide started out as an educational tool for my company's customers. The enthusiastic reception it received has caused the book to evolve into a rather complete survey of fiber optics. It is intended to be a very practical look at the technology and industry. In a sense, it is an attempt to present a very technical subject in the most straightforward and intuitive manner possible. Because of this approach, the text is not cluttered with lots of esoteric equations. The equations that are presented are those necessary to deal with fiber optic technology in real applications. There are many good, deep theoretical texts on fiber optics. You'll recognize them because they will tell you, for instance, that *Refractive Index (n) is identical to the square root of the relative dielectric constant* ε_r. Personally, explanations like that do not help me much. To me it seems sufficient to view refractive index as a property of a material that causes it to bend light and causes light to travel proportionally slower compared to its speed in a vacuum. That is something that I can relate to and can use in calculations about real systems.

This book will appeal to those who are newcomers to the field or those who cannot get past the techno-jargon that seems to go with every technology. The *Fiber Optic Reference Guide* presents the essential concepts of the fiber optics industry and will give the reader a good feeling of how the technology really works rather than present endless pages of mind-numbing equations. Insights into the history of fiber optics and its components are also included in this book. I think this is important because it helps clarify why things are done the way they are. By having some sense of history, what worked and what did not, it becomes easier to predict what the future will bring.

The book is laid out with fourteen chapters and six appendices. The chapters cover the fundamentals of fiber optics from the inception of the industry, through the basic components of a fiber optic system, to the elements of system design, fiber optic applications, testing and troubleshooting fiber optic installations, and the future of the industry. Italicized words within the text have accompanying glossary entries. Selected references and additional suggested readings are given at the end of each chapter. The appendices provide two basic types of information. General reference material is covered in the first three appendices. This information includes: specifying numbers, scientific constants, conversion factors, fiber optic symbols, and the glossary of fiber optics and electronics terms that is quite possibly the most comprehensive glossary of its kind. The last three appendices deal with fiber optics industry-specific information including a fiber optic system checklist, an extensive listing of industrial, military, and Bellcore standards for fiber optics, a list of the industry's tradeshow sponsors, conference sponsors, and leading industry magazines.

The content of this book has evolved over a period of several years and has benefited from the input from a great many people, especially the editor, Kimberly S. Hansen. I thank all of those who contributed to make this a useful reference guide for the fiber optics industry and its present and future customers.

David R. Goff

V.P. Engineering
Force, Incorporated
825 Park Street
Christiansburg, VA 24073
TEL: (540) 382-0462
E-Mail: Force.Inc@bev.net

A HISTORY OF FIBER OPTIC TECHNOLOGY

1

THE NINETEENTH CENTURY

In 1870, the British Royal Society in London, England witnessed a thought-provoking demonstration given by natural philosopher, John Tyndall. Tyndall, using a jet of water that flowed from one container to another and a beam of light, demonstrated that light used internal reflection to follow a specific path. As water poured out through the spout of the first container, Tyndall directed a beam of light at the path of the water. The light, as seen by the audience, followed a zigzag path inside the curved path of the water. This simple experiment, illustrated in Figure 1.1, marked the first research into the guided transmission of light.

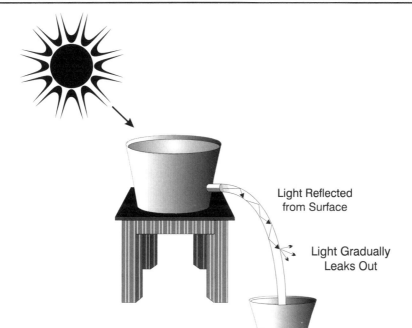

FIGURE 1.1

John Tyndall's Experiment

Light Reflected from Surface

Light Gradually Leaks Out

Water Flowing Out of Basin

William Wheeling expanded upon Tyndall's experiment when, in 1880, he patented a method of light transfer he called "piping light." Wheeling believed that by using mirrored pipes branching off from a single source of illumination, i.e. a bright electric arc, he could send the light to many different rooms in the same way that water, through plumbing, is carried throughout buildings today. Due to the ineffectiveness of Wheeling's idea and to the concurrent introduction of Edison's highly successful incandescent light bulb, the concept of piping light never took off. Interestingly, several Japanese companies are still pursuing Wheeling's century old idea because of improvements in technology and interest in energy efficiency.

That same year, Alexander Graham Bell developed an optical voice transmission system he called a photophone. Decades ahead of its time, the photophone used free-space light to carry the human voice 200 meters. Specially placed mirrors reflected sunlight onto a diaphragm attached within the mouthpiece of the photophone. At the other end, mounted within a parabolic reflector, was a light-sensitive selenium resistor. This resistor was connected to a battery that was, in turn, wired to a telephone receiver. As one spoke into the photophone, the illuminated diaphragm vibrated, casting various intensities of light onto the selenium resistor. The changing intensity of light altered the current that passed through the telephone receiver which then converted the light back into speech. The technology to support this invention would not be available for many years, but Bell believed this invention was superior to the telephone because it did not need wires to connect the transmitter and receiver. It was, in fact, the world's first optical *Amplitude Modulation (AM)* audio link.

THE TWENTIETH CENTURY

Fiber optic technology experienced a phenomenal rate of progress in the second half of the twentieth century. Early success came during the 1950's with the development of the fiber-scope. This image-transmitting device, which used the first practical all-glass fiber, was concurrently devised by Brian O'Brien at the American Optical Company and Narinder Kapany and colleagues at the Imperial College of Science and Technology in London. (In fact, Narinder Kapany first coined the term "fiber optics" in 1956.)

The development of glass-coated glass fibers was motivated by the optical loss experienced when using uncoated glass fibers. The inner fiber, or *core*, was used to transmit the light, while the glass coating, or *cladding*, prevented the light from leaking out of the core by reflecting the light within the boundaries of the core. This concept is explained by Snell's Law which states that the angle at which light is reflected is dependent on the refractive indices of the two materials — in this case, the core and the cladding. The lower refractive index of the cladding (with respect to the core) causes the light to be angled back into the core as illustrated in Figure 1.2.

FIGURE 1.2

Optical Fiber with Cladding

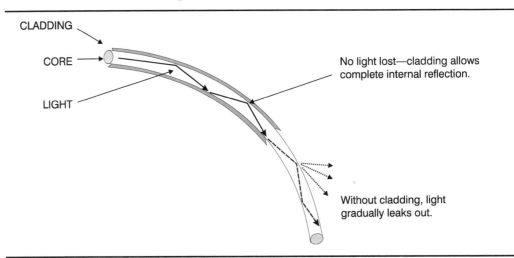

CLADDING

CORE

LIGHT

No light lost—cladding allows complete internal reflection.

Without cladding, light gradually leaks out.

The fiberscope quickly found application inspecting welds inside reactor vessels and combustion chambers of jet aircraft engines as well as in the medical field. Fiberscope technology has evolved over the years to make laparoscopic surgery one of the great medical advances of the twentieth century.

The development of laser technology was the next important step in the establishment of the industry of fiber optics. Only the *laser diode* or its lower-power cousin, the *light-emitting diode (LED)*, had the potential to generate large amounts of light in a spot tiny enough to be useful for fiber optics. In 1957, Gordon Gould popularized the idea of using lasers when, as a graduate student at Columbia University, he described the laser as an intense light source. Shortly after, Charles Townes and Arthur Schawlow at Bell Laboratories supported the laser in scientific circles. Lasers went through several generations including the development of the ruby laser and the helium-neon laser in 1960. Semiconductor lasers were first realized in 1962, marking the beginning of the laser most widely used today in fiber optics.

Because of their higher frequency, the importance of lasers as a means of carrying information did not go unnoticed by communications engineers. Light has an information-carrying capacity 10,000 times that of the highest radio frequencies being used. However, the laser was unsuited for open-air transmission because it was adversely affected by environmental conditions such as weather and smog. Faced with the challenge of finding a transmission medium other than air, Charles Kao and Charles Hockham, working at the Standard Telecommunication Laboratory in England in 1966, published a landmark paper proposing that optical fiber might be a suitable transmission medium if *attenuation* (loss of signal strength) could be kept under 20 dB/km. At the time of this proposal, optical fibers exhibited losses of 1,000 dB/km or more. The problem was that at a loss of only 20 dB/km, 99% of the light would be lost over only 3,300 feet. In other words, only 1/100th of the optical power reached the receiver. Intuitively, they postulated that these restrictive optical losses were the result of impurities in the glass, and not the glass itself.

Intrigued by Kao and Hockham's proposal, glass researchers began to work on the problem of purifying glass. In 1970, Drs. Robert Maurer, Donald Keck, and Peter Schultz of Corning developed a glass fiber that measured attenuation at less than 20 dB/km, the threshold for making fiber optics a viable technology. It was the purest glass ever made. Concurrent advances in laser technology, semiconductor chips, detectors, and connectors, combined with the optical fiber ushered in the true beginnings of the fiber optics industry.

The early work on fiber optic light sources and *detectors* was slow and often had to borrow technology developed for other reasons. For example, the first fiber optic light sources were derived from visible indicator LED's. As demand grew, light sources were developed for fiber optics that offered higher switching speed, more appropriate wavelengths, and higher output power.

Fiber optics developed over the years in a series of generations that can be closely tied to *wavelength*. The earliest fiber optic systems were developed at an operating wavelength of about 850 nm. This wavelength corresponds to the so-called "first window" in a silica-based optical fiber. This window is a wavelength region that offers low optical loss. It sits between several large absorption peaks caused primarily by moisture in the fiber glass and *Rayleigh scattering* at shorter wavelengths.

The 850 nm region was initially attractive because the technology for light emitters at this wavelength had already been perfected in visible indicator LED's. Low-cost silicon detectors could also be used at the 850 nm wavelength. As technology progressed, the first window became less attractive because of the approximately 3 dB/km loss limit.

Most companies jumped to the "second window" at 1300 nm, but a few companies, notably ITT, spent effort developing the wavelength region near 1060 nm. The 1060 nm region still allowed low-cost silicon detectors to be used; however, the light emitter technology was more complex. This region did offer lower attenuation, about 1.7 dB/km. Ultimately however, the second window at 1300 nm won out with lower attenuation of below 1 dB/km. Later, the "third window", at 1550 nm, was developed. It offered the theoretical minimum optical loss for silica-based fibers, about 0.2 dB/km.

Today, 850 nm, 1300 nm, and 1550 nm systems are all manufactured and deployed along with very low-end systems using visible wavelengths near 660 nm. Each has its advantage. Longer wavelengths offer higher performance, but always come with higher cost. The shortest link lengths can be handled with wavelengths of 660 nm or 850 nm. The longest link lengths require 1550 nm wavelength systems. Today, there is talk about exploiting a "fourth window" at 1650 nm. While it is not lower loss than 1550 nm, the loss is comparable, and it might simplify long-length, multiple-wavelength communications systems. Figure 1.3 illustrates the absorption peaks and the optical loss in silica glass as it relates to wavelength.

FIGURE 1.3

Three Wavelength Regions of Optical Fiber

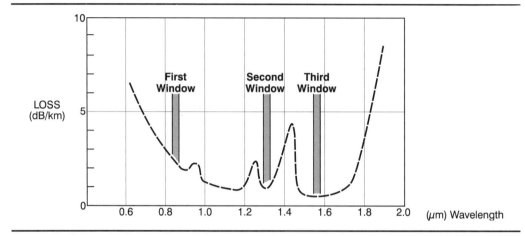

Most optical fiber is based on silicon, like most of today's electronics. There have been numerous attempts over the last few decades to develop alternate materials to either reduce cost or improve performance. To date, none have shown any real promise in dethroning silicon-based fiber. Much effort has gone into developing plastic fiber, but its impact on the marketplace has been minimal. It is suited for very short distances, typically around tens of meters only. As the cost of glass fiber has plunged over the last decade, the advantage of plastic fiber has faded.

The other big push has been to develop heavy metal halide fibers for use at long wavelengths in the $3 \mu m$ to $5 \mu m$ region (two to three times longer than the third window). The attraction here is that, in theory, such a fiber could have optical losses as low as 0.001 dB/km. However, after much effort, the losses of such fibers cannot even match the best losses of silica-based fibers. It seems likely that silicon will continue to be the basis of most of today's high technology.

REAL WORLD APPLICATIONS

The U.S. Military was quick to recognize the potential of fiber optics for improving its communications and tactical systems. In the early 1970's, the U.S. Navy installed a fiber optic telephone link aboard the U.S.S. Little Rock. The Air Force followed suit by developing its Airborne Light Optical Fiber Technology (ALOFT) program in 1976. Encouraged by the success of these applications, military R&D programs were funded to develop stronger fibers, tactical cables, ruggedized, high-performance components, and numerous demonstration systems ranging from aircraft to undersea applications.

Commercial applications quickly followed their military predecessors. In 1977, AT&T and GTE installed the first fiber optic telephone system, an application that had immediate success in meeting stringent standards for reliability and performance. As a result, fiber optic telephone networks are commonplace today. While computers, information networks, and data communications have been slower to embrace fiber, they too find application in a transmission system that is immune to lightning and carries more information faster and over longer distances using lighter weight cable.

In 1990, Bell Labs researcher, Linn Mollenauer, transmitted a 2.5 Gb/s signal over 7,500 km without regeneration. His system used a *soliton* laser and an *erbium-doped fiber amplifier* that allowed the light wave to maintain its shape and density. These "hero" experiments push the envelope further each year. As glass becomes more pure and lasers become more powerful, the bandwidth and distance capabilities for fiber optic transmission will increase.

THE FUTURE

Because of fiber optic technology's immense potential bandwidth (over 1 teraHertz, 10^{12} Hz), there are extraordinary possibilities for future fiber optic applications. And with millions of miles of fiber optic cable already installed for smaller network applications, it may be an easy transition to bring *broadband* services into the home. Broadband service available to a mass market would open up a wide variety of interactive communications for both consumers and businesses, bringing to reality interactive video networks, interactive banking and shopping from the home, and even interactive distance learning.

It is predicted that broadband networks will be commonplace by the turn of the twenty-first century. As a result, sales of fiber optic cables, connectors, other related components, and systems are expected to continue to be growth areas for many years. However, the deciding factor for the success of broadband distribution systems will ultimately be the cost and utility to the consumer. Fiber optic researchers must be able to devise a way to integrate existing technology and equipment with inexpensive new technology and equipment.

The industry continues to mature, and the future holds promise for optical fiber. Teleconferencing, distance learning, security and surveillance, high-speed data communication, digitized video, cable television—the list of applications that will benefit from future fiber optic technology grows longer each day.

Chapter Summary

- John Tyndall demonstrated in 1870 that light used internal reflection to follow a specific path.
- William Wheeling patented his method of light transfer, "piping light," in 1880.
- The photophone, the world's first optical AM audio link, was developed in 1880 by Alexander Graham Bell.
- The fiberscope, which used the first practical glass-coated fiber was concurrently developed in the 1950's by Narinder Kapany and colleagues and Brian O'Brien.
- The term "fiber optics" was coined by Narinder Kapany in 1956.
- The laser gained acceptance in scientific circles during the late 1950's and early 1960's.
- Optical fiber was proposed as a medium which could carry laser signals by Charles Kao and Charles Hockham, in 1966, at the Standard Telecommunication Laboratory in England.
- Fiber optic telephone systems were installed by AT&T and GTE in 1977.
- Linn Mollenauer transmitted a 2.5 Gb/s signal a distance of 7,500 km without regeneration in 1990 using a soliton laser and an erbium-doped fiber amplifier.
- Broadband networks offering interactive communication services are predicted to be commonplace by the 21st century.

Selected References and Additional Reading

—. 1982. *Designers Guide to Fiber Optics.* Harrisburg, PA: AMP, Incorporated.

Hecht, Jeff. 1993. *Understanding Fiber Optics.* 2nd edition. Indianapolis, IN: Sams Publishing.

—. 1992. *Just the Facts.* New Jersey: Corning, Incorporated.

Palladino, John R. 1990. *Fiber Optics: Communicating By Light.* Piscataway, NJ: Bellcore.

FIBER OPTIC FUNDAMENTALS

In the growing world of improved communication, fiber optics offers a method of transmission that allows for clearer, faster, more efficient communications than copper. A fiber optic system holds many advantages over a copper wire system. For example, while a simple two-strand wire can carry a low-speed signal over a long distance, it cannot send high-speed signals very far. Coaxial cables can better handle high-speed signals but only over a short distance. Fiber optics holds a great advantage over the two transfer media as it can handle high-speed signals over extended distances. Other advantages of fiber include:

Immunity from Electromagnetic (EM) Radiation and Lightning: This refers to the effect of EM radiation on the fiber itself. Because fiber is made from *dielectric* (nonconductive) materials, it is unaffected by EM radiation. The electronics required at the end of each fiber, however, are still susceptible to EM radiation and require shielding. Immunity from EM radiation and lightning was initially most important to the military. First of all, it makes secure communications easier. Fiber itself is not inherently secure, but it does not normally emit any EM radiation that can be readily detected. Second, immunity to EM radiation is important in aircraft designs that have composite skins; these nonconductive skins do not shield the electronics or wiring from EM fields or radiation. In recent years, lightning immunity has been one of the key drivers for fiber optics being used in commercial security systems. Because security systems are usually dispersed over a wide area, they are susceptible to lightning strikes and interference. Immunity from EM radiation has recently become an important factor in choosing fiber to upgrade communication systems in existing buildings. This is because fiber can often be run in the same conduits that currently carry power lines, simplifying installation.

Lighter Weight: This feature refers to the optical fiber itself. In real world applications, copper cables can often be replaced by fiber optic cables that weigh at least ten times less. For long distances, a complete fiber optic system (optical fiber and cable, plus the supporting electronics) also has a strong weight advantage over copper systems. This is often not true for short systems, however, because fiber optic systems almost always require more elaborate, and thus larger and heavier electronics than copper systems.

Higher Bandwidth: Fiber has higher bandwidth than any alternative available. The CATV industry in the past required amplifiers every thousand feet or so on their supertrunks when copper cable was used. This is due to the limited bandwidth of the copper cable. A modern fiber optic system could carry the same signals with similar or superior signal quality for 50 miles or more without needing amplification (repeaters). Even at that, modern fiber optic communication systems use less than one percent of fiber's inherent bandwidth.

Better Signal Quality: Because fiber is immune to EM interference, has lower loss per unit distance, and wider bandwidth, signal quality is usually substantially better compared to copper.

Lower Cost: This has to be qualified. Fiber certainly costs less for long distance applications. However, for signal transmission requirements over distances of a few feet, copper is cheaper and probably always will be. The cost of fiber itself is cheaper per unit distance than copper if bandwidth and transmission distance requirements are high; however, the cost of the electronics and electro-optics at the end of the fiber can be substantial. The price of copper is such today that for very long distance links, converting to a fiber optic system can be completely paid for by the salvage value of the copper that is removed. Fiber and copper systems can be compared by finding a break-even distance. At distances shorter than the break-even distance, copper is cheaper and vice versa. In the mid 1980's the break-even distance was 10 km or more. Today, the break-even distance is often less than 100 meters. Most of this gain has come from the reduced cost of fiber optic systems and components. Some of the gain is also due to the fact that the cost of copper has increased over the same period of time.

Easily Upgraded: The limitation of fiber optic systems today, and for many years to come, is the electronics and electro-optics used on each end of the fiber. The fiber itself usually has much more transmission capability, especially higher bandwidth, than is being utilized. Once fiber is installed, particularly a single-mode fiber, advances in electronics and electro-optics can be readily incorporated using the installed fiber.

Ease of Installation: Many newcomers to fiber optics are often concerned about glass being very brittle and prone to breakage. In fact, glass can be many times stronger than steel, and optical fibers are so small that they are very flexible. A good quality fiber optic cable incorporates strain relief materials as well as bend limiters that make the cable very hardy. A copper coax cable is in fact many more times fragile than a fiber optic cable. Copper coax cables are prone to kinks and deformities that will permanently degrade the performance of the cable. Glass however, will not deform or kink. Thus it can usually take much more abuse than copper.

PRINCIPLES OF FIBER OPTIC TRANSMISSION

Fiber optic components transmit information by turning electronic signals into light. *Light* refers to more than the portion of the *electromagnetic spectrum* that is visible to the human eye. The electromagnetic spectrum is composed of visible and near-infrared light like that transmitted by fiber, and all other wavelengths used to transmit signals such as AM and FM radio and television. Figure 2.1 illustrates the electromagnetic spectrum. As one can see, only a very small part of it is perceived by the human eye as light.

FIGURE 2.1

Electromagnetic Spectrum

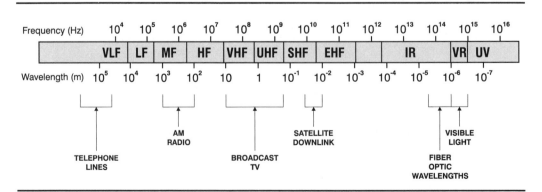

The term wavelength refers to the wavelike property of light, a characteristic shared by all forms of electromagnetic radiation. Wavelength is the measurement of the distance a single cycle of an electromagnetic wave covers as it travels through a complete cycle. Wavelengths for fiber optics are measured in nanometers (the prefix *nano* meaning one-billionth) or microns (the prefix *micro* meaning one-millionth). Wavelengths of light used in fiber optic applications can be broken into two main categories: near-infrared and visible. Visible light, as defined by the human eye, ranges in wavelengths from 400 to 700 nanometers (nm) and has very limited uses in fiber optic applications, due to the high optical loss. Near-infrared wavelengths range from 700 to 2,000 nanometers; these wavelengths are almost always used in modern fiber optic systems.

The principles behind fiber optic systems are relatively simple. As shown in Figure 2.2, fiber optic links contain three basic elements: the *transmitter* that allows for data input and outputs an optical signal, the *optical fiber* that carries the data, and the *receiver* that decodes the optical signal to output the data.

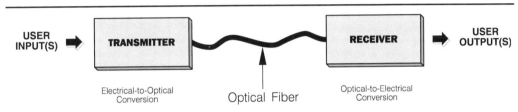

FIGURE 2.2

Elements of a Fiber Optic Link

The transmitter (Tx), shown in Figure 2.3, uses an electrical interface, either video, audio, data, or other forms of electrical input, to encode the user's information through *modulation*. Three forms of modulation are typically used: amplitude modulation (AM), *frequency modulation (FM)*, and digital modulation. The electrical output of the modulator is usually transformed into light either by means of a light-emitting diode (LED) or a laser diode (LD). The wavelengths of these light sources range from 660 nm to 1550 nm for most fiber optic applications.

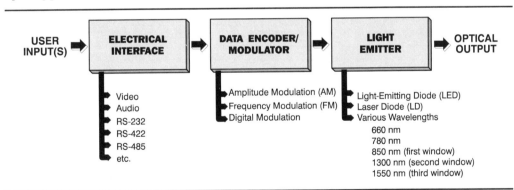

FIGURE 2.3

Fiber Optic Transmitter

The receiver (Rx), illustrated in Figure 2.4, decodes the light signal back into electrical signals. Two types of light detectors are typically used: the *PIN photodiode* or the *avalanche photodiode (APD)*. Typically, these detectors are made from *silicon (Si)*, *indium gallium arsenide (InGaAs)*, or *germanium (Ge)*. The detected and amplified electrical signal is then sent through a data decoder or demodulator that converts the electrical signals back into video, audio, data, or other forms of user input.

FIGURE 2.4

Fiber Optic
Receiver

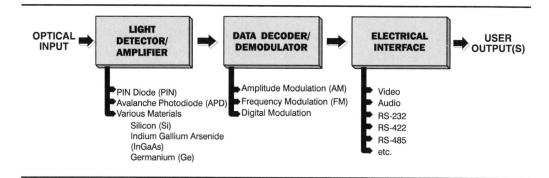

Why is fiber important? Because it provides a private pipeline that can carry huge amounts of data. Alternatives are over-the-air broadcast or hardwired copper wires carrying electrons. Figure 2.5 explains three schemes for transmitting information from one point to another.

FIGURE 2.5

Transmission
Schemes

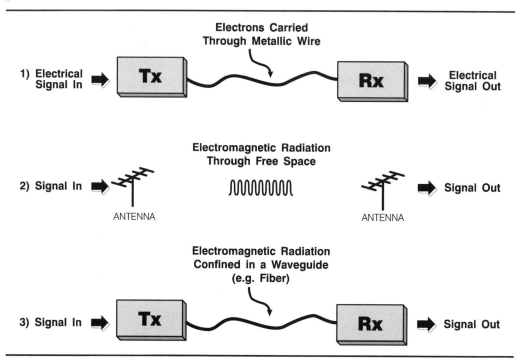

The first scheme is metallic transmission which uses a copper wire or coaxial cable to carry a modulated electrical signal containing information. This method allows for a limitless number of private channels (assuming you have that much copper cable), but each channel has limited information and distance capability due to the inherent characteristics of copper cable. The second scheme for moving information between two locations is free-space transmission. This is how radio signals and over-the-air TV signals are received. Free-space transmission has the advantage of providing very large bandwidth capability as well as long distance capability, but it does not provide a private channel. Also, the free-space spectrum is a finite, limited, and costly commodity. Witness the recent FCC spectrum auctions for proof. Free space cannot provide the millions of high-speed communication channels required by tomorrow's information age. The last scheme is waveguide transmission. This describes optical fiber transmission. A waveguide (optical fiber) confines the electromagnetic radiation (light) and moves it along a prescribed path. Optical fiber offers the best of both metallic and free-space transmission. It has the key advantage of metallic transmission, the ability to carry a signal from point A to point B without cluttering the limited free-space electromagnetic spectrum; however, fiber does not have the disadvantage of metallic transmission: very limited bandwidth and data rate.

FIBER OPTIC COMPONENTS

Fiber: Optical fibers are extremely thin strands of ultra-pure glass designed to transmit light from a transmitter to a receiver. These light signals represent electrical signals that include video, audio, or data information in any combination. Figure 2.6 shows the general cross-section of an optical fiber. The fiber consists of three main regions. The center of the fiber is the core. This region actually carries the light. It ranges in diameter from 9 microns to 100 microns in the most commonly used fibers. Surrounding the core is a region called the cladding. This part of the fiber reflects the light back into the core. The cladding typically has a diameter of 125 microns or 140 microns. A key design feature of all optical fibers is that the refractive index of the core is higher than the refractive index of the cladding. Both the core and cladding are usually *doped* glass materials. Other fiber types incorporate quartz or pure fused silica and plastic, but these are not used in mainstream high-performance applications. The outer region of the optical fiber is called the coating or *buffer*. The buffer, typically a plastic material, provides protection and preserves the strength of the glass fiber. Typical diameters for the buffer are 250 microns, 500 microns, and 900 microns. Optical fiber is treated more extensively in Chapter 3.

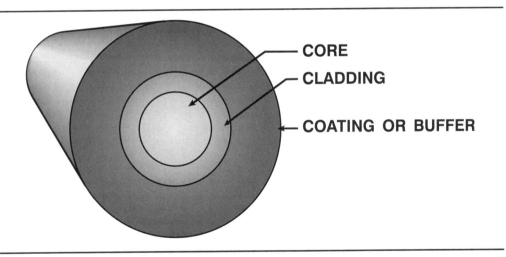

CORE
CLADDING
COATING OR BUFFER

FIGURE 2.6

Cross-Section of an Optical Fiber

Light Emitters: The two types of light sources used in fiber optics are light-emitting diodes (LED's) and laser diodes (LD's). LED's may be *surface-emitting* or *edge-emitting* sources.

The surface-emitting LED (SLED) emits light over a very wide angle. This type of light source is often called a *Lambertian emitter* because of the nature of the emission pattern. This broad emission angle is attractive for use as an indicating LED because of the wide viewing angle, but is a detriment for fiber optic uses. Because the emitting angle is so large, it is difficult to focus more than a few percent of the total light output into the fiber core. The key advantage of surface-emitting LED's is that they are low cost, making these light emitters the dominate type in use. The second type of light emitter is the edge-emitting LED (ELED). This LED type has a much narrower angle of light emission and also has a smaller emitting area. This allows a larger percentage of the total light output to be focused into the fiber core. ELED's are also generally faster than their surface-emitting cousins. The last light emitter type is the laser diode. A laser has a very narrow emission angle, and the emitting spot is very small, usually only a few microns in diameter. Because of the small emission angle and emitting spot, a very high percentage, often more than 50%, of the output light can be focused into the fiber core. The laser diode is the fastest of these three emitter types. Figure 2.7 illustrates the emission patterns of these sources.

FIGURE 2.7

Sources and
Emission Patterns

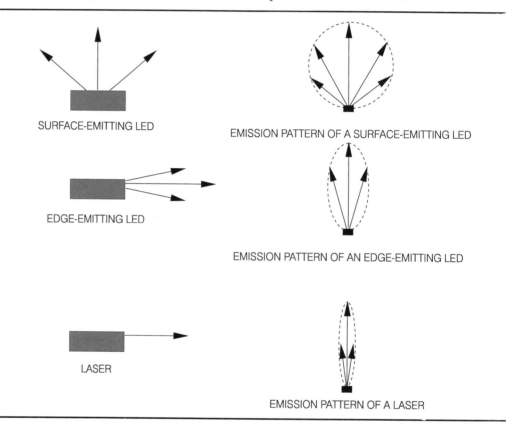

Many characteristics are considered when making the decision to use LED's or LD's in fiber optic systems. Some basic considerations include center wavelength, range of the wavelength (often called the *FWHM, Full Width Half Maximum*), and power of the light source. The source must be at the proper wavelength (780 nm, 850 nm, 1300 nm, or 1550 nm for glass fibers). The range of the wavelength is important as larger wavelength spreads bring with them an increased possibility of dispersion problems, limiting bandwidth. The power of the light source is also important in that it must be neither too weak, nor too strong. A weak source will not provide enough power to transmit a light signal through an optical fiber, while a source that is too strong could cause distortion of the signal by overloading the receiver with optical input. These and other characteristics will be explored at length in Chapter 5.

Detectors: The function of the detector is to convert optical power into an electrical signal. Detectors may be *PIN photodiodes* or *avalanche photodiodes (APD's)*. The basic difference between the two is that the APD provides a good deal of internal gain, simplifying the task of signal amplification. As is almost always the case with engineering trade-offs, with the advantage comes a disadvantage. APD's are, not surprisingly, more expensive and require extensive additional support electronics because of the need to provide a high-voltage biased signal that must be temperature compensated. PIN diodes on the other hand, do not require this additional support circuitry and are very economical but require more complex amplifier stages. PIN detectors are by far the most economical option which gives them the largest share of the market. Light detectors are discussed in Chapter 6.

Interconnection Devices: An interconnection device is any component or technique used to connect a fiber or a fiber optic component to another component or another fiber. Interconnection devices serve a twofold role in fiber optics; they provide both light junctions and mechanical junctions for interconnecting fiber optic systems. As light junctions, they provide outputs or inputs for signal sources. Mechanically, they hold the connections in place. Interconnection devices include connectors, splices, couplers, splitters, switches and wavelength-division multiplexers (WDM's). Some examples of uses of interconnection devices are:

- Interfaces between local area networks and devices
- Patch panels
- Portable military communication links
- Network-to-terminal connections
- Connections between recording equipment, cameras, and sound equipment in portable studios

Interconnection devices are treated in detail in Chapters 7 and 8.

In addition to these fiber optic components, this guide will discuss system design (Chapter 9), field applications for optical fiber (Chapter 10), video transmission (Chapter 11), data transmission (Chapter 12), testing and measurement techniques (Chapter 13), and the future of the fiber optics industry (Chapter 14).

Chapter Summary

- Fiber optics holds many advantages over conventional copper wire and coax cable, including EMI immunity, lighter weight, higher bandwidth, lower cost, and better signal quality.
- Fiber optic components transmit information by turning electronic signals into light.
- Wavelengths of light used in fiber optic applications include near-infrared and visible.
- The three basic elements of a fiber optic link are the transmitter, the receiver, and the optical fiber.
- The most common data transmission schemes are metallic transmission, free-space transmission, and waveguide transmission.
- Optical fibers are extremely thin strands of ultra-pure glass designed to transmit light signals from a transmitter to a receiver.

- The two types of light sources used in fiber optics are light-emitting diodes (LED's) and laser diodes (LD's).
- Two types of detectors associated with fiber optics are the PIN photodiode, and the avalanche photodiode.
- An interconnection device is any component or technique used to connect a fiber or fiber optic component to another component or another fiber. These devices include connectors, splices, couplers, splitters, switches and wavelength-division multiplexers (WDM's).

Selected References and Additional Reading

Baack, Clemens. 1986. *Optical Wideband Transmission Systems.* Florida: CRC Press, Inc.

Hecht, Jeff. 1993. *Understanding Fiber Optics.* 2nd edition. Indianapolis, IN: Sams Publishing.

Sterling, Donald J. 1993. *Amp Technician's Guide to Fiber Optics*, 2nd Edition. New York: Delmar Publishers.

OPTICAL FIBER

By 1950, the challenge to scientists studying optical fiber transmission was not whether light could carry information, but whether a glass conduit could be developed that was pure enough to keep losses below 20 dB/km. A flexible glass-coated glass fiber served as a suitable transmission medium for the fiberscope, but losses remained unworkably high. Scientists persevered.

In 1970, Corning scientists Robert Maurer, Donald Keck, and Peter Schultz developed a fiber with a measured attenuation of less than 20 dB/km. It was the purest glass ever made, and the breakthrough led to the commercialization of fiber optic technology. Corning's success was the result of a new process for manufacturing optical fiber which they called inside vapor deposition (IVD). Instead of melting the raw silica the way most glass is made, they formed the glass from vaporized chemicals which were deposited inside a silica tube. The outer tube became the cladding, and the core of the fiber formed within. An inherent disadvantage to this method was that the cladding was formed using the traditional method of making glass. Thus impurities still existed in the cladding and reduced the purity of the overall fiber. Back to the drawing board they went, and before long, Corning scientists had developed a method of outside vapor deposition (OVD) which formed the entire fiber from ultra-pure, vapor deposited chemicals. Today scientists still experiment with ever purer forms of optical fiber, and losses as low as 0.2 dB/km at 1550 nm are not uncommon.

MANUFACTURE OF OPTICAL FIBER

Modified chemical vapor deposition (MCVD), another term for the IVD method, and outside vapor deposition (OVD) are the two predominant methods for manufacturing optical fiber. The MCVD process involves depositing ultra-fine, vaporized raw materials into a pre-made silica tube. The soot that develops from this deposition is consolidated by heating. The resulting *preform* or blank is heated and drawn into a hair-thin optical fiber. Figure 3.1 illustrates the MCVD process.

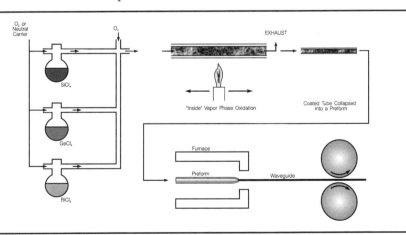

FIGURE 3.1

Manufacturing Optical Fiber (MCVD Process)

The optical fiber is encased in several protective layers to ensure integrity under various conditions. The first layer is applied to the glass fiber as it is drawn from the preform. This coating is generally made of ultraviolet-curable acrylate or silicone, and it serves as a moisture shield and as mechanical protection during the early stages of cable production. A secondary buffer is often extruded over the primary coating to further improve the fiber's strength.

The OVD process involves deposition of raw materials onto a rotating rod. This occurs in three steps: laydown, consolidation, and draw. During the laydown step, a soot preform is made from ultra-pure vapors of silicon tetrachloride and germanium tetrachloride. The vapors move through a traversing burner and react in the flame to form soot particles of silica and germanium oxide. These particles are deposited on the surface of the rotating target rod. When the deposition is complete, the rod is removed, and the deposited material is placed into a consolidation furnace. The water vapor is removed, and the preform is collapsed to become dense, transparent glass. This preform is drawn into a continuous strand of glass fiber in much the same way that taffy is stretched and thinned. By varying the mixture of gases throughout the process, the preform has a step-index or a graded-index of refraction. Figure 3.2 displays the method of drawing optical fiber.

FIGURE 3.2

Optical Fiber Draw Process

(Illustration courtesy of Corning, Incorporated.)

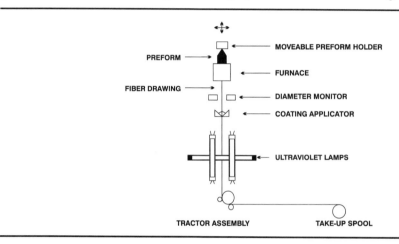

PRINCIPLES OF OPERATION

The operation of optical fiber is based on the principle of *total internal reflection*. This is a process that we do not encounter in our everyday lives. When we look in a mirror each morning, we see an image that is created by reflecting perhaps 90% of the light that strikes it. Total internal reflection reflects 100% of the light. Figure 3.3 demonstrates the equations involved in this principle.

FIGURE 3.3

Snell's Law—Equations for Total Internal Reflection

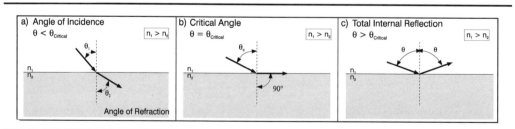

In Figure 3.3, the upper, lighter region of each frame always has a higher refractive index than the lower, darker region. The refractive index of the upper region is designated n_1 while the lower region refractive index is n_2. Figure 3.3a shows a case where the angle of incidence is less than the critical angle. Note that the angle the light is travelling changes at the interface between the higher refractive index n_1 region and the lower refractive index n_2 region. In Figure 3.3b, the angle of incidence has increased to the critical angle. At this angle the light ray travels parallel to the interface region. In Figure 3.3c, the incidence angle has increased to a value greater than the critical angle. In this case 100% of the light reflects at the interface region.

One can observe this principle when viewing a fish tank. Figure 3.4 shows the top view of a fish tank complete with an occupant and an observer. From a viewpoint behind the position of the fish, both the observer and the fish can see light bulb "A". However, light bulb "B" cannot be seen because of the total internal reflection that occurs. In this example, the water has a refractive index of 1.33 and the air has a refractive index of about 1.00. (Actually there is a glass wall between the water and air with a refractive index of 1.45, but this doesn't materially affect the experiment.)

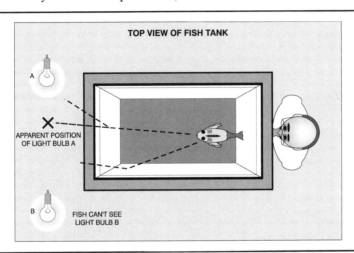

FIGURE 3.4

Principle of Total Internal Reflection

Total internal reflection occurs because light travels at different speeds in different materials. A dimensionless number called the *index of refraction* is used to characterize the different mediums through which the light is traveling. The index of refraction is the ratio of the velocity of light in a specific medium (v) to its velocity in a vacuum (c).

Eq. 3.1
$$n = \frac{c}{v}$$

As light passes from one medium to a medium with a different index of refraction, it is bent, or refracted. If light passes from a medium with a lower index of refraction to one with a higher index of refraction the light is bent toward the normal, and if the light passes from a higher to lower index of refraction the light is bent away from the normal. Snell's Law, in Figure 3.3 determines the amount the light is bent and is given by:

Eq. 3.2
$$n_1 \sin\Theta_1 = n_2 \sin\Theta_2$$

As the angle of incidence increases, the angle of refraction approaches 90° (see Figure 3.3). The angle of incidence that produces an angle of refraction of 90° is the *critical angle*. Increasing the angle of incidence past the critical angle results in total internal reflection. In total internal reflection the angle of incidence is equal to the angle of reflection. This is the basis for the operation of optical fiber. The critical angle is calculated as follows:

Eq. 3.3
$$\Theta_c = \sin^{-1}\left(\frac{n_2}{n_1}\right)$$

where:

n_1=Refractive index of the core.
n_2=Refractive index of the cladding.

The core of an optical fiber has a higher index of refraction than the cladding ($n_1 > n_2$), allowing for total internal reflection. Light entering the core of a fiber at an angle sufficient for total internal reflection travels down the core reflecting off the interface between the core and the cladding. Light entering at an angle less than the critical angle is refracted into the cladding and lost.

There is an imaginary cone of acceptance with an angle α determined by the critical angle. This is related to a parameter called the *numerical aperture (NA)* of the fiber. NA describes the light gathering capability of fiber and is given as:

Eq. 3.4

$$NA = \sin\alpha = \sqrt{n_1^2 - n_2^2}$$
$$\text{and}$$
$$\alpha = \sin^{-1}\left(\sqrt{n_1^2 - n_2^2}\right)$$

Figure 3.5 illustrates numerical aperture, and shows the location of angle α.

FIGURE 3.5

Numerical
Aperture

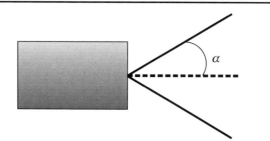

Here is an example of the above parameters for a real fiber. Let's assume that the core refractive index is 1.47 and the cladding refractive index is 1.45. So:

$n_1 = 1.47$
$n_2 = 1.45$

From equation 3.1 we can find the speed at which light travels in the fiber.

$$v = \frac{c}{n} = \frac{(2.998 \cdot 10^8 \text{ m/s})}{1.47} = 2.039 \cdot 10^8 \text{ m/s}$$

This can be inverted to find that it takes light 4.903 ns to travel 1 meter in an optical fiber.

From equation 3.3 we can calculate the critical angle, Θ_C:

$$\Theta_c = \sin^{-1}\left(\frac{1.45}{1.47}\right) = 80.5°$$

From equation 3.4 we can calculate the numerical aperture to be:

$$NA = \sqrt{(1.47^2 - 1.45^2)} = 0.2417$$

Finally, also from equation 3.4 we can calculate α.

$$\alpha = \sin^{-1}(0.2417) = 13.98°$$

Table 3.1 lists the refractive indices and propagation times of a variety of mediums.

Medium	Refractive Index	Propagation Time
Vacuum	1.000	3.336 ns/m
Air	1.003	3.346 ns/m
Water	1.333	4.446 ns/m
Fused Silica	1.458	4.863 ns/m
Belden Cable (RG-59/U)	N/A	5.051 ns/m
Refractive Index = C/V where: C = Speed of light in a vacuum. V = Speed of light in a medium.		

TABLE 3.1

Refractive Indices and Propagation Times

In this table, the propagation times of optical fiber and Belden copper cable appear almost identical. While the propagation times are similar, they are determined by different factors. In metallic cables, propagation delays are dependent on the cable dimensions and the frequency. In optical fiber, propagation delays tend to be related to the material. Most applications are insensitive to the absolute propagation time delay caused by fiber, but there are some applications where it is very critical. System designs that involve sending many synchronous digital signals over separate fibers require that all signals arrive about the same time. Another application involves sending multiple signals over the same fiber using different wavelengths of light. Both of these applications require careful analysis of the propagation time through the fiber. Propagation time through a fiber is calculated as follows:

Eq. 3.5
$$t = \frac{L \bullet n}{c}$$

Where:

t = Propagation time in seconds.
L = Fiber length in meters.
n = Refractive index of the fiber core (approximately 1.45).
c = Speed of light (2.998×10^8 meters/second).

In doing a detailed analysis one must consider that the fiber length is slightly temperature dependent, and the refractive index of the fiber is dependent on the wavelength, as shown in Figure 3.6 and is also slightly dependent on temperature. It is difficult to obtain exact data for these variables, so consulting the fiber manufacturer is recommended.

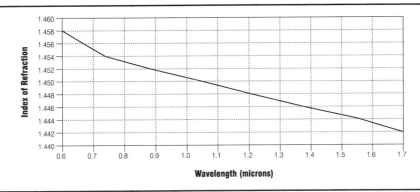

FIGURE 3.6

Refractive Index of Pure Fused Silica

Figure 3.7 shows how the principle of total internal reflection applies to optical fiber. Because the core's index of refraction is higher than the cladding's index of refraction, the light that enters at a certain angle is guided along the fiber.

FIGURE 3.7

Total Internal Reflection in Step-Index Multimode Fiber

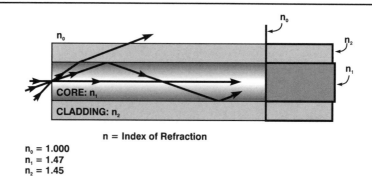

n = Index of Refraction

$n_0 = 1.000$
$n_1 = 1.47$
$n_2 = 1.45$

The illustration shows a step-index multimode fiber. Three different rays of light, also called *modes*, are pictured traveling down the fiber. One mode travels straight down the center of the core. A second mode travels at a steep angle and bounces back and forth by total internal reflection. The third mode exceeds the critical angle and is refracted into the cladding and lost as it escapes into the air. Intuitively, it can be seen that the second mode travels a longer distance than the first mode causing the two modes to arrive at two separate times. This creates a disparity between arrival times of the different light rays that is known as *dispersion*, and the result is a muddied signal at the receiving end. Dispersion will be treated in detail later in this chapter; however, it is important to note that high dispersion is an unavoidable characteristic of multimode step-index fiber. To compensate for the dispersion inherent in multimode step-index fiber, graded-index fiber was developed. Graded-index refers to the fact that the refractive index of the core gradually decreases farther from the center of the core. The increased refraction in the center of the core slows the speed of some light rays, allowing all the light rays to reach the receiving end at approximately the same time, reducing dispersion.

For an understanding of how multimode graded-index fibers work, refer back to Figure 3.7. In Figure 3.7, the second mode would still travel a longer distance, but since it spends most of its time in the outer region of the core where the refractive index is lower, it travels faster. The first mode that travels straight down the center of the fiber travels the shortest distance, but since the refractive index is higher there, it travels slower. A modern multimode graded-index fiber is designed so that all modes travel at nearly the same speed. This is accomplished by designing the fiber's core to have a parabolic index of refraction profile.

TYPES OF OPTICAL FIBER

There are two basic types of optical fiber: *multimode fiber* and *single-mode fiber*. Multimode fiber was the first type to be commercialized. Its core is much larger than that of single-mode fiber, allowing hundreds of rays (modes) of light to move through the fiber simultaneously. Single-mode fiber, on the other hand, has a much smaller core. While it would seem that a larger core would allow for a higher *bandwidth* or higher capacity to transmit information, this is not true. Single-mode fibers are better at retaining the fidelity of each light pulse over longer distances, and they exhibit less dispersion caused multiple rays or modes. Thus, more information is transmitted per unit of time. This gives single-mode fiber higher bandwidth compared to multimode fiber. Single-mode fiber is generally characterized as *step-index fiber* meaning the refractive index of the fiber core is a step above that of the cladding rather than graduated as it is in *graded-index fiber*. Single-mode fiber also enjoys lower fiber attenuation than multimode fiber. (Attenuation will be addressed further in the next section of this chapter.)

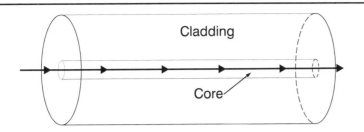

FIGURE 3.8

Single-mode Fiber

Single-mode fiber has disadvantages. The smaller core diameter makes coupling light into the core more difficult. The tolerances for single-mode connectors and splices are also much more demanding.

Multimode fiber may be categorized as step-index or graded-index fiber. The term multimode simply refers to the fact that numerous modes or light rays are carried simultaneously through the waveguide. The larger core diameter increases coupling ease and generally multimode fiber can be coupled to lower cost light sources. However, *multimode dispersion* is a drawback, and fiber attenuation is also higher.

Figure 3.9 shows the principle of graded-index fiber. The core's central refractive index n_A is greater than that of the outer core's refractive index n_B. As discussed earlier, the core's refractive index is parabolic, being higher at the center. As Figure 3.8 shows, the light rays no longer follow straight lines. Rather, they follow a serpentine path being gradually bent back toward the center by the continuously declining refractive index.

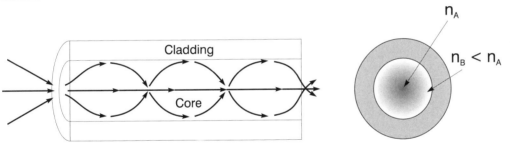

FIGURE 3.9

Multimode Graded-Index Fiber

Now that the general differences in fiber types have been examined, it is time to explain what is meant by such terms as attenuation and dispersion and relate these loss mechanisms to the fiber's information-carrying capacity, bandwidth.

ATTENUATION

Attenuation loss is a logarithmic relationship between the optical output power and the optical input power in a fiber optic system. It is a measure of the decay of signal strength, or loss of light power that occurs as light pulses propagate through the length of the fiber. The decay along the fiber is exponential and can be expressed as:

Eq. 3.6 $$P(z) = P_0 \cdot \exp(-\alpha'z)$$

where:

$P(z)$=Optical power at distance z from the input.
P_0=Optical power at fiber input.
α'=Fiber attenuation coefficient, [1/km].

Engineers usually think of attenuation in terms of decibels; therefore, the equation may be rewritten using $\alpha = 4.343\,\alpha'$, from conversion of base e to base 10.

$$P(z) = P_0 \bullet 10^{-\alpha z}$$

Eq. 3.7
$$\log P(z) = -\alpha z/\log 10 + \log P_0$$

$$\alpha = \alpha_{scattering} + \alpha_{absorption} + \alpha_{bending}$$

where:

α = Fiber loss, [dB/km].

The decibel is treated more extensively in Chapter 9.

Attenuation in optical fiber is caused by several intrinsic and extrinsic factors. Two intrinsic factors are scattering and *absorption*. The most common form of scattering, Rayleigh Scattering (see Figure 3.10), is caused by microscopic non-uniformities in the optical fiber. These non-uniformities cause rays of light to partially scatter as they travel along the fiber, thus, some light energy is lost. Rayleigh scattering represents the strongest attenuation mechanism in most modern optical fibers; nearly 90% of the total attenuation can be attributed to it. It becomes important when the size of the structures in the glass itself are comparable in size to the wavelengths of light traveling through the glass. Thus, longer wavelengths are less affected than short wavelengths. The attenuation coefficient (α) decreases as the wavelength (λ) increases and is proportional to λ^{-4}. Rayleigh scattering increases sharply at short wavelengths. (Rayleigh scattering causes the sky to be blue. Only the short wavelength blue colors are significantly scattered by air molecules.)

FIGURE 3.10

Scattering

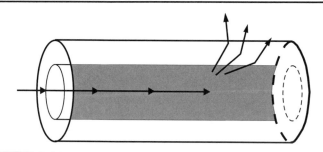

Absorption, illustrated in Figure 3.11, can be caused by the molecular structure of the material, impurities in the fiber such as metal ions and OH⁻ ions (water), and atomic defects such as unwanted oxidized elements in the glass composition. These impurities absorb the optical energy and dissipate it as a small amount of heat. As this energy dissipates, the light becomes dimmer. At 1.25 and 1.39 μm wavelengths, optical loss occurs because of the presence of OH⁻ ions in the fiber. Above a wavelength of 1.7 μm, glass starts absorbing light energy due to the molecular resonance of the SiO_2 molecule.

FIGURE 3.11

Absorption

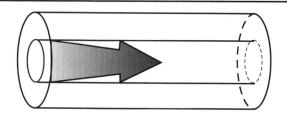

Extrinsic causes of attenuation include cable manufacturing stresses, environmental effects, and physical bends in the fiber. Physical bends break down into two categories: *microbending* and *macrobending* (see Figure 3.12). Microbending is the result of microscopic imperfections in the geometry of the fiber. These imperfections could be rotational asymmetry, changes of the core diameter, rough boundaries between the core and cladding, a result of the manufacturing process itself, or mechanical stress, pressure, tension, or twisting. Macrobending describes fiber curvatures with diameters on the order of centimeters. The loss of optical power is the result of less-than-total reflection at the core-to-cladding boundary. In single-mode fiber, the fundamental mode is partially converted to a radiating mode due to the bends in the fiber. Because of the increase in mode field diameter with wavelength, in a given single-mode fiber, macrobending loss will be higher at longer wavelengths. Bending loss is usually unnoticeable if the radius of the bend is larger than 10 cm.

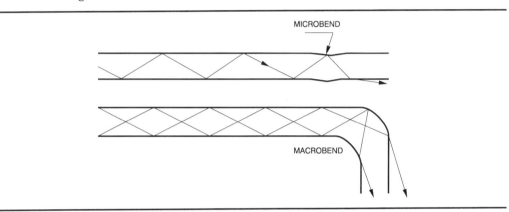

FIGURE 3.12

Bending

(Illustration courtesy of AMP, Inc.)

The amount of attenuation caused by an optical fiber is primarily determined by its length and the wavelength of the light traveling through the fiber. There are also many secondary and tertiary factors that contribute. Figure 3.13 shows the loss per unit length of a typical modern optical fiber. The plot covers wavelengths from 0.5 μm to 1.9 μm. As a point of reference, the human eye sees light in the range from 0.4 μm (blue) to 0.7 μm (red). Most modern fiber optic transmission takes place at wavelengths longer than red, in the infrared region. There are three important fiber optic wavelength regions, 850 nm, 1300 nm, and 1550 nm. These particular wavelengths were chosen because the loss of the fiber is lowest at these wavelengths. There are three primary mechanisms that influence the fiber's loss at a given wavelength. At shorter wavelengths, Rayleigh scattering is important, increasing as λ^{-4}. At longer wavelengths, absorption becomes dominant as the molecules in the glass start to resonate. In between, absorption by impurities is important. The dotted line in Figure 3.13 shows the approximate location of the absorption caused by the OH$^-$ ions. This is often the most harmful impurity in fiber. When these three loss mechanisms are considered together, there are only a few dips. The plot shows that there are really four dips. The 1060 nm region is a low-spot that was skipped over and never became significant although a few companies, notably ITT, did produce fiber links in the early 1980's that used this region.

The 850 nm region, called the first window, was the first to be widely exploited because of the LED and detector technology that was available in fiber's early days. The 1300 nm region, the second window, is very popular today because of its dramatically lower loss and, as will be shown subsequently, the lower dispersion at this wavelength. The 1550 nm region, the third window, is generally used only in cases where the use of repeaters might otherwise be required or in conjunction with other wavelengths as in a wavelength-division multiplexed system. A good rule thumb is that performance and cost increase as wavelength increases. A fourth wavelength, 780 nm, is also used. Low-cost "CD" lasers in this wavelength are manufactured in high volume, making them very economical.

FIGURE 3.13

Optical Loss versus Wavelength

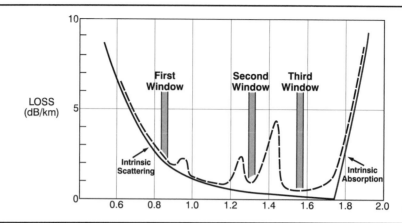

DISPERSION

Dispersion is the technical term for the spreading of the pulses of light as they travel down the optical fiber. Dispersion limits bandwidth in optical fiber, reducing the amount of information that the optical fiber can carry. The most important types of dispersion are multimode dispersion and *chromatic dispersion* which includes *material dispersion*, *waveguide dispersion*, and *profile dispersion*.

MULTIMODE DISPERSION

Multimode dispersion describes the pulse broadening in multimode fibers caused by different modes traveling at different speeds through the fiber. This form of dispersion is sometimes called modal dispersion because it is characteristic of multimode fiber only. Multimode dispersion can be reduced three ways:

1. Use a smaller core diameter fiber to reduce the number of modes traveling through the fiber.
2. Use graded-index fiber. As described earlier, graded-index fiber uses different refractive indices within the fiber to move the modes along so that they arrive at the end of the fiber together.
3. Use single-mode fiber. Under certain circumstances, this option eliminates multi-mode dispersion altogether.

To qualify number three, single-mode fiber is only single-mode at wavelengths longer than the *cutoff wavelength*. For a typical 1300 nm single-mode fiber (9 μm core diameter), this cutoff wavelength is between 1150 nm and 1200 nm. At wavelengths below this threshold, the single-mode fiber will become dual mode, then tri-mode and so on. Increasingly, 9 μm single-mode fiber is being used with short-wavelength lasers at 780 nm to 850 nm. At these wavelengths, 9 μm core fiber is actually dual-mode. There is a specialty size single-mode fiber with a 5 μm core diameter, but it is rarely used because of its cost and the difficulty coupling a light source to such a small core size. The point is, multimode dispersion is also a factor in short wavelength systems that use 9 μm core diameter single-mode fiber. When multimode dispersion is present, it dominates to the point that other types of dispersion can typically be ignored.

CHROMATIC DISPERSION

Chromatic dispersion represents the fact that different colors or wavelengths propagate (travel) at different speeds, even within the same mode. Chromatic dispersion is the result of material dispersion, waveguide dispersion, or profile dispersion.

Figure 3.14 below shows chromatic dispersion along with its key components, waveguide dispersion and material dispersion. In this example, chromatic dispersion goes to zero at a wavelength near 1550 nm. This is characteristic of dispersion-shifted fiber. Standard fiber, single-mode or multimode, has zero dispersion at a wavelength of 1300 nm. At this wavelength, waveguide dispersion is small. In fact, special steps must be taken to increase waveguide dispersion in order to shift the zero-dispersion wavelength to 1550 nm.

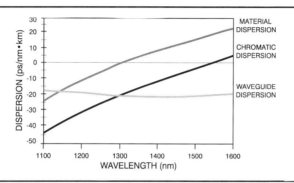

FIGURE 3.14

Chromatic Dispersion

If you can operate a fiber at the zero-dispersion wavelength with a monochromatic light source, the bandwidth of the fiber will be very large. Figure 3.15 shows the bandwidth-distance product for a hypothetical single-mode fiber. The x-axis is the center wavelength for the light source. Three curves are shown for light sources with FWHM (Full Width Half Maximum) values of 2 nm, 5 nm and 10 nm. FWHM is the width of the spectral emission at the 50% amplitude points.

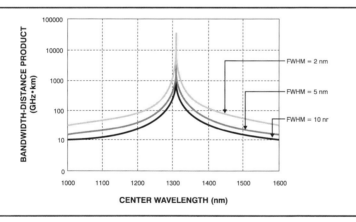

FIGURE 3.15

Single-mode Fiber Bandwidth

For the most narrow source shown, FWHM = 2 nm, the bandwidth-distance product for the fiber is over 30,000 GHz•km at a center wavelength of 1300 nm, the fiber's zero-dispersion wavelength. As the center wavelength moves even a few nanometers from 1300 nm, the fiber's bandwidth-distance product drops dramatically. At a center wavelength of 1310 nm, the fiber's bandwidth-distance product has dropped by a factor of 30, to 1,000 GHz•km. For wider optical sources, the bandwidth is even lower, as can be seen in the plot. If one used a very narrow light source tuned exactly for the fiber's zero-dispersion wavelength, the peak would be much higher than those shown in the plot.

Material Dispersion

Different wavelengths travel at different velocities through a fiber, even in the same mode. We know that the index of refraction (n) is given as:

Eq. 3.12 $$n = \frac{c}{v}$$

where

c = The speed of light in a vacuum.

v = The speed of the same wavelength in the material.

Each wavelength travels at a different speed through the material. This changes the value of v in the equation at each wavelength. Dispersion from this phenomenon is called material dispersion. Two factors determine the amount of material dispersion that will occur.

1. The range of light wavelengths injected into the fiber: a source emits several wavelengths rather than a single wavelength. This is the spectral width of a source. LED's have a much wider spectral width (about 35-170 nm) than lasers (0.1-5 nm).

2. The center operating wavelength of the source: around 850 nm, longer wavelengths (red) travel faster than shorter wavelengths (blue). However, at 1550 nm the situation is reversed, and the shorter blue wavelengths travel faster than the longer red ones. The crossover where the wavelengths travel at the same speed occurs around 1300 nm, the zero-dispersion wavelength.

Material dispersion greatly affects single-mode fibers. In multimode fibers, multimode dispersion usually supersedes material dispersion in terms of its impact on the system. Figure 3.16 shows the refractive index versus wavelength for pure fused silica core fiber. The group refractive index, also called *group index*, is the speed of light in a vacuum (c) divided by the *group velocity* of the mode. This group velocity is the reciprocal of the rate change of the *phase constant* with respect to the angular frequency. Keep in mind that this relationship varies according to the composition of the fiber core.

FIGURE 3.16

Material Dispersion

(Illustration courtesy of Hewlett-Packard.)

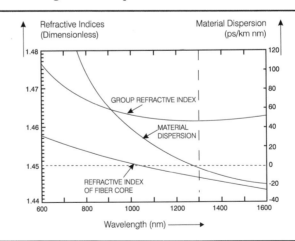

Waveguide Dispersion

Like material dispersion, waveguide dispersion is a greater concern for single-mode fibers than for multimode fibers. Waveguide dispersion occurs because optical energy travels in both the core and the cladding at slightly different speeds. This is a result of the difference in the indices of refraction between the core and the cladding. Practical single-mode fiber is designed so that material dispersion and waveguide dispersion cancel one another at the wavelength of interest.

Profile Dispersion

The refractive indices of both the core and the cladding affect the group velocity in a fiber. These differences are usually stated in terms of the *refractive index profile*. This is a description of the value of the refractive index as a function of distance from the optical axis along an optical fiber diameter. Profile dispersion is caused by the different wavelength dependencies of the refractive indices of the core and cladding. These differences are caused by the different materials involved. This parameter is more important in multimode fibers than in single-mode fibers because the profile can be optimized for only one wavelength.

FIBER COMPARISONS

This section of the guide offers a comparison of various sizes of optical fiber. Figure 3.17 has three sections labeled Attenuation, Bandwidth-Distance Products, and Power, that give a quick visual comparison between three important fiber parameters.

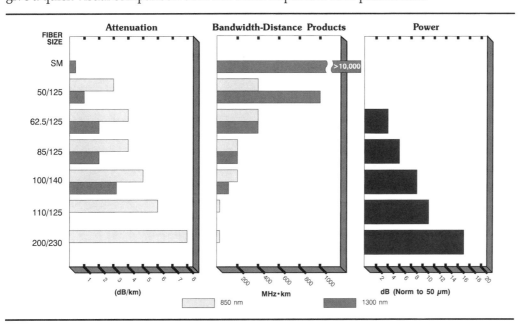

FIGURE 3.17

Comparison of Optical Fiber Sizes

The Attenuation section shows attenuation by different size fibers. Smaller numbers indicate the better fiber. The Bandwidth-Distance Products section tells how much information can be carried per second on a single fiber. Higher numbers show the greater information carrying capacity. The Power section shows the relative optical power that can be coupled into an optical fiber using an LED. In this graph, 50/125 μm fiber is the reference. With 62.5/125 μm fiber, one can typically couple 4.5 dB more power into the fiber. That amounts to 181% more power compared to 50/125 μm fiber. If single-mode fiber was shown on the same chart it would be about -13 dB relative to 50/125 μm fiber. As is the case with almost all engineering trade-offs, you do not get something for nothing. Fibers that are easy to couple into lots of power tend to have poor bandwidth-distance products and attenuation rates.

It is impossible to say which fiber type is best without examining the specific problem to be solved. There are applications where 200/230 μm fiber is the best choice and others where single-mode fiber is the optimum choice. Which fiber is best is akin to asking which car is best. If economy is the only consideration, one might choose a Hyundai. If speed is the goal, one might pick a Porsche or Lotus. If luxury is the goal, then a Lincoln Town Car or a Cadillac might be the best choice.

Despite all of the sizes of fibers described, only a few fiber types are important. These are:

Single-mode: Widely used for high data rate and long distance applications.

62.5/125 μm: Very popular in most commercial applications; it has wide uses with low- to moderate-speed data links and video links.

50/125 μm: This fiber type is mainly used by military customers.

100/140 μm: Once a very popular size, there are only a few remaining applications. The only major application is use in aircraft.

The first number listed (e.g., 50 or 62.5) represents the fiber core diameter in microns. The number after the slash is the cladding diameter. This is the key dimension when picking a fiber optic connector. For instance, 50/125 μm and 62.5/125 μm size fibers can use the same size connector. A connector with a larger hole would be required for 100/140 μm fiber. Figure 3.18 illustrates the most popular fiber sizes. By studying the figure, one can get a visual feel for the relative core and cladding size that is useful for identifying an unknown fiber's core and cladding using a microscope.

FIGURE 3.18

Popular Fiber
Sizes—Magnified
Drawings

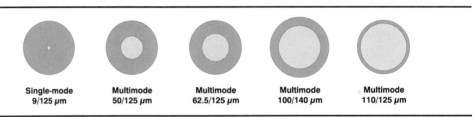

Several factors must be considered when selecting optical fiber for a given application:

- Fiber core/cladding size (e.g., 50/125 μm)
- Fiber material and construction (e.g., glass core/glass cladding)
- Fiber attenuation measured in dB/km
- Fiber bandwidth-distance product measured in MHz•km
- Environmental considerations (e.g., temperature)

Most fiber optic applications utilize fibers made from glass cores and glass cladding. This combination yields excellent performance at a reasonable cost. Other combinations include plastic clad silica (plastic cladding/silica core, called PCS) and all-plastic fiber. PCS fiber is used mainly in applications where high nuclear radiation levels may be present. It is also widely used for imaging fibers because of its superior transmission at visible wavelengths. All-plastic fibers can cost less, but they are typically limited to very short distances (a few meters) because attenuation is very high and bandwidth-distance product is low.

Fiber loss is an important parameter in picking an optical fiber. Fiber loss depends heavily on the operating wavelength. Practical fibers have the lowest loss at 1550 nm and the highest loss at 780-850 nm. (Only four wavelengths are important: 780 nm, 850 nm, 1300 nm, and 1550 nm.)

Fiber bandwidth is another critical parameter. It determines the maximum rate at which data can be transmitted. It is very important to understand that this commonly used parameter only gives the bandwidth limitation due to multimode dispersion. Quite often, material dispersion is much more limiting, especially at shorter wavelengths. The discussion earlier in this chapter gives a better description of the dispersion mechanisms that must be considered in a full systems analysis.

Table 3.2 shows the typical fiber optical loss at the four key fiber optic wavelengths. The data is taken from a number of fiber manufacturers. Many manufacturers offer multiple levels of performance for each fiber size. The table lists the range of values seen with mainstream fibers.

Fiber		Optical Loss (dB/km)			
Size	Type	780 nm	850 nm	1300 nm	1550 nm
9/125 μm	SM	⁓	⁓	0.5-0.8	0.2-0.3
50/125 μm		4.0-8.0	3.0-7.0	1.0-3.0	1.0-3.0
62.5/125 μm		4.0-8.0	3.0-7.0	1.0-4.0	1.0-4.0
100/140 μm	MM	4.5-8.0	3.5-7.0	1.5-5.0	1.5-5.0
110/125 μm		⁓	15.0	⁓	⁓
200/230 μm		⁓	12.0	⁓	⁓

TABLE 3.2

Typical Fiber Loss

Table 3.3 presents the bandwidth-distance product of the common fiber sizes at the four fiber optic wavelengths. Table 3.4 shows several other miscellaneous fiber parameters.

Fiber		Bandwidth-Distance Product (MHz•km)			
Size	Type	780 nm	850 nm	1300 nm	1550 nm
9/125 μm	SM	<800	2,000	20,000+	4,000-20,000+
50/125 μm		150-700	200-800	400-1,500	300-1,500
62.5/125 μm		100-400	100-400	200-1,000	150-500
100/140 μm	MM	100-400	100-400	100-400	10-300
110/125 μm		⁓	17	⁓	⁓
200/230 μm		⁓	17	⁓	⁓

TABLE 3.3

Typical Fiber Bandwidth

Fiber Size	Numerical Aperture	Temperature Range	Min. Bend Radius
9/125 μm	⁓	-60 to +85°C	12 mm
50/125 μm	0.20	-60 to +85°C	12 mm
62.5/125 μm	0.275	-60 to +85°C	12 mm
100/140 μm	0.29	-60 to +85°C	12 mm
110/125 μm	0.37	-65 to +125°C	15 mm
200/230 μm	0.37	-65 to +125°C	16 mm

TABLE 3.4

Miscellaneous Fiber Parameters

There are also environmental factors to consider in selecting a fiber. Temperature is often the most demanding parameter. Surprisingly, most optical fibers have more difficulty with low temperatures than with high temperatures.

Some generalizations that can be made about fiber types are as follows:

- Larger core size fibers are generally more expensive (for the fiber).
- Larger core size fibers have higher loss (attenuation) per unit distance.
- Larger core size fibers have lower bandwidth.
- Larger core size fibers allow lower cost light sources to be used.
- Larger core size fibers typically use lower cost connectors.
- Larger core size fibers typically yield the lowest system cost (for short distance systems).

The first three items tend to drive the selection to smaller fibers while the last three items drive the selection to larger fibers. Many factors will dictate the choice of fiber type. Transmission bandwidth, maximizing distance between repeaters/amplifiers, cost of splicing or connectorizing, tolerance of temperature fluctuations, strength and flexibility are just some of these factors. In most applications, cabling the fiber is required to meet the application requirements. Fiber optic cables are discussed in detail in the next chapter.

Chapter Summary

- Inside vapor deposition and outside vapor deposition are the two predominant methods for manufacturing optical fiber.
- The operation of optical fiber is based on the principle of total internal reflection which occurs because light travels at different speeds in different materials.
- As light passes from one medium to a medium with a different index of refraction, the light is bent, or refracted.
- The core of an optical fiber has a higher index of refraction than the cladding, allowing for total internal reflection.
- Multimode fiber and single-mode fiber are the two basic types of optical fiber.
- Attenuation is a logarithmic relationship between the optical output power of the transmitter and the optical input power of the receiver in a fiber optic system.
- Attenuation in optical fibers is caused by scattering, absorption, and factors such as cable manufacturing stresses, environmental effects, and physical bends in the fiber.
- The amount of attenuation caused by an optical fiber is primarily determined by its length and the wavelength of the light traveling through the fiber.
- Dispersion is the technical term for the spreading of the pulses of light as they travel down the optical fiber.
- Multimode dispersion describes the pulse broadening in multimode fibers caused by different modes traveling at different speeds through the fiber.
- Chromatic dispersion represents the fact that different colors or wavelengths propagate at different speeds, even within the same mode.
- Waveguide dispersion occurs because optical energy travels in both the core and the cladding at slightly different speeds.
- Profile dispersion is caused by the different wavelength dependencies of the refractive indices of the core and cladding.

Selected References and Additional Reading

Buck, John. 1995. *Fundamentals of Optical Fibers.* New York, NY: Wiley Interscience.

Hecht, Jeff. 1993. *Understanding Fiber Optics.* 2nd edition. Indianapolis, IN: Sams Publishing.

Hentschel, Christian. 1988. *Fiber Optics Handbook.* 2nd edition. Germany: Hewlett-Packard Company.

—. 1992. *Just the Facts.* New Jersey: Corning, Incorporated.

Sterling, Donald J. 1993. *Amp Technician's Guide to Fiber Optics,* 2nd Edition. New York: Delmar Publishers.

FIBER OPTIC CABLES

Chapter 3 discussed the methods for manufacturing optical fiber and detailed many of its important characteristics. Another characteristic is that bare optical fiber is quite fine, making the handling of bare optical fibers difficult, especially in multifiber situations. To make handling optical fiber easier, it is first "buffered" or coated with a thin primary coating by the fiber manufacturer. It is then most often cabled. Fiber optic cables also function to protect the optical fiber from mechanical damage such as bends, breaks, or cuts both during and after installation. The exact structure and properties of a fiber optic cable are very often directly related to the application for which the cable is constructed as well as the environment in which the cable will be installed.

FIBER OPTIC CABLE CONSTRUCTION

Although cable construction is usually application-specific, there are elements common to all types of fiber optic cables. The first element is the fiber housing. The fiber housing is either a *loose-tube* or a *tight-buffer* construction. Figure 4.1 shows the differences in these two types of construction in both cutaway and cross-section form. Loose-tube construc-

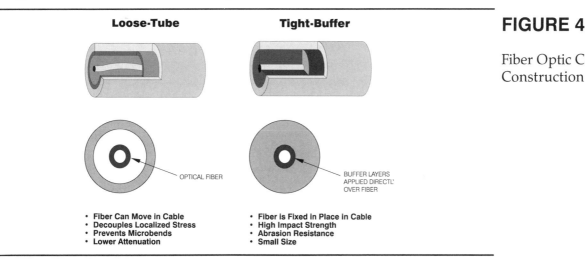

FIGURE 4.1

Fiber Optic Cable Construction

tion allows for lower attenuation but the tubes may permanently kink if the cable is bent beyond its limits, breaking the fiber. In loose-tube cable construction, the optical fibers are enclosed in plastic buffer tubes that are filled with a gel to impede water penetration. The buffer tubes are stranded around a dielectric or steel central member, which serves as an anti-buckling element. Loose-tube cable construction is used mainly for long distance applications and permanent installations. To save space, several fibers may be placed in the same tube before it is filled with a water-impeding gel. In high fiber counts, this is cost-effective, producing a lower cost per cabled fiber. A tight-buffer cable is smaller, more flexible, and more resistant to impact. In this cable design, the buffering material is in direct contact with the fiber. It is preferred when small size and high mechanical integrity are a main concern. In tight-buffer cable designs, the amount of material and effort used is more nearly proportional to the number of fibers; therefore, this design is more cost-effective in cables with lower fiber counts.

A key difference between loose-tube and tight-buffer cables is how they react to fluctuating temperatures. Since the fiber in a loose-tube cable is not mechanically coupled to the cable structure, it tends to do well in widely varying temperatures. The converse is true for most tight-buffer cable designs. The problem with the tight-buffer construction is that the materials used in the cable often have a higher thermal coefficient of expansion than the glass in the fiber itself. As the cable is heated, the fiber is stretched, and as the cable is cooled, the fiber is compressed. These stresses on the fiber cause the attenuation of the fiber to increase. This is usually referred to as cabling excess loss. This loss can range from a few hundredths of a dB per kilometer to several tenths of dB's per kilometer or more. Some very sophisticated fiber optic cable designs incorporate central strength members that have negative temperature coefficients of expansion so that the positive expansion caused by the outer jacket material is neutralized. Still, tight-buffer cable designs have some problems with temperature changes. Usually cold temperatures cause the most problems for fiber optic cables.

Ribbon cable is a variation of the loose-tube construction. A group of coated fibers is arranged so that the fibers are parallel to each other then coated with plastic to form a multifiber ribbon. Typically five to twelve fibers are encased in this manner. This set of fibers is then placed in one tube in the cable jacket and surrounded with gel, similar to other loose-tube designs. The simple structure of the ribbon cable makes it possible to splice all fibers of a ribbon simultaneously. This is a complex operation, but if done correctly, it can make splicing fast. The ribbon cable design allows for high packing density. However, mishandling during installation could cause uneven strain on different fibers, introducing the potential for uneven fiber losses.

The number of fibers used in the cable will affect the cable's current and future usability. The fiber count is usually based on three things: the intended end-user applications, both present and future; the level of multiplexing and use of bridges/routers; and the physical topology of the network. Simple applications require a minimum of one fiber to establish basic communication; however, fiber count in a fiber optic cable ranges from the simplest single-fiber construction to complex multifiber cables. In the more complex, multifiber cable constructions, a common structure would have a number of buffered fibers or buffered fiber bundles loosely wound around a central member. Table 4.1 describes the fiber count requirements for many of today's most often used applications.

TABLE 4.1

Cable Fiber Count Required for Specific Applications

Application	Number of Fibers
Voice; Two-Way	1 or 2
Voice; Intercom	1
Video; Security	1
Video; Interactive	1 or 2
Telemetry	1 or 2
Data Multiplexing	2
Channel Extension	2
Ethernet	2
Token Ring	2 or 4
FDDI	2 or 4

The central member described above may be steel, coated steel, coated glass, a fiber glass rod, or simply filler. Small, indoor cables rarely include a central member, but central members are very common in outdoor cables and cables with high fiber counts.

Depending on the application, a variety of *strength members* can be designed into the cable construction. A common example is the use of Kevlar®, an extremely strong aramid yarn. This yarn can be applied longitudinally around each buffered fiber or used under the outer jacket of the cable. The use of Kevlar allows the cable to withstand tension from 50-600 pounds for an extended period. It is common practice in cable installation to pull the cable through the duct work by the strength members to avoid putting stress on the fiber itself. In some cable designs, the strength members are integrated into the structure so that the entire cable may be pulled without concern over locating and attaching the strength members.

The final component of a fiber optic cable is the cable *jacket*. The selection of jacket material requires consideration of such factors as mechanical properties, attenuation, environmental stress to be placed on the cable, and flammability. Variations in the outer jacket design allow the jacket to be interlocked to the cable core by filling the core's outer interstices. Strength members in the cable reinforce the jacket, preventing any movement of the jacket along the axis. Figure 4.2 shows two types of cable jackets, and Table 4.2 lists the properties of common cable jacket materials.

FIGURE 4.2

Two Types of Cable Jackets

(Photos courtesy of Optical Cable Corporation.)

TABLE 4.2

Properties of Cable Jacket Material

Jacket Material	Properties
Polyvinyl Chloride (PVC)	Normal mechanical protection. Many different grades of PVC offer flame retardancy and outdoor use. Also for indoor and general purpose applications.
Hypalon ®	Can withstand extreme environments; flame retardant; good thermal stability; resistant to oxidation, ozone, and radiation.
Polyethylene	Used for telephone cables. Resistant to chemicals and moisture; low-cost. Polyethylene is flammable, so it is not used in electronic applications.
Thermoplastic Elastomer (TPE)	Low-cost; excellent mechanical and chemical properties.
Nylon	Used over single conductors to improve physical properties.
Kynar® (Polyvinylidene Fluoride)	Resistant to abrasions, cuts; thermally stable; resistant to most chemicals; low smoke emission; self-extinguishing. Used in highly flame retardant plenum cables.
Teflon® FEP	Zero smoke emission, even when exposed to direct flame. Suitable to temperatures of 200°C; chemically inert. Used in highly flame retardant plenum cables.
Tefzel®	Many of the same properties as Teflon; rated for 150°C; self-extinguishing.
Irradiated Cross-Linked Polyolefin (XLPE)	Rated for 150°C; high resistance to environmental stress, cracking, cut-through, ozone, solvents, and soldering.
Zero Halogen Thermoplastic	Low toxicity makes it usable in any enclosed environment.
Kevlar, Hyplon, Tefzel, and Teflon are registered trademarks of E.I. Du Pont Nemours & Company. Kynar is a registered trademark of Pennwalt, Inc.	

TYPES OF FIBER OPTIC CABLES

The simplex cable, illustrated in Figure 4.3 is round with a single fiber in the center. Duplex cables, cables with two fibers within, may be circular, oval, or arranged zipcord fashion like an electrical cable. A zipcord cable is illustrated in Figure 4.4.

FIGURE 4.3

Simplex Cable
Construction

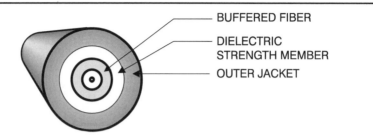

FIGURE 4.4

Duplex Zipcord
Cable Construction

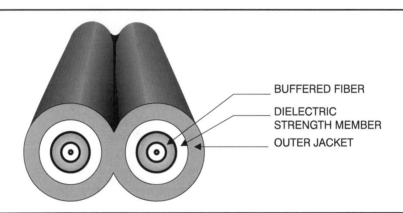

More complex cable assemblies fall into several categories including breakout (fanout) cables, composite cables, and hybrid cables. Breakout cables, also called fanout cables, are so called because the fibers are packaged in the cable as single-fiber or multifiber subcables. This allows the individual fibers to be accessed without the need for patch panels to terminate the multiple fibers. Figure 4.5 shows a cross-section of a typical breakout cable.

FIGURE 4.5

Multifiber
Breakout Cable
Construction

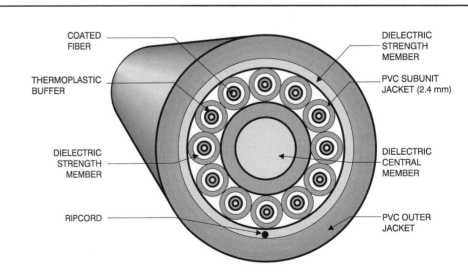

There is some disagreement in the cable industry as to the exact definitions of composite cables and hybrid cables. For the sake of this discussion, composite cables will be defined as cables mixing both single-mode and multimode optical fiber in a single cable assembly. The term hybrid cable will be used to describe a cable that incorporates mixed optical fiber with copper cable. Both composite cables and hybrid cables offer the advantage of time and cost savings. Multiple cables pulled separately would take longer to install. However, these cables are specialty items, custom-manufactured for a specific application, but the increase in demand for the convergence of video, audio, and data transmission make these types of cables a very attractive alternative to separate cables. Hybrid cables can also carry electrical signals and are getting wide use in broadband CATV networks. Figure 4.6 shows a typical hybrid cable.

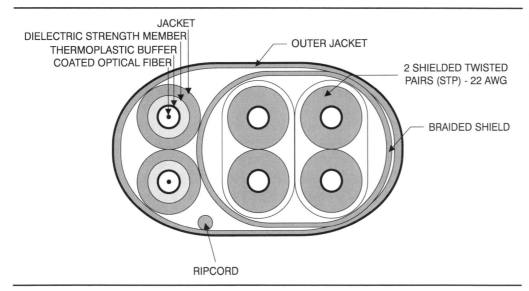

FIGURE 4.6

Hybrid Cable Construction

FIBER OPTIC CABLE VERSUS COPPER COAX CABLE

Copper coax cables and other conventional metal cables are the predecessors of fiber optic cables. Both fulfill the same basic function of signal transmission. Fiber optic cables resemble conventional cables and incorporate some of the same materials such as polyvinyl chloride (PVC) sheathing and polyethylene (PE) which is used for environmental protection, especially on buried and aerial cables. Optical fiber has many advantages however. One major advantage is the nonconductive nature of optical fiber. While some fiber optic cables may contain conductive material such as a steel central member or steel outer sheath, most fiber optic cables are dielectric, meaning they contain no conductive materials. This gives the cable complete EMI/RFI immunity and reduces the impact of lightning strikes.

Figure 4.7 illustrates another advantage of optical fiber. The graph shows a comparison of the attenuation of a low-loss optical fiber and four popular types of copper coax cable. Note the very high levels of attenuation for the coax cable at higher frequencies. RG-6/U has an attenuation rate of 70 dB/km at a frequency of only 100 MHz. That means that only 0.00001% of the signal strength will reach the other end of a 1 km length of cable. Even the best coax shown, RG-8, has an attenuation rate of 45 dB/km at a frequency of 100 MHz. At a frequency of 40 MHz, RG-58/U has an attenuation rate of 100 dB/km. Only 0.00000001% of the signal strength would remain after 1 km of cable. By comparison, the single-mode fiber optic cable has an essentially flat 1 dB/km of attenuation, even at 500 MHz. This means that 79% of the signal strength will remain after a 1 km length of fiber. The advan-

FIGURE 4.7

Comparison of
Copper Coax
versus Optical Fiber
Cable

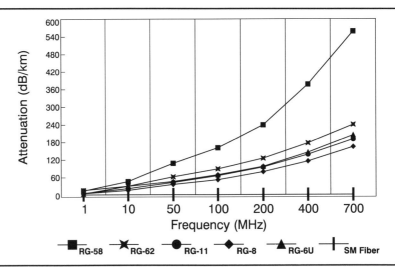

tages of fiber at high frequencies, high data rates, and long distances are obvious. Even multimode fiber only has 8 dB of loss at 500 MHz, so 16% of the power remains after 1 km of fiber.

TENSILE STRENGTH

Fiber optic cables offer a higher mean tensile breaking strength than other types of cable. This value considers the pounds per square inch (psi) that can be applied before a given cable will snap. The depth of the microscopic flaws inherent on the surface determine the strength of an optical fiber. The theoretical mean tensile breaking strength of optical fiber is about 600,000 pounds per square inch or 600 kpsi, enough strength for a fiber with a diameter equivalent to one square inch to suspend fifty elephants in an elevator. Figure 4.8 shows the comparative tensile breaking strengths of various cable materials.

FIGURE 4.8

Mean Tensile
Strength of Cable
Materials

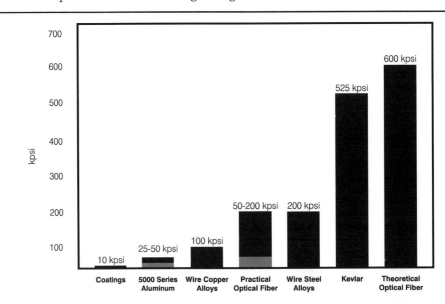

Usually the theoretical fiber strength of 600 kpsi can only be achieved for very short lengths of fiber. Just as a chain is only as strong as its weakest link, a length of optical fiber is no stronger than the weakest point along its length. Microscopic cracks and flaws in the surface of the fiber limit the strength of the fiber to values typically less than 200 kpsi. All modern fibers are proof-tested as they are manufactured. Most fibers are proof-tested at either 50 kpsi or 100 kpsi. In special cases, fiber can be proof-tested to 200 kpsi or even higher for high-reliability applications.

One very important result of a fiber's proof-tested strength is the allowable bend radius. A fiber with a proof-tested strength of 50 kpsi can have a minimum bend radius of 1.5". A fiber with a proof-tested strength of 100 kpsi can bend tighter to a 1" radius, and 200 kpsi fiber can bend as tight as a 0.75" radius.

Fiber with a higher proof-test will withstand more abuse during installation and throughout its lifetime. A fiber optic cable is generally assumed to have a 20 to 40 year life after it is installed. Stripping the fiber with mechanical strippers can degrade the life expectancy of a fiber. Moisture and excessive stress are also enemies of optical fiber. Moisture can invade cracks in the outer surface of the fiber and can cause the cracks to propagate. Stress on a fiber can also cause those microcracks to propagate, ultimately causing the fiber to break. The proof-test level to which a fiber is subjected determines the maximum possible flaw size in the fiber. The maximum flaw size in 100 kpsi proof-tested fiber is about 1/3 the size of possible flaws in 50 kpsi proof-tested fiber. Flaw size greatly affects the lifetime of a fiber under given stress conditions. The 100 kpsi proof-tested fiber survives many times longer that the 50 kpsi proof-tested fiber.

SPECIFYING FIBER OPTIC CABLES

As mentioned, fiber optic cables are often custom manufactured from application to application. Cable environment is a critical parameter in determining the cable construction. There are many types of environments for fiber optic cables. Some of these are illustrated in Figure 4.9.

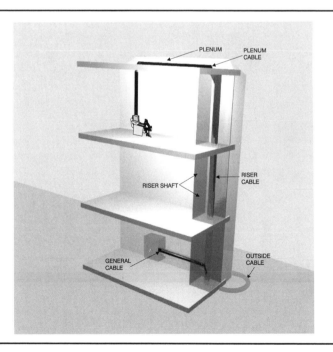

FIGURE 4.9

Cable Environments

The least strenuous cable environment can occur within devices such as computers, telephone switching systems, or distribution and splice organizers although high stresses can occur due to small bend radii. These cables are generally small and simple, and they tend to be low-cost. The device itself protects the cable from outside forces.

Intraoffice and intrabuilding cables are installed across a room, under a floor, between walls or above suspended ceilings, areas called plenum. These cables must meet fire and electrical safety standards before they are specified for installation in an office or an office building. The National Fire Protection Association publishes the National Electrical Code which defines safety considerations for using fiber optic cables within a building. There are three listings for cables: nonconductive optical fiber general purpose cable (OFN), nonconductive optical fiber riser cable (OFNR), and nonconductive optical fiber plenum cable (OFNP). Cable manufacturers must submit the cable design to an independent test laboratory such as Underwriter's Laboratories in order to obtain a listing for their cables. OFN cable is tested per UL 1581, a general test to determine the cable's resistance to generating smoke and spreading flames. OFNR cable is tested per UL 1666-1986, the riser flame test. This vertical flame test ensures that the fiber optic cable does not facilitate the spread of fire from floor to floor within the riser shaft of a building. OFNP cable is tested per UL 910, the plenum flame and smoke test. This horizontal flame test verifies that the cable will not spread flame or generate too much smoke within the plenum, ducts, or other spaces in a room or building.

Intraoffice and intrabuilding cables that meet these safety standards are often called plenum cables because that is usually where they are installed in the office or building. Intrabuilding cables are often constructed as breakout cables. This subcable design eliminates the need for patch panels in terminal closets. The subcables, which are color coded to simplify identification, allow the cable to be divided into individual fibers for distribution to separate end points in the office. Intraoffice cables tend to be of simpler construction, usually simplex or duplex cables.

Direct-burial cables are usually laid into deep trenches or plowed into the ground. This environment is very demanding and requires extra protection against moisture and temperature extremes. Direct-burial cables usually incorporate an extremely strong outer jacket to protect the cable from damage caused by digging or chewing rodents. (In fact, military standards for fiber optic cables include rodent protection tests.) Moisture protection is often achieved by filling the cable with a gel that impedes the infiltration of water. Moisture is a concern because long-term exposure to moisture will degrade the fiber's optical characteristics. Extremely cold temperatures compound this problem when the moisture expands as it freezes, causing forces to be applied to the fiber that could cause microbends and increased optical loss.

FIGURE 4.10

Aerial Cable
Installation

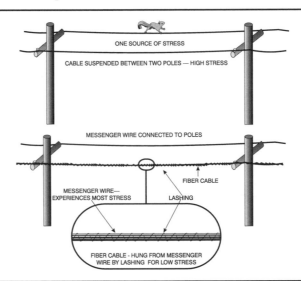

Fiber optic cables intended for aerial installation must also be constructed to handle environmental extremes. Aerial installation involves stringing the cable from utility poles as illustrated in Figure 4.10. Generally, these cables are all-dielectric; that is, they contain no metal. This prevents ground loops and provides the cable with lightning immunity. Suspension between two poles is the classic installation method. Internal strength members must be strong enough in this installation to prevent sagging that would put excess stress on the fibers. In other cases, the cable is wound or lashed to a parallel strength member. This supports the cable at more frequent intervals, reducing the stress along the length of the cable. Regardless of the actual installation method, all aerial cables have strength members and structures that isolate the fibers from the stress on the cable, and the outer jacket material offers UV protection from the sun in addition to protection from moisture and temperature extremes.

In situations where the cable must be used both within a building and outside from building to building, for example in a college campus network, indoor/outdoor cable may be specified. This cable type features the materials needed to meet fire and electrical safety standards encased in a removable layer of outdoor jacket material. Prior to the availability of indoor/outdoor cable, mixed environment cabling was handled in one of three ways: by using only tight-buffer cables (typically specified for indoor use), by using only loose-tube cables (typically specified for outdoor use), or by splicing combinations of tight-buffer cable inside the building to loose-tube cable outside the building. Loose-tube cables may be used in a building provided they are properly enclosed in metallic tubing per local building codes, and tight-buffer cables may be installed outside provided they are placed in a duct below the frost line. By splicing both types of cables together, each cable type is used in the most suitable environment. While the splice points will affect the *optical link loss budget* of the application and require additional labor for the splice, the cost differences between loose-tube cable and tight-buffer cable may make this an attractive solution in some applications when long outdoor runs are required.

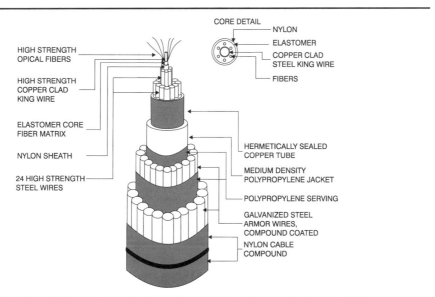

FIGURE 4.11

Submarine Cable Construction

Another very demanding environment for fiber optic cable is the submarine environment where the cable is completely submerged in fresh or salt water. Submarine cables being used for short distances are usually no more than ruggedized, waterproofed versions of direct-burial cable. Long-distance cables such as the one illustrated in Figure 4.11 are more complex. Damage from boat anchors, trawlers, fishermen, and sharks are a concern as is the water pressure in applications where the cable is laid in very deep water. There are three types of submarine cable systems that impact the actual construction of the cable. Short systems, usually no more than 100 km, link islands with nearby continents or other islands. Moderate-distance systems run over 1,000 km. These systems require the use of repeaters or optical amplifiers. Transoceanic cables, as the name suggests, run thousands of kilometers across oceans from continent to continent. These are hybrid cables that use copper to power the repeaters. Higher data rates and exploitation of the 1550 nm window in optical fiber are allowing for longer repeater distances; however, a new technology known as erbium-doped fiber amplifiers, discussed in detail in Chapter 10, will provide even greater distance capability for the optical signal and further reduce the need for repeaters in submarine systems.

While this chapter discusses many types of cables, new applications will bring new variety. The implementation of a global network has already created design challenges for fiber optic cable manufacturers. Still, the cable is but a conduit. Now it is time to turn the discussion to what this conduit carries.

Chapter Summary

- Optical fiber is buffered to protect the fiber from damage during and after installation.
- Fiber cables can be either loose-tube or tight-buffer construction.
- Central members add rigidity to the cable.
- Strength members such as Kevlar can be designed into the cable construction.
- The final component of a fiber optic cable is the cable jacket.
- The fiber count of a cable will affect the cable's present and future usability.
- Simplex cables have a single fiber in the center, while duplex cables contain two fibers.
- Complex cable assembly types include breakout cables, composite cables, and hybrid cables.
- Composite cables incorporate both single-mode and multimode optical fiber.
- Hybrid cables incorporate mixed optical fiber with copper cable.
- The least strenuous cable environment occurs within devices such as computers, telephone switching systems, or distribution and splice organizers.
- Direct-burial cables are usually laid into deep trenches or plowed into the ground and must be constructed to handle environmental extremes.
- Submarine cables being used for short distances are usually no more than ruggedized, waterproofed versions of direct-burial cable.
- Transoceanic cables run thousands of kilometers from continent to continent.

Selected References and Additional Reading

—. 1992. *Just the Facts.* New Jersey: Corning, Incorporated.

Kachmar, Wayne M. 1990. *An Overview of Datacom Cable Installation: Installing the Fiber Network.* From the Proceedings of the Fiber Optic and Computer Networking Conference & Exhibition. 24-26 Oct. 1990. Boston: World Trade Center.

Palladino, John R. 1990. *Fiber Optics: Communicating By Light.* Piscataway, NJ: Bellcore.

Sterling, Donald J. 1993. *Amp Technician's Guide to Fiber Optics*, 2nd Edition. New York: Delmar Publishers.

—. 1991. *Universal Transport System Design Guide.* Hickory, NC: Siecor Corporation.

LIGHT EMITTERS

5

Light emitters are a key element in any fiber optic system. These components convert the electrical signal into a corresponding light signal that can be injected into the fiber. The light emitter is important because it is often one of the most costly elements in the system, and its characteristics often strongly influence the final performance limits of a given fiber optic link.

There are two types of light emitters in widespread use in modern fiber optic systems: laser diodes (LD's) and light-emitting diodes (LED's). Laser diodes may be Fabry-Perot or *distributed feedback* (DFB) types while LED's are usually specified as surface-emitting diodes or edge-emitting diodes. These different classifications will be discussed in detail later in the chapter. All light emitters are complex semiconductors that convert an electrical current into light. The conversion process is fairly efficient in that it creates very little heat compared to the heat generated by incandescent lights. LED's and laser diodes are of interest for fiber optics because of five inherent characteristics:

- Small size
- High radiance (i.e., They emit a lot of light in a small area.)
- Small emitting area (The area is comparable to the dimensions of optical fiber cores.)
- Very long life (i.e. They offer high reliability.)
- Can be modulated (turned on and off) at high speeds

LED's and laser diodes are found in a variety of consumer electronics products. LED's are used as visible indicators in most electronics equipment, and laser diodes are most widely used in compact disk players. The LED's used in fiber optics differ from the more common indicator LED's in two ways:

1. The wavelength is generally in the near infrared because the optical loss of fiber is the lowest at these wavelengths.
2. The LED emitting area is generally much smaller to allow the highest possible modulation bandwidth and improve the coupling efficiency with small core optical fibers.

THEORY OF OPERATION

Laser diodes and LED's operate on the same basic principle. They use the principle of the p-n semiconductor junction found in transistors and diodes. A p-n junction is comprised of a group IV element (Si, Ge, etc.) doped with a group III element (Al, Ga, In) and a group V element (P, As). (These element groups are based on the standard periodic table of elements, a copy of which can be found in most dictionaries.) The group III, IV, and V elements form a similar crystal lattice structure so the impurities added (group III and V) replace the group IV atoms on a one-to-one basis. The added group V atoms have one more valence electron than the group IV atoms, so they form an area with excess electrons called an n-type semiconductor. The added group III atoms have one less valence electron, forming an area called a p-type semiconductor. Although the doped semiconductor area has excess and deficient electrons, it is still overall electrically neutral.

When an n-type and a p-type semiconductor are placed together, the excess electrons from the n region move over into the p region to fill the holes left by the deficiency of electrons in the p-type material. The electrons stop moving into the p-type area when enough of them have built up to begin to repel any more electrons moving over (electrons with the same charge repel each other). This buildup of charges creates a potential that prevents a current from flowing through the junction.

Placing a potential across the p-n junction counteracts the internal potential enough to allow current to pass. In direct semiconductors (semiconductors designed for optics) the electrons lose an amount of energy corresponding to a property of the semiconductor material. This loss is called the *bandgap energy*. This energy is released as light (photons) that has a wavelength related to the bandgap energy by the formula:

Eq. 5.1
$$E_g = \frac{hc}{\lambda} = \frac{1240 eV \cdot nm}{\lambda}$$

where:

λ=Photon wavelength (nm).
E_g=Energy gap (eV).

The bandgap energy and wavelength of various semiconductors are listed in Table 5.1.

TABLE 5.1

Bandgap Energy & Wavelengths of Various Semiconductors

Material	Formula	Energy Gap	Wavelength
Gallium Phosphide	GaP	2.24 eV	550 nm
Aluminum Arsenide	AlAs	2.09 eV	590 nm
Gallium Arsenide	GaAs	1.42 eV	870 nm
Indium Phosphide	InP	1.33 eV	930 nm
Aluminum Gallium Arsenide	AlGaAs	1.42-1.61 eV	770-870 nm
Indium Gallium Arsenide Phosphide	InGaAsP	0.74-1.13 eV	1100-1670 nm

LIGHT EMITTER PERFORMANCE CHARACTERISTICS

Several key characteristics of LED's and lasers determine their usefulness in a given application. These are:

Peak Wavelength: This is the wavelength at which the source emits the most power. It should be matched to the wavelengths that are transmitted with the least attenuation through optical fiber. The most common peak wavelengths are 780, 850, 1300, and 1550 nm.

Spectral Width: Ideally, all the light emitted from an LED or a laser would be at the peak wavelength, but in practice the light is emitted in a range of wavelengths centered at the peak wavelength. This range is called the spectral width of the source.

Emission Pattern: The pattern of emitted light affects the amount of light that can be coupled into the optical fiber. The size of the emitting region should be similar to the diameter of the fiber core.

Power: The best results are usually achieved by coupling as much of a source's power into the fiber as possible. The key requirement is that the output power of the source be strong enough to provide sufficient power to the detector at the receiving end, considering fiber attenuation, coupling losses and other system constraints. In general, lasers are more powerful than LED's.

Speed: A source should turn on and off fast enough to meet the bandwidth limits of the system. The speed is given according to a source's rise or fall time, the time required to go from 10% to 90% of peak power. Lasers have faster rise and fall times than LED's.

SPECTRAL CHARACTERISTICS

Spectral characteristics in LED's and lasers can be important. Figure 5.1 shows the typical optical spectra for LED's.

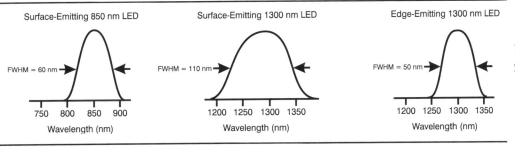

FIGURE 5.1

LED Optical Spectra

An 850 nm surface-emitting LED has a FWHM of 60 nm. FWHM stands for full width half maximum. It is a measure of the width of the optical spectrum taken at the point where the intensity falls to half of the maximum value. A surface-emitting 1300 nm LED has an even wider spectrum with a FWHM of 110 nm. This wider spectrum causes increased dispersion and also poses difficulties in WDM systems. A 1550 nm surface-emitting LED would even be wider. The last figure shows an edge-emitting 1300 nm LED. It has a much more compact spectrum with a FWHM of about 50 nm.

Figure 5.2 shows similar information for laser diodes. There are two different types of lasers shown. First, a 1300 nm Fabry-Perot laser is shown. The spectrum consists of nine discrete lines. This would properly be called a multimode laser, not referring to multimode fiber, but to the fact that the laser emits light at a number of discrete frequencies. Unlike a helium-neon (HeNe) laser which emits a single line at 633 nm, Fabry-Perot lasers emit at multiple, close-spaced wavelengths simultaneously. It can be said that light from diode lasers is less coherent than the light from a HeNe laser. This wider spectrum does cause some additional dispersion in the fiber, but it minimizes a nasty problem that may occur when using multimode fiber called *modal noise.*

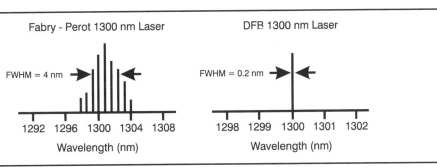

FIGURE 5.2

Laser Optical Spectra

A HeNe laser spot has a curious sparkle appearance to it. This is modal noise. It is caused by constructive and destructive interference of the light, producing dark and light spots. Highly coherent diode lasers will produce a similar speckle pattern inside an optical fiber. This speckle pattern is very easily disturbed by connectors and discontinuities in the fiber. The result is a sharp increase in system noise. Generally multimode lasers are a better choice when used with multimode fiber since they are less coherent and produce a lower contrast speckle pattern. The DFB laser, by contrast, is almost perfectly coherent emitting light at a single frequency like the HeNe laser.

LIGHT-EMITTING DIODES (LED's)

LED's are made of several layers of p-type and n-type semiconductors. A p-n junction generates the photons, and several p-p and n-n junctions direct the photons to create a focused emission of light. The p-p and n-n junctions direct the light by providing energy barriers and changes in the index of refraction. There are two main types of LED's currently being used, surface-emitters and edge-emitters.

FIGURE 5.3

Typical Packaged LED

(Photo courtesy of ABB Hafo, Inc.)

Surface-emitters are made of layers of semiconducting material that emit light in a 180° arc. They are relatively inexpensive and very reliable, but the emission pattern limits the coupling efficiency with the fiber, and therefore, the power that can be transmitted. Surface-emitters are the most economical of the two types of LED's, but they have low output power and are generally slower devices.

FIGURE 5.4

Surface-Emitting LED

(Illustration courtesy of AMP, Inc.)

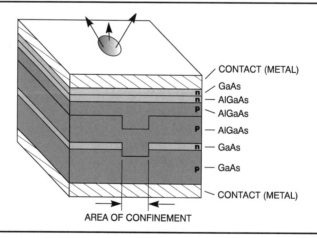

Edge-emitters are designed to confine the light to a narrow path directed out the side of the emitter. This focusing of the light means more power is emitted, and more power can be coupled to the fiber because the path is comparable to the size of the optical fiber core. Edge-emitters can provide high optical power levels (even into smaller core fibers) and are generally faster devices than surface-emitting LED's.

LED's can survive and operate in extreme temperature ranges (up to full military specifications, -55°C to +125°C), although their optical output power can drift considerably as temperature varies. In general, shorter wavelength LED's are less prone to drift with temperature. Also, surface-emitting LED's are almost always more stable over temperature than the edge-emitting type. An 850 nm LED may only drift -0.03 dB/°C, while a 1300 nm

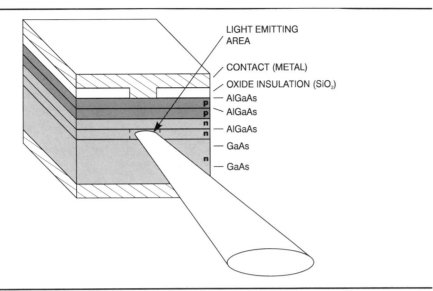

FIGURE 5.5

Edge-Emitting LED

(Illustration courtesy of AMP, Inc.)

LED may drift three to five times as much. Always, the LED's optical power drops as the temperature increases. Temperature also affects the peak emission wavelength. Most LED's exhibit a 0.3 nm/°C to 0.6 nm/°C drift in the peak emission wavelength as temperature varies. This can be important in multi-wavelength systems where there is a possibility of *crosstalk* between channels if the wavelength of the LED drifts.

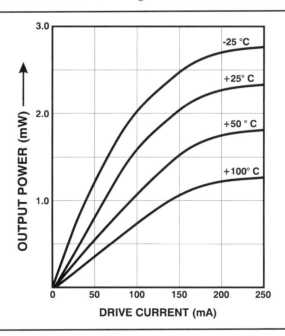

FIGURE 5.6

Typical LED
Behavior versus
Temperature

Of the two light source types, LED's are the most widely used for short system fiber optic applications. In general, LED's tend to cost less than laser diodes, so they find wider application.

A common trait of all LED types is that they turn on much faster than they turn off. This is due to the long carrier recombination lifetimes in LED's. Often, this carrier recombination will be in the tens of nanoseconds range. Most LED driver circuits employ special tricks to overcome this deficiency. These tricks generally involve applying quick reverse bias to the

LED during turnoff to more quickly sweep the carriers out of the active region. Even with these circuit tricks, LED's are limited to lower speed applications than laser diodes. Most commercial LED's have a top-end bandwidth of 100-200 MHz. A few specialty LED's are available with bandwidths as high as 600 MHz, according to manufacturer's claims, but still, LED's are only widely used for data rates below 266 Mb/s. Part of the reason is that the wider spectral bandwidth of LED's causes more dispersion in the optical fiber. This increased dispersion severely limits the maximum distance that the optical signal can be sent.

The other factor that has limited the use of LED's in high data rate applications has been the availability of low-cost "CD" lasers. These lasers are manufactured in very high volumes (over one million a month) making them very economical. While these lasers were not specifically developed for data applications, they often do very well at data rates of one gigabit per second or more.

LED Driver Circuits

LED optical output is approximately proportional to drive current. Other factors, such as temperature, also affect the optical output. Figure 5.7 shows the typical behavior of an LED. Two curves are shown. The top curve represents a 0.1% duty cycle with the peak current as shown on the horizontal axis. The bottom curve shows the output with 100% duty cycle. The drop is primarily due to the heating of the LED chip. Most LED's have light versus current curves that droop or fall below a linear curve. A few, such as super-radiant LED's (near lasers) can have curves that bow upward rather than downward.

FIGURE 5.7

Optical Output versus Current in an InGaAsP Light-Emitting Diode

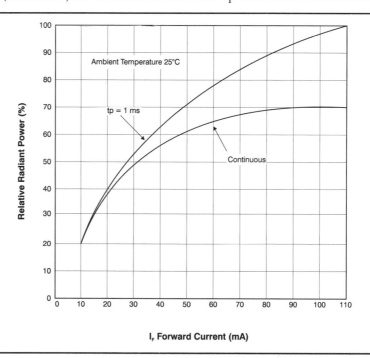

LED's are usually driven with either a digital signal or an analog signal. When the drive signal is digital, there is no concern about LED linearity. The LED is either on or off. There are special problems that need to be addressed when designing an LED driver. The key concern is driving the LED so that the maximum speed is achieved. Figures 5.8a, 5.8b and 5.8c show three popular digital LED driver circuits. The first circuit, shown in Figure 5.8a,

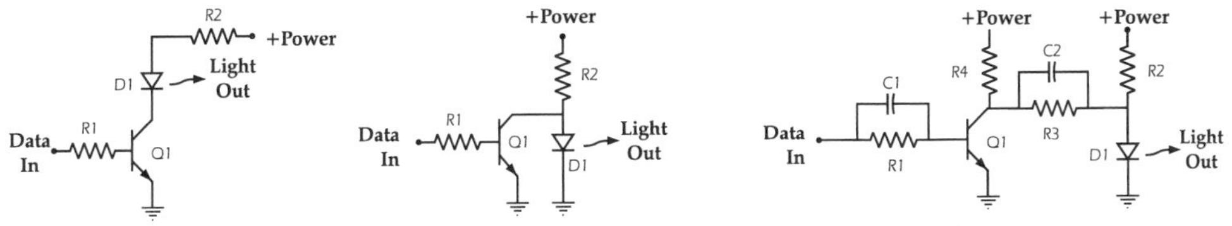

FIGURE 5.8

LED Driver Circuits

(a) Series   (b) Shunt   (c) Faster

is a simple series driver circuit. The input voltage is applied to the base of transistor Q1 through resistor R1. The transistor will either be off or on. When transistor Q1 is off, no current will flow through the LED, and no light will be emitted. When transistor Q1 is on, the cathode (bottom) of the LED will be pulled low. Transistor Q1 will pull its collector down to about 0.25 Volts. The current is equal to the voltage across resistor R2 divided by the resistance of R2. The voltage across R2 is equal to the power supply voltage less the LED forward voltage drop and the saturation voltage of the drive transistor. The key advantage of the series driver shown in Figure 5.8a is its low average power supply current. If one defines the peak LED drive current as I_{LEDmax} and assumes that the LED duty cycle is 50%, then the average power supply current is only $I_{LEDmax}/2$. Further, the power dissipated is $(I_{LEDmax}/2) \bullet V_{SUPPLY}$ where V_{SUPPLY} is the power supply voltage. The power dissipated by the individual components, the LED, transistor and resistor R1, is equal to the voltage drop across each component multiplied by $(I_{LEDmax}/2)$. The key disadvantage of the circuit shown in Figure 5.8a is low speed. This type of driver circuit is rarely used at data rates above 30-50 Mb/s. In general, there are two ways to design an LED drive circuit for low power dissipation. The first is to use a high-efficiency LED and reduce I_{LEDmax} to the lowest possible value. The second is to reduce the duty cycle of the LED to a low value. Usually larger gains can be made with the second method.

The second LED driver circuit, shown in Figure 5.8b, offers much higher speed capability. It uses transistor Q1 to quickly discharge the LED to turn it off. This circuit will drive the LED several times faster than the series drive circuit shown in Figure 5.8a. The key advantage of the shunt drive circuit is that it gives much better drive symmetry. LED's are easy to turn on quickly, but are difficult to turn off because of the relatively long carrier lifetime. In the shunt driver circuit in Figure 5.8b, resistor R2 provides a positive current to turn on the LED. Typically, R2 would be in the 40 Ω range. This makes the turn-on current about 100 mA peak. Transistor Q1 provides the turn-off current. When saturated, transistor Q1 will have an impedance of a few Ohms. This provides a much larger discharging current allowing the LED to turn off quickly. The key disadvantage of the shunt driver is the power dissipation. It is typically more than double that of the series driver. In fact, the circuit draws more current and power when the LED is off than when the LED is on! The exact power dissipation can be computed by first analyzing the off and on state currents and then combining the two values using information about the operating duty cycle.

The last driver circuit, shown in Figure 5.8c, is a variation on the shunt driver shown in Figure 5.8b. Two additional resistors and two capacitors have been added to the basic circuit. The purpose of these additional components is to further improve the operating speed. Capacitor C1 serves to improve the turn-on and turn-off characteristics of transistor Q1 itself. One has to be careful that C1 is not made too large. If this occurs, the transistor base may be overdriven and damaged. The additional components, resistors R3 and R4, and capacitor C2 provide overdrive when the LED is turned on and underdrive when the transistor is turned off. The overdrive and underdrive accelerates the LED transitions. Typically, the RC time constant of R3 and C2 is made approximately equal to the rise or fall time of the LED itself when driven with a square wave. All of these tricks together can increase the operating speed of the LED and driver circuit to about 270 Mb/s. There have been numerous laboratory tests and prototype circuits that have achieved rates to 500-1000 Mb/s, but none of these have ever made it into mass production. Typically these levels of performance require a great deal of custom tweaking on each part to achieve the high data rates.

FIGURE 5.9

Response of an LED to a Digital Modulation Signal

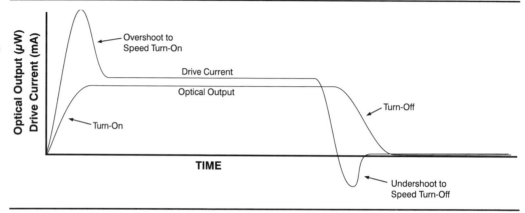

Figure 5.9 shows the response of an LED to a digital modulation signal. The electrical signal shown is the type generated by more sophisticated LED driver circuits such as that shown in Figure 5.8c. Starting at time zero, we first see the digital signal go to a logic level *1*. The most remarkable part of this event is the strong overshoot seen on the electrical drive signal. This overshoot may be two times the steady state logic *1* drive current. This overshoot accelerates the turn-on time or rise time of the LED. Even so, we see that the optical output lags behind the electrical signal. Typical values for very high-performance LED's and driver circuits would be 0.7 ns rise time of the electrical signal and 1.5 ns optical rise time. Later, when the digital signal goes back to a logic *0*, we see the same process repeated. The electrical signal has a strong undershoot component which acts to accelerate the turn-off of the LED. LED's are much harder to turn off than they are to turn on. This is because of the relatively long carrier lifetime in the active region. The undershoot serves to reverse bias the LED, sweeping out the carriers. Even so, the turn-off time of most LED's is always slower than the turn-on time. Typical values for turn-off times are 0.7 ns for the electrical signal and 2.5 ns for the optical signal. Note that while in a logic *0* state, the drive current does not quite go to zero. It is common to provide a small amount of pre-bias current, typically a few percent of the peak drive current, to keep the LED forward biased and improve dynamic response.

LASER DIODES (LD's)

There are two main types of laser diode structures, Fabry-Perot (FP) and distributed feedback (DFB). DFB lasers offer the highest performance levels and also the highest cost of the two types. They are nearly monochromatic while FP lasers emit light at a number of discrete wavelengths. DFB lasers tend to be used for the highest speed digital applications and for most analog applications because of their faster speed, lower noise, and superior linearity. Fabry-Perot lasers further break down into buried hetero (BH) and multi-quantum well (MQW) types. BH and related styles ruled for many years, but now MQW types are becoming very widespread. MQW lasers offer significant advantages over all former types of Fabry-Perot lasers. They offer lower threshold current, higher slope efficiency, lower noise, better linearity, and much greater stability over temperature. As a bonus, the performance margins of MQW lasers are so great, laser manufacturers get better yields, so laser cost is reduced. One disadvantage of MQW lasers is their tendency to be more susceptible to backreflections.

All laser diodes used for fiber optic communications incorporate a rear facet photodiode to provide a real-time means of monitoring the output of the laser. This is necessary because the threshold current of the laser changes with temperature as does slope efficiency. (Figure 5.15, later in this chapter, shows these effects.)

Figure 5.10 shows the typical optical construction of a laser diode. The laser diode chip emits light in two directions. The light from one end of the laser chip is focused onto the fiber and provides the useful output. The light from the other end falls on a large area photodiode. Usually, this photodiode is mounted some distance from the laser chip and is angled to reduce backreflections into the laser cavity.

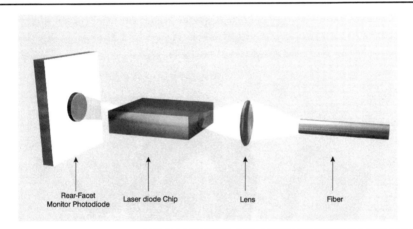

Rear-Facet Monitor Photodiode | Laser diode Chip | Lens | Fiber

FIGURE 5.10

Laser Construction

Fabry-Perot lasers are the most economical, but they are generally noisy, slower devices. DFB lasers are quieter devices (e.g., exhibit high signal-to-noise) have narrower spectral width, and are usually faster devices. Laser noise is expressed as relative intensity noise (RIN) in units of dB/Hz. Typical RIN values for good Fabry-Perot devices are -125 to -130 dB/Hz while a good DFB laser can have RIN values below -155 dB/Hz. To convert RIN values into a resultant signal-to-noise ratio, take the absolute RIN value and subtract $10\log_{10}$ (BW) where BW is the bandwidth of interest in Hertz. For instance, if the laser has a RIN of -130 dB/Hz and the system has a bandwidth of 1 MHz, the signal-to-noise ratio of the laser's optical signal would be 70 dB.

All lasers are susceptible to backreflections. Backreflections disturb the standing-wave oscillation in the laser cavity, and the net effect is an increase in the effective noise floor of the laser. A strong backreflection can cause some lasers to become wildly unstable and completely unusable in some applications. Backreflection can also generate nonlinearities in the laser response. These nonlinearities are often described as kinks. Most analog applications and some digital applications cannot tolerate these degradations. In these cases, an optical isolator is used to prevent reflections from returning into the laser cavity.

The importance of controlling backreflections depends on the type of information being sent and the particular laser. Some lasers are very susceptible to backreflections due to the design of the laser chip itself. Most often the determining factor is how tightly the fiber is coupled to the laser chip. A low-power laser generally has weak coupling to the fiber.

Perhaps only 5-10% of the laser power is coupled into the fiber. This means that only 5-10% of the backreflection would be coupled into the laser cavity, making the laser relatively immune to backreflections. On the other hand, a high-power laser may have 50-70% of the laser chip output coupled to the fiber. This also means that 50-70% of the backreflection will be coupled back into the laser cavity. This makes high-power lasers more susceptible to backreflections.

One strategy to reduce backreflections is to place an optical isolator at the laser output. An optical isolator will reduce backreflections by 20 to 45 dB by rotating the polarization of the light 45° each time it passes through the isolator and then blocking the returned light (which is now rotated 90°) with a polarizer. Unfortunately, optical isolators are not a substitute for properly polished, low-backreflection connectors. The amount of rejection offered by an optical isolator will improve problems caused by backreflections but often won't eliminate them. Optical isolators require the polarization of the light to be maintained for proper operation. Dirty or scratched connectors or short lengths of multimode fiber can destroy the polarization, making the optical isolator worthless.

One type of isolator is the Faraday rotator, a device based on the *Faraday effect,* that is used to rotate the plane of polarization of a light wave. Figure 5.11 illustrates this isolator.

FIGURE 5.11

Isolator Based on
Faraday Rotator

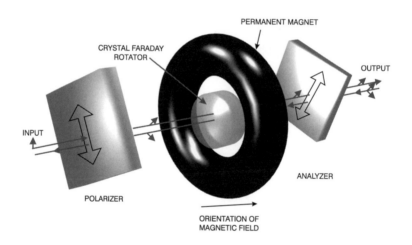

Before entering the Faraday rotator, which is usually an yttrium-iron-garnet (YIG) material, the light beam passes through a polarizer which is oriented parallel to the incoming state of polarization. The Faraday rotator then rotates the polarization by 45°. At the output, the beam passes an analyzer which is oriented at an angle of 45° relative to the first polarizer. Of all possible reflected beams, only those with a 45° orientation of the polarization are allowed to pass backwards. The polarization of the reflected beam is rotated by another 45° which results in a total rotation of 90°. This way, the reflected beam is blocked by the polarizer. In order not to disturb the proper function of the isolator, all its surfaces should be antireflection-coated.

Backreflection

What are the effects of backreflections? Figure 5.12 shows a clean laser waveform and Figure 5.14 shows a laser waveform with a strong backreflection. First, the waveform in Figure 5.12 needs some explanation.

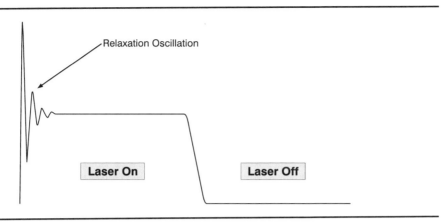

FIGURE 5.12

Laser Optical
Output With No
Backreflection

As seen in the figure, the rising edge is followed by a damped oscillation. This overshoot and the subsequent oscillation is called the relaxation oscillation. Most lasers exhibit this phenomenon. It can be understood by looking at the frequency versus amplitude response of the laser as illustrated in Figure 5.13.

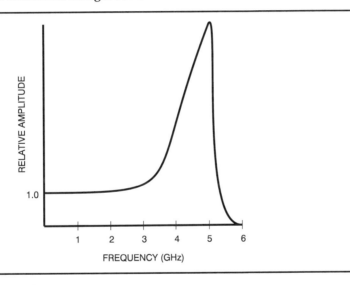

FIGURE 5.13

Frequency versus
Amplitude
Response of a Laser

The laser exhibits a resonance (high gain point) near 5 GHz in this case. This will be the approximate frequency of the relaxation oscillation. The frequency of the resonance peak is one of the factors that limits the maximum data rate that can be transmitted by a given laser. When the maximum frequency component of the data stream gets within a factor of two or three of the laser resonance frequency, performance usually degrades quickly. The frequency of the resonance and the magnitude of the overshoot depends on the drive levels applied to the laser. Overshoot is generally most severe when the laser is turned completely off and then back on. This condition is avoided in most practical data links.

The backreflection illustrated in Figure 5.14 shows the same characteristics as the initial relaxation oscillation. The time, T_1, can be precisely measured to determine the distance to the reflecting point. Often, the frequency has to be greatly reduced to observe such reflections. A fallacy is that only close reflections matter. Closer reflections are often a bit stronger, but at 1300 nm or 1550 nm, fiber loss is so low that reflections from several kilometers distance can be significant. Another useful trick is to monitor the photodiode servo loop to

FIGURE 5.14

Laser Optical
Output With
Backreflection

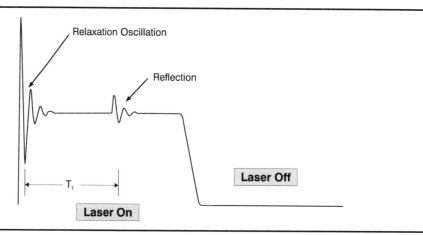

look for disturbances. A starting state is to place the end of the laser pigtail in some glycerin. This will eliminate virtually all backreflections. Note the output of the servo loop at that time. Then connect the laser pigtail into the system. If any significant perturbations are noted, there are backreflections of sufficient amplitude to disturb the standing wave in the laser cavity. This is a relatively direct method of observing laser backreflections. A less direct way to test for laser backreflections is to test at frequencies where backreflections will occur at an exact multiple of the bit time. Basically, the procedure is to calculate the round trip transit time to the potential reflection interfaces in the system.

It is generally easiest to measure the spacing between the high interference points when using this method. Often, the laser pigtail is made very short so that the first reflection occurs at a frequency higher than any frequency being transmitted by the system. However, longer fiber segments in the system will yield a low fundamental interference frequency and harmonics that will clutter the spectrum. The only practical strategy with laser-based systems is to design for low backreflections.

Temperature Effects on Lasers

Lasers can survive wide temperature ranges (up to full industrial specifications, -40°C to 85°C). Temperature affects the peak emission wavelength as well as the threshold current and the slope efficiency of the laser. Most lasers exhibit a 0.3-0.6 nm/°C drift in the peak emission wavelength as temperature varies. This can be important in multi-wavelength systems where there is a possibility of crosstalk between channels if the laser's wavelength drifts. Optical output power can be relatively stable over a wide temperature range because of the servo-loop that typically stabilizes the output based on the internal monitor photodiode. The internal photodiode monitors the rear facet of the laser chip while the front facet output is coupled to the optical fiber. If the front and rear facet outputs are perfectly correlated and the mechanical assembly is unaffected by temperature, then the optical output power would be perfectly stable as temperature varied. Neither is true, however, so the output power is affected by temperature. This effect is called tracking error. It is generally less than a few dB's over the temperature range, although its direction and magnitude vary widely from part to part.

Generally, laser optical output is approximately proportional to the drive current above the threshold current. Below the threshold current, the output is from the LED action of the device. Above the threshold, the output dramatically increases as the laser gain increases. Figure 5.15 shows the typical behavior of a laser diode. As operating temperature changes, several effects can occur. First, the threshold current changes. The threshold current is always lower at lower temperatures and vice versa. The second change that can be important is the slope efficiency. The slope efficiency is the number of milliwatts or microwatts of light output per milliampere of increased drive current above threshold. Most lasers show a drop in slope efficiency as temperature increases. Figure 5.15 shows that the 50°C curve is not as steep as the curve at the lower temperatures.

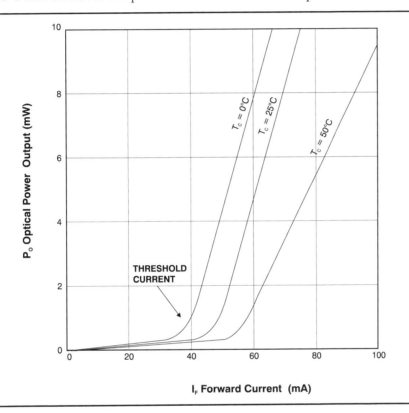

FIGURE 5.15

Laser Optical Power Output versus Forward Current

Peltier Coolers

A decade ago, most lasers used in fiber optic links were cooled devices. Peltier coolers were used to cool the lasers. These devices are specialized semiconductors that act like miniature heat pumps, similar in function to those found in many homes. They are built from stacks of p and n elements made from doped bismuth telluride ($BiTe_2$). Peltier coolers have the property that when current passes through them, they pump heat from one end to the other. The process is fully reversible. If one reverses the direction of current flow through the elements, the direction of heat transfer is reversed as well.

Small Peltier coolers were used to heat or cool the laser chip so that it could be maintained at a constant temperature. The laser package also generally included a thermistor so that the laser temperature could be accurately measured along with the customary rear-facet monitor photodiode. Cooled lasers were prevalent in those early days because long-term laser stability and life could only be achieved by maintaining the laser at tightly controlled temperatures. Today, the most reliable lasers are uncooled. Cooled lasers are now much less reliable than uncooled lasers because of the fragile nature of the Peltier cooler elements. Curiously, cooled lasers cannot cope with as wide a temperature range as uncooled lasers. This is true for two reasons. First, single-stage Peltier coolers, the type most often used with lasers, can only maintain about a 40°C temperature delta. Second, the Peltier cooler elements are usually assembled using low melting point solder which limits their top end temperature range.

Today cooled lasers are used for two main purposes. Used in conjunction with very high-power lasers, they remove heat from the laser chip, and they assure wavelength stability of the laser. It was mentioned earlier that the center wavelength of a laser (or LED) varies by about 0.3-0.6 nm/°C. This wavelength variation can cause three problems. First, it can affect the bandwidth that can be achieved over very long lengths of fiber. Long-haul systems often try to operate as close to the zero-dispersion wavelength of the fiber as possible. A drift of a few nanometers can cause the fiber's bandwidth to drop by a factor of ten or more. Second, it can affect the attenuation of the fiber over very long distances. This is usually only a problem at the 1550 nm wavelength. Last, laser center wavelength drift can cause unacceptable levels of crosstalk in wavelength-division multiplexed (WDM) systems with catastrophic results.

Figure 5.16 shows a functional drawing of a single-stage, multi-element Peltier cooler. The

FIGURE 5.16

Cross-Section of a Single-stage, Multi-element Peltier Cooler

(Illustration courtesy of Melcor Corp.)

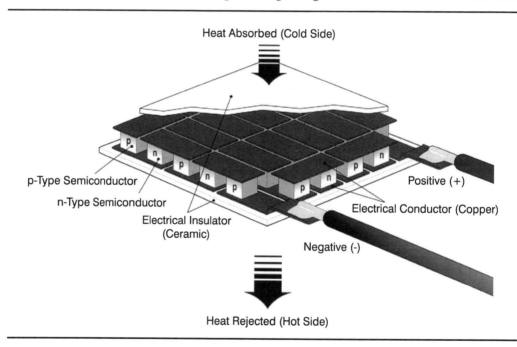

heat flow will be in the direction shown as long as the electrical polarity is as shown. If the electrical polarity is reversed, the direction of heat flow will reverse as well. The amount of heat pumped from one surface to the other depends on the amount of current flowing through the cooler. Typical laser and detector Peltier coolers require ±1 Amp at 1 to 2 Volts maximum.

Because of the way Peltier coolers work, good thermal design is essential. There has to be adequate heat sinking available on the non-laser side of the device, otherwise the Peltier cooler will overheat and destroy itself. Figure 5.17 shows a Peltier cooler mounted in a 14-pin laser package.

FIGURE 5.17

A Peltier Cooler
Mounted in a 14-
Pin Laser Package

(Photo courtesy of Melcor Corp.)

LASER SAFETY

While laser-based fiber optic products are in widespread use and present no hazard to personnel when handled properly, there are a few basic rules that must be observed to limit exposure to laser radiation. Laser radiation will damage eyesight under certain conditions. The following guidelines are important in laser safety:

- **Always** read the product data sheet and the laser safety label before applying power. Note the operating wavelength, optical output power, and safety classification.

- If safety goggles or other eye protection are used, be certain that the protection is effective at the wavelength(s) emitted by the device under test before applying power.

- **Always** connect a fiber to the output of the device **before** power is applied. Under no circumstances should the device ever be powered if the fiber output is unattached. If the device has a connector output, a connector that is connected to a fiber should be attached. This ensures that all light is confined within the fiber waveguide, virtually eliminating all potential hazard.

- **NEVER** look in the end of a fiber to see if light is coming out. Most fiber optic laser wavelengths (1300 nm and 1550 nm) are totally invisible to the unaided eye and will cause permanent damage. Shorter wavelength lasers (e.g. 780 nm) are visible and are potentially very damaging. Always use instrumentation, such as an optical power meter, to verify light output.

- **NEVER NEVER** look into the end of a fiber on a powered device with any sort of magnifying device. This includes microscopes, eye loupes, and magnifying glasses. This **will** cause a permanent, irreversible burn on the eye's retina. Always double check that power is disconnected before using such devices. If possible, completely disconnect the unit from any power source.

Laser safety classes are as follows in Table 5.2:

Class	Wavelength Range	Optical Power Accession Limits
I	180 nm to 10^6 nm	Varies with λ and exposure time.
IIa	400 nm to 710 nm	3.9×10^{-6} W (3.9 microwatts)
II	400 nm to 710 nm	1.0×10^{-3} (1.0 milliwatt)
IIIa	400 nm to 710 nm	5.0×10^{-3} (5.0 milliwatts)
IIIb	180 nm to 400 nm	Varies with λ and exposure time.
	400 nm to 10^6 nm	0.5 Watts

TABLE 5.2

Laser Safety Classes

Figure 5.18 illustrates typical laser warning labels used on the product packaging to identify laser-based products.

FIGURE 5.18

Typical Laser
Warning Labels

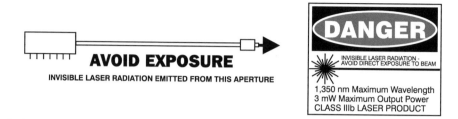

COMPARISON OF LED'S AND LASER DIODES

Table 5.3 offers a quick comparison of some of the characteristics for lasers and LED's.

TABLE 5.3

Comparison of
Light Emitters

Parameter	Light-Emitting Diode (LED)	Laser Diode (LD)
Output Power	Linearly proportional to drive current.	Proportional to current above the threshold.
Current	Drive Current: 50 to 100 mA Peak	Threshold Current: 5 to 40 mA
Coupled Power	Moderate	High
Bandwidth	Moderate	High
Wavelengths Available	0.66 to 1.55 μm	0.78 to 1.55 μm
Emission Spectrum	40 nm to 190 nm FWHM	0.1 nm to 10 nm FWHM
Cost	$5 to $300	$5 to $3,000

Both types of light sources use the same key materials. GaAlAs (gallium aluminum arsenide) is commonly used for short-wavelength devices. Long-wavelength devices generally incorporate InGaAsP (indium gallium arsenide phosphide).

Linearity is an important characteristic to both types of light sources for some applications. Linearity represents the degree to which the optical output is directly proportional to the electrical current input. Most light sources give little or no attention to linearity, making them usable only for digital applications. Analog applications require close attention to linearity. Nonlinearity in LED's and lasers causes harmonic distortion in the analog signal that is transmitted over an analog fiber optic link.

LED's are generally more reliable than lasers, but both sources will degrade over time. This degradation can be caused by heat generated by the source and uneven current densities. In addition, LED's are easier to use than lasers. Lasers are temperature sensitive; the lasing threshold will change with the temperature. Thus, lasers require a method of stabilizing the threshold to achieve maximum performance. Often, a photodiode is used to monitor the light output on the rear facet of the laser. The current from the photodiode changes with variations in light output and provides feedback to adjust the laser drive current.

Another issue in selecting LED's or lasers is the packaging. The main concern with packaging is how the light source couples with the fiber. One method uses a microlensed device. A small drop of epoxy is placed directly on the chip. The epoxy bead focuses the light in a uniform spot. The fiber, which usually has a smaller diameter than the bead, can be placed anywhere on the epoxy drop and will receive the same amount of optical energy. This allows for more efficient coupling of the fiber to the source. Pigtails use a short length of fiber as part of the optical device. Pigtailing can improve coupling efficiency by bringing the fiber end closer to the emitting area of the chip. The light is coupled into the fiber before it has a chance to spread out.

LIGHT EMITTERS AS DETECTORS

All LED's and lasers have the ability to act as detectors, but a few perform this task much better than others. The key parameter to look for is very efficient coupling between the light emitter and the fiber. This allows good performance in both modes. It is also important that the LED's have consistent spectral characteristics. In ping-pong LED's or full-duplex LED's the LED is used intermittently as a light emitter then as a light detector. In this way, information can be sent in either direction over the fiber. Some sophisticated techniques exist that combine this simple idea with time compression and an elaborate synchronizing protocol that appears to allow information to be sent both ways at the same time. In fact, information only travels in one direction at a time. The interchange back and forth is so quick that it seems to be simultaneous.

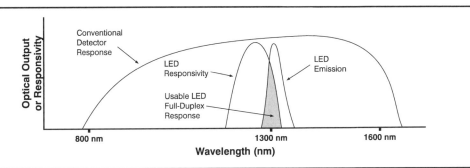

FIGURE 5.19

Ping-Pong LED Operation

Now that we've discussed light emitters, it is time to move to the components at the other end of the fiber, light detectors.

Chapter Summary

• Light emitters convert electrical signals into corresponding light signals that can be injected into a fiber.

• Laser diodes (LD's) and light-emitting diodes (LED's) are the two types of light emitters in widespread use in modern fiber optic systems.

• LD's and LED's both use the principle of the p-n semiconductor junction found in transistors and diodes.

• Some characteristics that determine the usefulness of LD's and LED's are peak wavelength, spectral width, emission pattern, power, and speed,.

• The two main types of LED's currently being used are surface-emitters and edge-emitters.

• Surface-emitters are the most economical of the two types of LED's, but they have low output power and are generally slower devices.

• Edge-emitters can produce high optical power levels and are generally faster than surface-emitters.

• A common trait of all LED types is that they turn on much faster than they turn off.

• The two main types of laser diode structures are the Fabry-Perot (FP) and distributed feedback (DFB).

• Fabry-Perot lasers emit at multiple, close-spaced wavelengths simultaneously.

• All lasers are susceptible to backreflections.

• The importance of controlling backreflections depends on the type of information being sent and the particular laser.

- Peltier coolers are specialized semiconductors that act like miniature heat pumps.
- Today, Peltier coolers are used in conjunction with very high-power lasers to remove heat from the laser chip and assure wavelength stability.
- Laser radiation can damage eyesight under certain conditions.
- LED's and LD's can survive and operate in extreme temperature ranges.
- Of LED's and LD's, LED's are the most widely used for fiber optic systems, mainly due to their lower cost.
- LED's and LD's are usually driven with either a digital signal or an analog signal.
- Ping-pong, or full-duplex LED's are used intermittently as light emitters, then light detectors, enabling information to be sent both ways over the fiber.

Selected References and Additional Reading

Baack, Clemens. 1986. *Optical Wideband Transmission Systems.* Florida: CRC Press, Inc.

—. 1982. *Designers Guide to Fiber Optics.* Harrisburg, PA: AMP, Incorporated.

Kopp, Greg and Laura A. Pagano. 1995. "Polarization Put in Perspective." *Photonics Spectra.* February 1995: 103-107.

Palladino, John R. 1990. *Fiber Optics: Communicating By Light.* Piscataway, NJ: Bellcore.

Rockwell, R. James, James F. Smith, and William J. Ertle. 1995. "Playing it Safe with Industrial Lasers." *Photonics Spectra.* February 1995: 118-124.

Sterling, Donald J. 1993. *Amp Technician's Guide to Fiber Optics*, 2nd Edition. New York: Delmar Publishers.

Yeh, Chai. 1990. *Handbook of Fiber Optics: Theory and Applications.* New York: Academic Press, Inc.

LIGHT DETECTORS

Light detectors perform the opposite function of light emitters. Emitters, as we already know, are electro-optic devices. They convert electrical pulses into light pulses. Detectors are opto-electric devices. They enable the optical signal to be converted back into electrical impulses that are used by the receiving end of the fiber optic data, video, or audio link. The most common detector is the semiconductor photodiode, which produces current in response to incident light.

In an LED, the energy emitted during the recombination of electron-hole pairs is in the form of light. In a photodiode, the opposite phenomenon occurs. Light striking the photodiode creates a current in the external circuit. Absorbed photons excite the electrons and the result is the creation of an electron-hole pair. For each electron-hole pair created, an electron is set flowing as current in the external circuit. Like light emitters, detectors operate based on the principle of the p-n junction. An incident photon striking the diode gives an electron in the valence band sufficient energy to move to the conduction band, creating a free electron and a hole. If the creation of these carriers occurs in a depleted region, the carriers will quickly separate and create a current. As they reach the edge of the depleted area, the electrical forces diminish and current ceases. While the p-n diodes are insufficient detectors for fiber optic systems, both PIN photodiodes and avalanche photodiodes (APD's) are designed to compensate for the drawbacks of the p-n diode.

IMPORTANT PHOTODETECTOR PARAMETERS

Responsivity

The *responsivity* of a photodetector is the ratio of the current output to the light input. Other factors being equal, the higher the responsivity of the photodetector, the better the sensitivity of the receiver. Since responsivity varies with wavelength, it is specified either at the wavelength of peak responsivity or at a wavelength of interest. For most applications, responsivity is the most important characteristic of a detector because it defines the relationship between optical input and electrical output. The theoretical maximum responsivity is about 1.05 A/W at a wavelength of 1300 nm. Commercial InGaAs detectors provide typical responsivity of 0.8 to 0.9 A/W at a wavelength of 1300 nm.

The theoretical maximum responsivity of a photodetector occurs when the *quantum efficiency* of the detector is 100%. Responsivity and quantum efficiency (η) are related by:

Eq. 6.1
$$R = \frac{\eta \bullet \lambda}{1240}$$

where:

R = Theoretical maximum responsivity in Amps/Watt.
η = Quantum efficiency.
λ = Wavelength in nanometers.

Thus a detector at 1300 nm would have a theoretical maximum responsivity of 1.05 A/W and a detector at 850 nm would have a theoretical maximum responsivity of 0.68 A/W.

FIGURE 6.1

Typical Spectral
Response of Various
Detector Materials

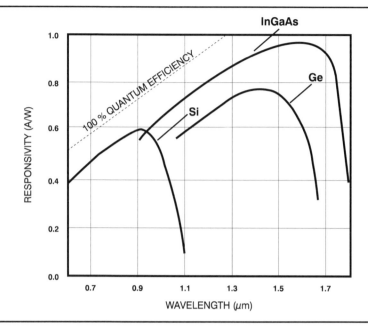

Quantum Efficiency

Quantum efficiency is the ratio of primary electron-hole pairs created by incident photons to the photons incident on the diode material. A quantum efficiency of 100% means that every absorbed photon creates an electron-hole pair. A typical quantum efficiency for a commercial detector is 70% to 90%. This quantum efficiency rating indicates that seven to nine out of ten photons will create carriers. Factors that prevent the quantum efficiency from being 100% include coupling losses from the fiber to the detector, absorption of light in the p or n region, and leakage currents in the detector.

Capacitance

The capacitance of a detector is dependent upon the active area of the device and the reverse voltage across the device. A small active diameter allows for lower capacitance. However, as the active diameter decreases, it becomes harder to align the fiber to the detector. This is complicated by the fact that photodiode response is slower at the edges of the active area. If the edges are illuminated, a slow response component will be present, increasing edge jitter. It is important to illuminate only the center region of the active area to minimize this effect.

Photodiode capacitance decreases with increasing reverse voltage. Figure 6.2 shows a typical C-V curve for a high-speed photodiode. This curve shows that as the reverse voltage is increased beyond 5 or 6 Volts, the decrease in capacitance becomes minimal. At this point the detector is said to be fully depleted. Higher reverse voltages also speed up the detector. Excessive reverse voltage may increase detector noise in some cases.

RESPONSE TIME

Response time represents the time needed for the photodiode to respond to optical inputs and produce an external current. The combination of the photodiode capacitance and the load resistance, along with the design of the photodiode, sets the response time. As with light emitters, detector response time is specified as rise time or fall time, and it is measured between 10% and 90% points of amplitude. The response time of a diode relates to its usable bandwidth.

FIGURE 6.2

Capacitance versus
Reverse Voltage

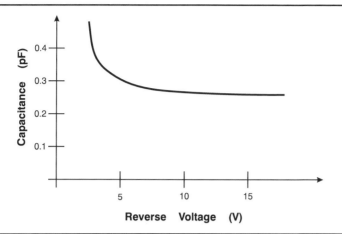

Response time is influenced by the design of the photodiode as well as its application parameters. For instance, the size of a detector (active area, usually expressed as a diameter in mm or μm) directly influences its capacitance. The applied reverse voltage decreases the capacitance and speeds response. The impedance that the detector operates into also affects the response time. The approximate -3 dB frequency of a detector is given in Equation 6.2.

$$\text{Eq. 6.2} \qquad f_{-3dB} = \frac{1}{(2 \bullet \pi \bullet R \bullet C)}$$

where:

R = Impedance that the detector operates into.
C = Capacitance of the detector.

The 10-90% rise or fall time of the detector can also be estimated from Equation 6.3.

$$\text{Eq. 6.3} \qquad \tau = 2.2 \bullet R \bullet C$$

Equation 6.4 relates rise and fall time to -3 dB bandwidth.

$$\text{Eq. 6.4} \qquad \tau = \frac{0.35}{f_{-3dB}} \quad \text{or} \quad f_{-3dB} = \frac{0.35}{\tau}$$

As an example, consider a high-speed detector with a capacitance (C) of 0.5 pF operating into an impedance (R) of 50 Ω. This detector would exhibit a rise time of 55 ps and a -3 dB frequency of 6.4 GHz. Keep in mind that other detector factors could limit the performance to lower values. The calculated values should be considered the best that could be achieved.

Dark Current

This is one of the worst terms that has been ever been conceived for a phenomenon. It implies that somehow the detector manages to put out a current when there is no light. What really happens is that a current flows through the detector in the absence of light because of the intrinsic resistance of the detector and the applied reverse voltage. The voltage acting on the bulk resistance of the detector causes a small current to flow. This current is very temperature sensitive and may double every 5°C to 10°C. *Dark current* contributes to the detector noise and also creates difficulties for DC coupled amplifier stages.

Linearity & Backreflection

All PIN diodes are inherently linear devices. However, for the most demanding applications, such as multichannel CATV links, special care must be taken to reduce distortion to very low levels. These so-called analog PIN detectors often have distortion products below -60 dB. Another factor that is very important for analog applications is the backreflection of the detector. Generally the fiber is coupled to the detector at a perpendicular angle. For low backreflection detectors, the detector may be tilted by 7° to 10° as shown in Figure 6.3 below.

FIGURE 6.3

Low Backreflection
Detector Alignment

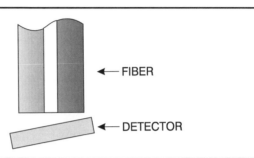

Noise

Noise is an ever-present phenomenon that limits a detector's performance. It is any electrical or optical energy other than the signal itself. Noise appears in all elements of a communication system; however, it is usually most critical to the receiver. This is because the receiver is trying to interpret an already weak signal. The same noise in a transmitter is usually insignificant because the signal at the transmitter is much stronger than the attenuated signal that the receiver picks up. *Shot noise* occurs because the process of creating the current is a set of discrete occurrences rather than a continuous flow. As more or less electron-hole pairs are created, the current fluctuates, creating shot noise. Shot noise occurs when no light falls on the detector. Even without light, a small current, dark current, is thermally generated. Noise increases with current and bandwidth. When dark current only is present, noise is at a minimum, but it increases with the current resulting from optical input.

A second type of noise, *thermal noise*, arises from fluctuations in the load resistance of the detector. The electrons in the resistor are not stationary, and their thermal energy allows them to move about. At any given moment, the net movement toward one electrode or the other generates random currents that add to and distort the signal current from the photodiode.

Shot noise and thermal noise exist in the receiver independent of the arriving optical power. They are a result of matter itself, but they can be minimized. A rule of thumb is that the signal power should be ten times that of the noise or the signal will not be adequately detected. This signal quality can be expressed as a *signal-to-noise ratio (SNR)*. A large SNR means that the signal is much larger than the noise. Different applications require different SNRs. This signal-to-noise ratio is treated further in Chapter 9, "System Design Considerations."

PIN PHOTODIODE

A p-n diode's deficiencies are related to the fact that the depletion area (active detection area) is small; many electron-hole pairs recombine before they can create a current in the external circuit. In the PIN photodiode, the depleted region is made as large as possible. A lightly doped intrinsic layer separates the more heavily doped p-types and n-types. The diode's name comes from the layering of these materials Positive, Intrinsic, Negative - PIN.

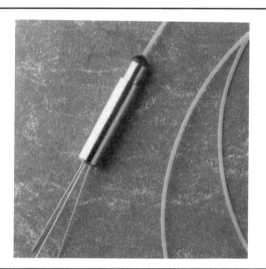

FIGURE 6.4

PIN Photodiode for CATV Applications

(Photo courtesy of Epitaxx.)

In the absence of light, PIN photodiodes behave electrically just like an ordinary rectifier diode. If forward biased, they conduct large amounts of current. The forward turn-on voltage is related to the energy gap of the detector. For silicon, this energy gap is around 1.1 eV (electron Volts). For InGaAs, the energy gap is 0.77 eV and for germanium, the energy gap is around 0.65 eV. Figure 6.5 shows the cross section and operation of a PIN photodiode.

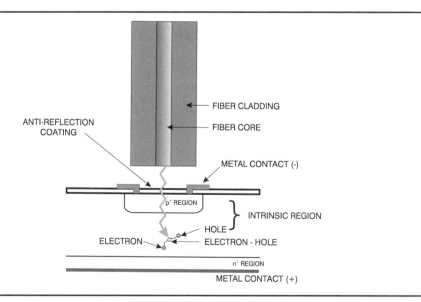

FIGURE 6.5

Cross-Section and Operation of a PIN Photodiode

PIN detectors can be operated in two modes: *photovoltaic* and *photoconductive*. In the photovoltaic mode, no bias is applied to the detector. In that case the detector will be very slow and the detector output is a voltage that is approximately logarithmic to the input light level. Real-world fiber optic receivers never use the photovoltaic mode. In the photoconductive mode, the detector is reverse biased. The output in this case is a current that is very linear with the input light power. PIN detectors can be linear over seven or more decades of input light intensity.

IDP DETECTORS

An alternative to the PIN photodiode is known as an integrated detector/preamplifier (IDP). Noise that limits the operation of the receiver can occur between the diode and the first receiver stage. To reduce this noise an IDP may be used. An IDP is an integrated circuit that has both a detector and a transimpedance amplifier (see discussion below). Characteristic specifications for IDP's are similar to those of PIN detectors or APD's. One difference is that the output of an IDP is voltage, so responsivity is specified in Volts/Watt (V/W). Typical responsivity for an IDP is 4000 V/W.

Transimpedance Amplifier

These devices are used to improve the performance of a detector in a fiber optic system. The transimpedance amplifier (TIA) takes the photodiode output current, multiplies it by the transimpedance gain, A_z, and outputs a voltage signal. The TIA must have sufficient bandwidth to handle the maximum system data rate. It should also have high gain and low noise to maximize receiver sensitivity. Of course, as the bandwidth or the gain increases, the noise floor will increase. A TIA with differential outputs is very beneficial to receiver sensitivity because of the differential input stage of the decision circuit. This effectively adds 3 dB of gain without a noise penalty. Table 6.1 shows some typical characteristics of a transimpedance amplifier.

TABLE 6.1

TIA Characteristics

Transimpedance Gain	10,000 Ω
Bandwidth	150 MHz
Input Noise Current Spectral Density	2.9 pA/$\sqrt{Hz}$
Output Stage	Single-Ended

AVALANCHE PHOTODIODE (APD)

Another type of detector is the avalanche photodiode (APD). For PIN diodes, each absorbed photon ideally creates one electron-hole pair that sets one electron flowing in the external circuit. In this one-to-one relation, PIN photodiodes resemble LED's. Lasers offer a higher-than-one-to-one ratio of photons to carriers, and current: one primary carrier can result in the emission of several photons. By this comparison, APD's are like lasers. The primary carriers, the free electrons and holes created by absorbed photons, accelerate, gaining several electron Volts of kinetic energy. A collision of these fast carriers with neutral atoms causes the accelerated carriers to use some of their own energy to move the bound electron out of the valence shell. Free electron-hole pairs appear; these pairs are known as the secondary carriers. Collision ionization is the name for the process that creates these secondary carriers. As primary carriers create secondary carriers, the secondary carriers themselves accelerate and create new carriers. Collectively, this process is known as photomultiplication. Typical multiplication ranges in the tens and hundreds. A multiplication factor of eighty means that, on average, eighty external electrons flow for every photon of light absorbed. This factor is a statistical average—the actual number of electrons generated per photon absorbed may be more or less than the multiplication factor.

APD's are of interest because of the inherent gain that they provide. Because they include gain, the electronics amplifier chain can be simplified. However, as in most engineering trade-offs, there is a price. For APD's the price is heavy. APD's require high-voltage power supplies for their operation. The voltage can range from 20 or 30 Volts for InGaAs APD's to over 300 Volts for Si APD's. This adds circuit complexity. Also, APD's are very temperature sensitive, further increasing circuit complexity. In general, APD's are only useful for digital systems because they possess very poor linearity. All of these factors combined weigh heavily against the use of APD's. Because of the added circuit complexity and the high voltages that the parts are subjected to, they are almost always less reliable than PIN detectors. This added to the fact that PIN detector-based receivers can almost match the performance of an APD make PIN detectors the first choice for most deployed systems.

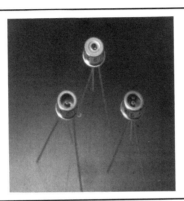

FIGURE 6.6

Typical Packaged APD's

(Photo courtesy of Epitaxx.)

COMPARISON OF PIN PHOTODIODES AND APD'S

Selecting a detector depends on the application involved. Other factors such as cost and bandwidth must also be taken into consideration. Table 6.2 offers a comparison of PIN photodiodes and APD's.

Parameter	PIN Photodiodes	APD's
Construction Materials	Silicon, Germanium, InGaAs	Silicon, Germanium, InGaAs
Bandwidth	10 MHz to 20 GHz	100 MHz to 3 GHz
Wavelength	0.6 to 1.8 μm	0.6 to 1.8 μm
Conversion Efficiency	0.5 to 1.0 Amps/Watt	0.5 to 100 Amps/Watt
Support Circuitry Required	None	High Voltage and Temperature Stabilization
Cost (Fiber Ready)	$1 to $500	$100 to $2000

TABLE 6.2

Comparison of PIN Photodiodes and APD's

As with light emitters, coupling the detector to the fiber is an important consideration. Detectors are packaged in the same ways as light emitters, using microlensed devices or pigtailing the detector.

Chapter Summary

- Light detectors enable an optical signal to be converted back into electrical impulses that are used by the receiving end of the fiber optic data, video or audio link.

- The most common detector is the semiconductor photodiode, which produces current in response to incident light.

- Like light emitters, detectors operate based on the principle of the p-n junction.

- The responsivity of a photodetector is the ratio of the current output to the light input.

- For most applications, responsivity is the most important characteristic of a detector because it defines the relationship between optical input and electrical output.

- Quantum efficiency is the ratio of primary electron-hole pairs created by incident photons to the number of photons incident on the diode material.

- The theoretical maximum responsivity of a photodetector occurs when the quantum efficiency of the detector is 100%.

- The capacitance of a detector is dependent upon the active area of the device, the design of the junction, and the reverse voltage across the device.

- Photodiode capacitance decreases with increasing reverse voltage.

- Dark current refers to the flow of current through a detector in the absence of light.

- The combination of the photodiode capacitance and the load resistance, along with the design of the photodiode, sets the time needed for the photodiode to respond to optical inputs and produce an external current.

- Noise is an ever-present phenomenon that limits a detector's and a system's performance.

- Both shot noise and thermal noise exist in the receiver independent of the arriving optical signal.

- PIN photodiodes and avalanche photodiodes (APD's) are designed to compensate for the drawbacks of the p-n diode.

- PIN photodiodes behave electrically just like an ordinary rectifier diode in the absence of light.

- An IDP is an integrated circuit that has both a detector and a transimpedance amplifier.

- When speaking of APD's, the term collision ionization is the name for the process that creates secondary carriers.

- APD's require high-voltage power supplies for their operation, ranging from 20 to 30 volts for InGaAs APD's, to over 300 volts for Si APD's.

- Because of the many shortcomings of APD's, PIN detectors are the first choice for most deployed systems. Selecting a detector depends on the application involved, cost, and bandwidth.

Selected References and Additional Reading

Baack, Clemens. 1986. *Optical Wideband Transmission Systems.* Florida: CRC Press, Inc.

—. 1982. *Designers Guide to Fiber Optics.* Harrisburg, PA: AMP, Incorporated.

Hentschel, Christian. 1988. *Fiber Optics Handbook.* 2nd edition. Germany: Hewlett-Packard Company.

Palladino, John R. 1990. *Fiber Optics: Communicating By Light.* Piscataway, NJ: Bellcore.

Yeh, Chai. 1990. *Handbook of Fiber Optics: Theory and Applications.* New York: Academic Press, Inc.

INTERCONNECTION DEVICES

Interconnection devices refer to any mechanism or technique used to join an optical fiber to another fiber or to a fiber optic component. The most common interconnection device is the *connector*. Connectors were once the most difficult aspect of the commercialization of fiber optics. Today nothing could be further from the truth. Fiber optic connectors have gone through several development generations in a few years and are now mature and highly reliable devices. The only drawback at this time is the almost bewildering number of connectors to choose from. Table 7.1 summarizes the evolution of fiber optic connectors.

Parameter/Feature	1st Generation	2nd Generation	3rd Generation
Coupling Method	Threaded	Bayonet	Push-Pull
Ferrule Material	Steel, Brass	Steel, Ceramic	Ceramic, Plastic
Alignment Sleeve	Often Loose	Captive	Captive
Sleeve Material	Plastic	Beryllium, Copper	Beryllium, Copper, Ceramic
Body Material	Metal	Metal, Plastic	Metal, Plastic
Rotation Prevention	No	Yes	Yes
Repeatability	Poor	Good	Very Good
Installation Ease	Poor	Good	Very Good
Insertion Loss	High	Moderate	Low
Backreflection	Not Addressed	Moderate	Low
Cost	High	Moderate	Low
Multimode Use	Very Good	Very Good	Very Good
Single-mode Use	Unusable	Good	Very Good
Example	SMA 906	ST™	SC

TABLE 7.1
Connector
Evolution

OPTICAL CONNECTOR BASICS

Fiber optic connector types are as various as the applications in which they are used. Different connector types have different characteristics, advantages, and disadvantages, and performance parameters. But all connectors have the same four basic components.

The Ferrule: The fiber is mounted in a long, thin cylinder, the *ferrule*, which acts as a fiber alignment mechanism. The ferrule is bored through the center at a diameter that is slightly larger than the diameter of the fiber cladding. The end of the fiber is located at the end of the ferrule. Ferrules are typically made of metal or ceramic, but they may also be constructed of plastic.

The Connector Body: Also called the connector housing, the connector body holds the ferrule. It is usually constructed of metal or plastic and includes one or more assembled pieces which hold the fiber in place. The details of these connector body assemblies vary among connectors, but bonding and/or crimping is commonly used to attach strength members and cable jackets to the connector body. The ferrule extends past the connector body to slip into the coupling device.

The Cable: The cable is attached to the connector body. It acts as the point of entry for the fiber. Typically, a strain-relief boot is added over the junction between the cable and the connector body, providing extra strength to the junction.

The Coupling Device: Most fiber optic connectors do not use the male-female configuration common to electronic connectors. Instead, a coupling device such as an alignment sleeve is used to mate the connectors. Similar devices may be installed in fiber optic transmitters and receivers to allow these devices to be mated via a connector. These devices are also known as feed-through bulkhead adapters.

FIGURE 7.1

Parts of a Fiber Optic Connector

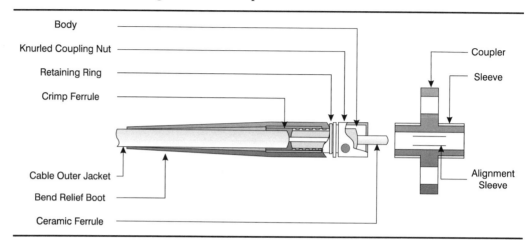

TYPES OF FIBER OPTIC CONNECTORS

There are many fiber optic connector types available, but only a few types are widespread. They are as follows:

SMA 906: SMA 906 connectors represent an old, first generation design now declining in use. Developed by the Amphenol Corporation, SMA's are an outgrowth of the SMA (Subminiature type A) electrical connector used in microwave applications. They are still used in some military applications because of their ability to withstand high temperatures. Overall, they are difficult to use, generally have poor performance and are suitable only for multimode fiber. They are typically used with 100/140 μm fiber or larger.

ST: ST connectors are very widespread in the U.S. and are used predominately with multimode fiber. AT&T introduced the ST connector in early 1985. The design features a spring loading twist-and-lock bayonet coupling that keeps the fiber and ferrule from rotating during multiple connections. This design provides more consistent insertion loss. The cylindrical ferrule may be made of plastic, ceramic, or stainless steel. ST connectors offer very good features, cost, and performance.

Biconic: Invented by Bell Laboratories in 1976, the biconic connector is a good first-to-second generation connector, and it was the first highly successful single-mode connector. The ferrule mates to a free-floating alignment sleeve, and a spring between the plug's threaded screw cap and ferrule ensures that a controlled longitudinal force seats the ferrule in the sleeve independent of the amount of tightening. The early models did not address rotation prevention or backreflection, but upgraded designs have addressed these issues. Biconic connectors are widely used by the telecommunications industry and are suitable for single-mode and multimode fiber.

FC: (Also available as FC/PC.) Nippon Telegraph and Telephone Corporation (NTT) designed the FC with a flat end face on the ferrule that provides "face contact" between joining connectors. The FC represents a good second generation connector design with very good performance and features but relatively high cost. It offers very good single-mode and multimode performance and was one of the first connectors to address backreflection. FC's are often used for analog systems or high bit-rate systems where backreflection management is important. FC/PC connectors incorporate a "physical contact" curved polished fiber end face that greatly reduces backreflections.

D4: Designed by Nippon Electric Corp. (NEC), the D4 connector is keyed, spring loaded, and uses a floating sleeve in its coupler. The ferrule has a diameter of 2.0 mm. It is usable for both single-mode and multimode applications, but at this time, NEC is its dominant user.

HMS-10: Developed by Diamond, Incorporated in Switzerland, the HMS-10 connectors offer precise construction and fiber alignment. They have low insertion loss, good return loss, and excellent repeatability, but they are expensive compared to other connectors and are only available completely assembled and terminated to a custom-ordered fiber specification.

SC: The SC (subscription channel) connector was also designed by NTT. The locking mechanism gives an audible click when pushed in or pulled out. This push-pull design prevents rotational misalignment and is intended to be pull-proof, meaning the ferrule is decoupled from the cable and the connector body. A slight pull on the cable will not cause the ferrule to lose optical contact at the interconnection. A duplex version of the SC connector is gaining popularity in networks and other applications requiring full-duplex transmission. The connector is suitable for single-mode and multimode fibers. The SC connector offers excellent packing density as well as exceptional performance and cost.

FDDI: The fiber distributed data interface (*FDDI*) is a duplex connector that has been in use since 1984. It is used primarily in duplex fiber operations, and it is endorsed by a standard of the same name. This standard describes duplex fiber optic networks. A fixed shroud protects the ferrule from damage. The FDDI connector may also be referred to as a media interface connector (MIC) or a fixed shroud duplex (FSD) connector.

ESCON: This connector is similar to the FDDI connector described above. Its principle difference from the FDDI connector is a retractable shroud which pulls back from the ferrule during mating and protects the ferrule when the connector is not in use. Like the FDDI connector, this connector allows the duplex transmission of information.

EC/RACE: This connector, developed by Radiall S.A. in France, uses a rectangular push-pull latching system that makes it nearly impossible to disconnect by pulling on the cable. Originally designed for use in the European community's research and development in advanced communications technologies in Europe (RACE) program, this low-cost, high-density connector is easy to install. High return loss is achieved by incorporating a silicone based index-matching membrane that negates the need for physical contact of the fibers.

Table 7.2 on the following page illustrates these connector types, offers some general specifications, and relates them to the applications in which they are most likely to find use.

TABLE 7.2

Connector Types

(**NOTE:** Connector sizes are not scaled and are not relative to one another.)

CONNECTOR	INSERTION LOSS	REPEATABILITY	FIBER TYPE	APPLICATIONS
BICONIC	0.60-1.00 dB	0.20 dB	SM, MM	Telecommunications
D4	0.20-0.50 dB	0.20 dB	SM, MM	Telecommunications
EC/RACE	0.10-0.30 dB	0.10 dB	SM	High-speed Datacom
ESCON	0.20-0.70 dB	0.20 dB	MM	Fiber Optic Networks
FC	0.50-1.00 dB	0.20 dB	SM, MM	Datacom, Telecommunications
FDDI	0.20-0.70 dB	0.20 dB	SM, MM	Fiber Optic Networks
HMS-10	0.10-0.30 dB	0.10 dB	SM	Test Equipment
SC	0.20-0.45 dB	0.10 dB	SM, MM	Telecommunications
SC DUPLEX	0.2-0.45 dB	0.10 dB	SM, MM	Datacom
SMA	0.40-0.80 dB	0.30 dB	MM	Military
ST	Typ. 0.40 dB (SM) Typ. 0.50 dB (MM)	Typ. 0.40 dB (SM) Typ. 0.20 dB (MM)	SM, MM	Inter-/Intra- Building, Security, Navy

The first optical connectors were little more than adaptations of existing electrical connector designs. This yielded functional, although far from optimum, optical connectors. Precision was low and user friendliness was not a consideration. To make matters worse, early fiber was not terribly accurate either. The fiber had considerable diameter variation and did not have good *concentricity*. Because they caused high connector loss, these fiber flaws were most often interpreted by users as poor connector performance, not poor fiber performance. These problems are no longer present in quality modern fibers.

Some examples of early connectors include the SMA 906 types and 38999 types. The SMA 906 connector, one of the earliest widespread types, contains many undesirable features. It is a threaded type, making connection and removal time consuming. Because of the threaded nature of the connector, no two individuals ever tightened the connector the same amount. One person might tighten only to the first resistance, while another might utilize a crescent wrench to grossly over-torque and possibly destroy the connector. Another bad feature of the SMA 906 is the requirement of a plastic alignment sleeve installed between two connectors. If this sleeve is omitted, the connector may be completely unusable or at best, intermittent. This is often called a non-captive alignment sleeve because the alignment sleeve is not permanently built into the connector adapter. The last problem with most first generation connectors is that they allow the connectors to be mated without controlling rotational alignment. This causes a large amount of variation in the insertion loss as the connector is unmated and remated.

To overcome some of the problems cited above, manufacturers like ITT developed connectors such as the FOT. This connector used a custom fit jewel to adapt to the varying outer diameter of the fiber. While effective, this approach was very costly and required a great deal of skilled labor for installation. Other unsuccessful approaches included a multitude of expanded beam connectors. Most have been discontinued because an expanded beam connector can never achieve the low loss of a physical contact connector. One clever approach to compensating for variations in fiber dimensions and concentricity is the AT&T rotary splice. This splice can be tuned to very low loss levels (<0.05 dB) with almost any fiber. As fiber quality has improved, the need for such measures has declined; however, the rotary splice is better than a fusion splice for average loss.

Fiber optic connectors developed in the first generation were mostly screw-type (threaded) connectors (e.g., SMA 905 and SMA 906). Connectors later evolved to second generation connectors such as the AT&T ST type. ST stands for straight tip. This connector solved most of the problems associated with the early connectors. Gone is the loose alignment sleeve; it is now captive in the bulkhead adapter. The ST is a bayonet type. This assures that the connector is always properly and consistently connected. With this connector, only one rotational alignment is allowed, greatly improving repeatability. Other notable second generation connectors are the biconic and FC types, although neither address as many first generation flaws as the ST type. The biconic connector is a threaded type and initially did not prevent rotation. Recent design upgrades to the biconic connector have addressed fixed rotational alignment with the keyed biconic. The FC still has the drawback of being a threaded connector. This is inconvenient and decreases packing density because of the need for more finger access while making the connection.

Some second generation connectors also incorporated a new polishing method called PC (physical contact). Earlier connectors used a flat polish on the connector end. This led to a large amount of backreflection, a highly detrimental quality for high bit-rate or analog laser-based fiber optic systems. Large amounts of backreflection in these systems seriously degrade the system's performance and margins. The PC uses a curved polish to dramatically reduce the reflection. Further reduction is achieved by forcing the connector ends together. This eliminates the glass-air-glass interface found in non-PC connectors and further reduces the backreflection. Other variations that achieve low backreflection include angle polishing (angle typically 5° to 15°). Angle polished connectors (APC) require special bulkhead couplings to properly align the connectors. The connectors are then aligned so that the surfaces touch. Names that imply low backreflection include *PC, Super-PC* and *Super Polish*.

Third generation connectors tend to be push-pull types. This makes for the fastest possible connection time and allows a significant increase in packing density because less finger access is required. Third generation connectors have also focused heavily on minimizing installation time (onto the fiber), reducing insertion loss, and have addressed backreflection more completely than in earlier generations.

INSTALLING FIBER OPTIC CONNECTORS

The method for attaching fiber optic connectors to optical fibers varies among connector types. While not intended to be a definitive guide, the following steps are given as a reference for the basics of optical fiber interconnection.

1. Cut the cable one inch longer than the required finished length.
2. Carefully strip the outer jacket of the fiber with "no nick" fiber strippers. Cut the exposed strength members, and remove the fiber coating. The fiber coating may be removed two ways: by soaking the fiber for two minutes in paint thinner and wiping the fiber clean with a soft, lint-free cloth, or by carefully stripping the fiber with fiber stripper. Be sure to use strippers made specifically for use with fiber rather than metal wire strippers as damage can occur, weakening the fiber.
3. Thoroughly clean the bared fiber with isopropyl alcohol poured onto a soft, lint-free cloth such as Kimwipes®. NEVER clean the fiber with a dry tissue.
4. The connector may be connected by applying epoxy or by crimping. If using epoxy, fill the connector with enough epoxy to allow a small bead of epoxy to form at the tip of the connector. Insert the clean, stripped fiber into the connector. Cure the epoxy according to the instructions provided by the epoxy manufacturer.
5. Anchor the cable strength members to the connector body. This prevents direct stress on the fiber. Slide the back end of the connector into place (where applicable).
6. Prepare the fiber face to achieve a good optical finish by cleaving and polishing the fiber end. Before connection is made, the end of each fiber must have a smooth finish that is free of defects such as hackles, lips, and fractures. These defects, as well as other impurities and dirt change the geometrical propagation patterns of light and cause scattering.

FIGURE 7.2

Fiber End Face
Defects

(Photo courtesy of AMP, Inc.)

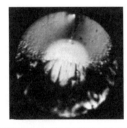

Cleaving

Cleaving involves cutting the fiber end flush with the end of the ferrule. Cleaving, also called the scribe-and-break method of fiber end face preparation, takes some skill to achieve optimum results. Properly done, the *cleave* produces a perpendicular, mirror-like finish. Incorrect cleaving will result in lips and hackles as seen in Figure 7.2. While cleaving may be done by hand, a cleaver tool, available from such manufacturers as Fujikura, allows for a more consistent finish and reduces the overall skill required. The steps listed below outline one procedure for producing good, consistent cleaves such as the one shown in Figure 7.3.

1. Place the blade of the cleaver tool at the tip of the ferrule.
2. Gently score the fiber across the cladding region in one direction. If the scoring is not done lightly, the fiber may break, making it necessary to reterminate the fiber.
3. Pull the excess, cleaved fiber up and away from the ferrule.
4. Carefully dress the nub of the fiber with a piece of 12-micron alumina-oxide paper.
5. Do the final polishing.

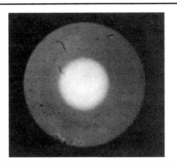

FIGURE 7.3

A Well-Cleaved Multimode Fiber

(Photo courtesy of AMP, Inc.)

Polishing

After a clean cleave has been achieved, the fiber end face is attached to a polishing bushing, and the fiber is ground and polished. The proper finish is achieved by rubbing the connectorized fiber end against polishing paper in a figure-eight pattern approximately sixty times. (See Figure 7.4.)

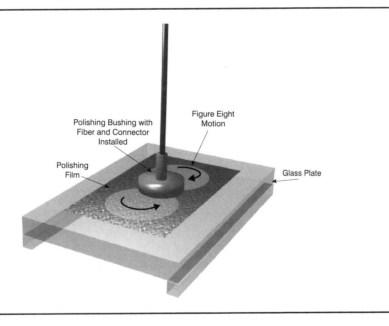

FIGURE 7.4

Polishing Technique

Polishing Bushing with Fiber and Connector Installed

Figure Eight Motion

Polishing Film

Glass Plate

To increase the ease and repeatability of connector installation, several companies offer connector kits. Some kits are specific to the type of connector to be installed while others supply the user with general tools and information for installing different types of connectors. Examples include Hewlett-Packard's Connector Assembly Tooling Kit, AMP, Incorporated's connector kits, or the line of field termination kits available from Fiber Instrument Sales.

As previously mentioned, many connectors require the use of an alignment sleeve. This sleeve increases repeatability from connection to connection. When using SMA connectors, alignment sleeves, also known as half-sleeves or full-sleeves (depending on the length), are required to reduce insertion loss and improve performance. An ST alignment sleeve is illustrated in Figure 7.5.

FIGURE 7.5

Alignment Sleeve

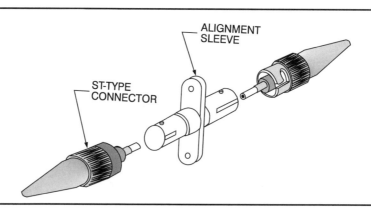

CARE OF FIBER OPTIC CONNECTORS

It is hard to conceive of the size of a fiber optic connector core. Single-mode fibers have cores that are only 8-9 μm in diameter. As a point of reference, a typical human hair is 50-75 μm in diameter, approximately 6-9 times larger! Dust particles can be 20 μm or larger in diameter. Dust particles smaller than 1 μm can be suspended almost indefinitely in the air. A 1 μm dust particle landing on the core of a single-mode fiber can cause up to 1 dB of loss. Larger dust particles (9 μm or larger) can completely obscure the core of a single-mode fiber. Fiber optic connectors need to be cleaned *every* time they are mated and unmated; it is essential that fiber optics users develop the necessary discipline to always clean the connectors before they are mated.

It is also important to cover a fiber optic connector when it is not in use. Unprotected connector ends are most often damaged by impact, such as hitting the floor. Most connector manufacturers provide some sort of protection boot. The best protectors cover the entire connector end, but they are generally simple closed-end plastic tubes that fit snugly over the ferrule only. These boots will protect the connector's polished ferrule end from impact damage that might crack or chip the polished surface. Many of the tight fitting plastic tubes contain jelly-like contamination (most likely mold release) that adheres to the sides of the ferrule. A blast of cleaning air or a quick dunk in alcohol will not remove this residue. This jelly-like residue can combine with common dirt to form a sticky mess that causes the connector ferrule to stick in the mating adapter. Often, the stuck ferrule will break off as one attempts to remove it. The moral of the story is always thoroughly clean the connector before mating, even if it was cleaned previously before the protection boot was installed.

Cleaning Technique

Required Equipment:

- Kimwipes or any lens-grade, lint-free tissue. The type sold for eyeglasses work quite well.
- Denatured alcohol.
- 30X microscope.
- Canned dry air.

Technique:

1. Fold the tissue twice so it is four layers thick.

2. Saturate the tissue with alcohol.

3. First clean the sides of the connector ferrule. Place the connector ferrule in the tissue, and apply pressure to the sides of the ferrule. Rotate the ferrule several times to remove all contamination from the ferrule sides.

4. Now move to a clean part of the tissue. Be sure it is still saturated with alcohol and that it is still four layers thick. Put the tissue against the end of the connector ferrule. Put your fingernail against the tissue so that it is directly over the ferrule. Now scrape the end of the connector until it squeaks. It will sound like a crystal glass that has been rubbed when it is wet.

5. Use the microscope to verify the quality of the cleaning. If it isn't completely clean, repeat the steps with a clean tissue. Repeat until you have a cleaning technique that yields good, reproducible results.

6. Mate the connector immediately! Do not let the connector lie around and collect dust before mating.

7. Air can be used to remove lint or loose dust from the port of a transmitter or receiver to be mated with the connector. Never insert any liquid into the ports.

Handling

1. Never touch the fiber end face of the connector.

2. Connectors not in use should be covered over the ferrule by a plastic dust cap. It is important to note that inside of the ferrule dust caps contain a sticky residue that is a by-product of making the dust cap. This residue will remain on the ferrule end after the cap is removed. It is critical to thoroughly clean the ferrule end BEFORE it is mated to the intended unit.

3. The use of index-matching gel, a gelatinous substance that has a refractive index close to that of the optical fiber, is a point of contention between connector manufacturers. Glycerin, available in any drug store, is a low-cost, effective index-matching gel. Using glycerin will reduce connector loss and backreflections, often dramatically. However, the index-matching gel may collect dust or abrasives that can damage the fiber end faces. It may also leak out over time, causing backreflections to increase.

SPLICING

Splices are permanent or semi-permanent connections between fibers. Typically, a *splice* is used to join lengths of cable outside buildings. Connectors are generally used at the ends of cables inside buildings. Splices offer lower attenuation and lower backreflection than connectors, and they are generally less expensive. There are two main types of splices: fusion splices and mechanical splices.

Fusion Splices

Fusion splicing involves butting two cleaved fiber end faces together and heating them until they melt together or fuse. This is normally done with a *fusion splicer* that carefully controls the alignment of the two fibers to keep losses as low as 0.05 dB. Fusion splicers are relatively expensive devices that usually include an electric arc welder to fuse the fibers, alignment mechanisms, a camera or binocular microscope to magnify the alignment by 50 times or more, and instruments to check the optical power through the fibers both before and after they are fused. The operation of a typical fusion splicer is illustrated in Figure 7.6.

FIGURE 7.6

Typical Fusion
Splicer

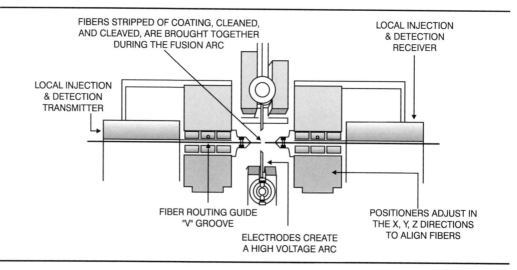

FIGURE 7.7

Capillary Splice

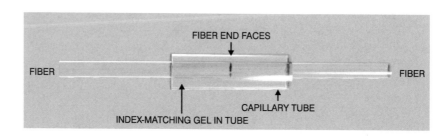

Mechanical Splices

Mechanical splices join two fibers together by clamping them within a structure or by epoxying the fibers together. Generally, the equipment needed to produce a *mechanical splice* is less expensive than the equipment for fusion splices. Mechanical splices may have a slightly higher loss, and backreflection can be a concern, but index-matching gel may be inserted in the mechanical splice to reduce this.

Capillary splicing is the simplest form of mechanical splicing. It involves inserting two fiber ends into a thin capillary tube as illustrated in Figure 7.7. Index-matching gel is typically used in this splice to keep backreflections at a minimum. Usually the fibers are held together by compression or friction although epoxy may be used to permanently secure the fibers.

The AT&T rotary splice, also called the polished-ferrule splice, involves inserting each clean fiber end into a glass ferrule. After the fiber is secured, the end is polished flush with the ferrule. The two polished ferrules are then butted together in an alignment sleeve that contains index-matching gel. The ferrules have slightly eccentric bores which mount the fibers slightly off-center. After the ferrules are inserted in the alignment sleeve, the ferrules are rotated while the splice loss is monitored. The ferrules are then fixed at the point where splice loss is lowest.

Multiple fiber array splices use grooved top and bottom plates to align multiple fibers with respect to one another. This is especially used to splice ribbon cables. Each fiber in the cable is placed in a separate groove in the bottom plate. When the top and bottom plates are put together, the fibers are automatically aligned.

INTERCONNECTION LOSSES

Connectors and splices both add loss to a system. These losses, like the losses in optical fiber, may be intrinsic or extrinsic. Intrinsic losses occur because of differences in the fibers being connected or spliced. These differences include variations in core and/or outer diameter, differences in the fibers' index profiles, and *ellipticity* and *concentricity* of the core. Extrinsic losses are the result of the splice or connector. These loss mechanisms include fiber end misalignment, the quality of the fiber end face finish, contamination on or within the fiber, refractive-index matching between fiber ends, spacing between fiber ends, imperfections in the fiber at the junction, and angular misalignment of the bonded fibers. Chapter 13, "Testing & Measurement Techniques," discusses these loss mechanisms and the methods for measuring these losses in greater detail.

Other Concerns

In addition to attenuation or loss mechanisms, other issues must be addressed to successfully interconnect two fibers. Fiber curl occurs when the ends of the fiber refuse to lie flat and straight. Although optical fiber curls only a very small degree, even a few micrometers of curl can cause problems in splicing, especially in single-mode fibers. Another concern is mismatched fibers. If the two fibers being spliced or connected are not identical, losses can increase dramatically. Strength of the fiber decreases with even a single splice. The act of stripping the fiber coating may cause microcracks that will weaken the fiber. In fusion splicing, contaminants can weaken the fusion zone, and thermal cycling can weaken the surrounding area. In connected fibers, the fiber is only as strong as the connectors that join it, but as mentioned, the later generation push-pull locking mechanisms can prevent the fibers from being accidentally pulled apart.

FIBER OPTIC CONNECTOR SELECTION GUIDE

The following table gives general information for selecting fiber optic connectors. Referenced notes are listed below the table.

Connector Type	Fiber Type	Generation[1]	Typ. Insertion Loss (dB)[2]	Cost[2]
ST	MM	2	0.20-0.60	$2.80-$6.35
ST	SM	2	0.20-0.60	$6.95-$11.85
SC	MM	3	0.20-0.30	$6.00-$9.45
SC	SM	3	0.15-0.30	$8.35-$13.95
D4	SM, MM	2	0.30	$11.00-$12.95
Biconic	MM	2	0.60	$9.35-$12.60
Biconic	SM	2	0.60	$21.95-$34.40
SMA	MM	1	0.60-0.80	$3.50-$6.50
FC	MM	2	0.20-0.50	$3.75-$8.95
FC	SM	2	0.20-0.50	$8.40-$16.00
ESCON	MM	3	0.15	$15.00
FDDI	MM	3	0.60	$16.20

TABLE 7.3

Fiber Optic Connector Selection Guide

Note 1: The Generation column is the estimated generation as described in Table 7.1 at the beginning of the chapter.

Note 2: Source is the 12th edition of the *Fiber Optic Supplies & Equipment* catalog published by Fiber Instrument Sales Inc., Oriskany, NY.

Splices and connectors are the two most common ways to join two fibers. The next chapter discusses some other methods.

Chapter Summary

- Fiber optic connectors have gone through several developmental generations.
- The four basic components of all optical connectors are the ferrule, the connector body, the cable, and the coupling device.
- The method for attaching fiber optic connectors to optical fiber varies from connector type to connector type.
- Cleaving involves cutting the fiber end flush with the end of the ferrule.
- After a cleave has been achieved, the connector is attached to a polishing bushing, and the fiber is ground and polished.
- Most connectors require the use of an alignment sleeve.
- Because single-mode fibers have cores that are only 8-9 μm in diameter, and because a dust particle anywhere from 9μm down to 1μm in diameter can greatly increase optical loss, connectors must always be cleaned before mating.
- Splices are permanent connections between fibers typically used to join lengths of cable outside buildings.
- Fusion splicing involves butting two cleaved fiber end faces together and heating them until they melt together.
- Mechanical splices join two fibers together by clamping them within a structure or by epoxying the fibers together.
- Different types of mechanical splices include the capillary splice and the AT&T rotary splice.
- Connectors and splices both add loss to a system, while other considerations such as fiber curl and mismatched fibers can further decrease loss.

Selected References and Additional Reading

Ajemian, Ronald G. 1995. "A Selection Guide for Fiber Optic Connectors." *Optics & Photonics News* June 1995: 31-36.

Hecht, Jeff. 1993. *Understanding Fiber Optics.* 2nd edition. Indianapolis, IN: Sams Publishing.

Miller, Mettler, and White. 1986. "Optical Fiber Splices and Connectors." NY: Maral Dekken, Inc.

Reed, Mike. 1995. "Making the Perfect Cleave." *Cabling Installation & Maintenance* April 1995: 38.

Sterling, Donald J. 1993. *Amp Technician's Guide to Fiber Optics,* 2nd Edition. New York: Delmar Publishers.

—. 1991. *Universal Transport System Design Guide.* Hickory, NC: Siecor Corporation.

OTHER PASSIVE DEVICES

Connectors and splices join two fiber ends together, but other passive devices must be used in applications where three or more fibers are to be joined. Types of passive devices include couplers, splitters, tap ports, switches, and wavelength-division multiplexers. These devices are used to divide or combine multiple optical signals.

FIBER OPTIC COUPLERS

Fiber optic couplers split optical signals into multiple paths or vice versa. Optical signals differ from electrical signals, making optical couplers trickier to design than their electrical counterparts. An optical signal is a flow of signal carriers (in this case, photons) similar to an electrical current. However, an optical signal does not flow through the receiver to the ground. Rather, it stops at the receiver where it is absorbed by the detector. If multiple fiber optic receivers were connected in a series, no signal would reach beyond the first receiver because the detector in that receiver would absorb the entire signal. Thus, multiple optical output ports must be parallel, and further, the signal must be divided between the ports, reducing its magnitude.

A coupler is identified by its characteristic number of input and output ports. This is expressed as an N x M configuration where N is the number of input fibers and M is the number of output fibers. Fused couplers can be made in any configuration, but even multiples of two are most common (2x2, 4x4, 8x8 etc.).

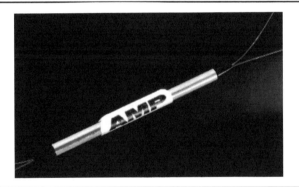

FIGURE 8.1

Typical Coupler

(Photo courtesy of AMP, Inc.)

The simplest couplers are fiber optic splitters. These devices have at least three ports but may have more than 32 for more complex devices. A simple 3-port device, also called a tee coupler, can be thought of as a directional coupler. One fiber is called the common fiber, while the other two fibers may be called input or output ports. A common application involves injecting light into the common port and splitting it into two independent legs (the output ports). The ratio of the distribution of light between the two output legs can be determined by the coupler manufacturer. Popular splitting ratios are 50%-50%, 90%-10%, 95%-5% and 99%-1%; however, almost any value can be achieved on a custom basis. (These values are sometimes specified in dB values.) If a 90%-10% splitter is used with a 50 μW light source, 45 μW and 5 μW would be at the output ports. Actually, the outputs never quite do that well. This is because of *excess loss*, a parameter that all couplers and splitters share. Excess loss assures that the total output is never as high as the input. Loss figures range from 0.05 dB to 2 dB for different coupler types. An interesting, and unexpected, property of splitters is that they are symmetrical. For instance, if the same coupler injected 50 μW into the 10% output leg, only 5 μW would reach the common port.

Common applications for couplers and splitters include:

- Local monitoring of a light source output (usually for control purposes).
- Distributing a common signal to several locations simultaneously. An 8-port coupler allows a single transmitter to drive eight receivers.
- Making a linear, tapped fiber optic bus. Here, each splitter would be a 95%-5% device that allows a small portion of the energy to be tapped while the bulk of the energy continues down the main trunk.

Optical couplers are used in applications where links other than point-to-point links are required. Examples are the bidirectional link and *local area networks* (LAN). In LAN applications, couplers are configured either as a star topology or a bus topology. In a star topology, stations branch off from a central hub, much like the spokes on a wheel. With a star coupler, the number of workstations can be easily expanded. For example, changing from a 4x4 to an 8x8 doubles the system capacity. Note that any input to the star coupler is divided to all outputs allowing every station to hear every other station. Star couplers have many ports (usually a power of two), and couplers with 32 or 64 ports are not uncommon. The key use of a star coupler is to make a large party-line circuit. Many transceivers can be connected to a star coupler and can freely communicate with all other transceivers assuming some sort of protocol is established to prevent two or more transceivers from communicating simultaneously. The key disadvantage of the star coupler is the rather large insertion loss (20 dB typically for a 64-port device) and the need for a complex collision-prevention protocol.

Bus topology utilizes a *tee coupler* and is constructed so that a series of stations can listen to a single backbone of cable. In a typical bus network, a coupler at each node splits off part of the power from the bus and carries it to a transceiver in the attached equipment. In a system with N terminals, a signal must pass through N-1 couplers before arriving at the receiver. Loss increases linearly as N increases. Bus topologies can be configured for a single direction or may have a duplex configuration. In a unidirectional setup a transmitter at one end of the bus communicates with a receiver at the other end. Each terminal also contains a receiver. A duplex network can be achieved by adding a second fiber bus or by using an additional directional coupler at each end and at each terminal. This second configuration allows signals to flow in both directions. Figure 8.2 illustrates the differences in a unidirectional bus and a duplex bus.

WDM's

A more complex coupler is the wavelength-division multiplexer (WDM). A WDM is a passive device that allows two or more different wavelengths of light to be split into multiple fibers or combined onto one fiber. WDM's allow the potential information-carrying capacity of an optical fiber to be increased significantly. Like the simple splitter, WDM's typically have a common leg and a number of input or output legs. Unlike the splitter, however, they have very little insertion loss. They do have the same range of excess loss. Two important considerations in a WDM device are *crosstalk* and channel separation. Crosstalk, also called directivity, refers to how well the demultiplexed channels are separated. Each *channel* should appear only at its intended port and not at any other output port. The crosstalk specification expresses how well a coupler maintains this port-to-port separation. Channel separation describes how well a coupler can distinguish

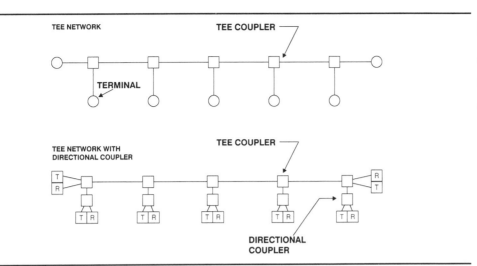

FIGURE 8.2

Tee Network
Configurations

(Illustration courtesy of AMP, Inc.)

wavelengths. In most couplers, the wavelengths must be widely separated. Light can travel in either direction without the penalty found in splitters. WDM's allow multiple, independent data streams to be sent over one fiber. The most common WDM system would use two wavelengths, although four or more-wavelength systems are available. Complex WDM systems employing more than twelve wavelengths have been described, but they are usually prohibitively expensive due to the extreme cost of the light sources and the required wavelength tolerances and stabilities.

Figure 8.3 shows two WDM's being used to allow bidirectional streams of data to be carried on a single fiber. It does not matter what the data streams are. One could be a video signal, the other an RS-232 data stream. Alternatively, both signals could be video signals or high-speed data signals at 1.26 Gb/s.

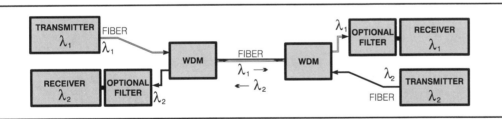

FIGURE 8.3

WDM—
Bidirectional

Figure 8.4 shows a second use of WDM's. In this example, two wavelengths of light are being used to transmit two independent signals in the same direction on one fiber. As in the previous example, the two signals can be anything.

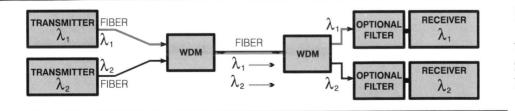

FIGURE 8.4

WDM—Dual
Signals, Same
Direction

The WDM shown in the Figure 8.5 is called a bulk optics WDM because it is constructed from discrete lenses and filters. The heart of this type of WDM is the *dichroic filter*. Dichroic filters are based on interferometric techniques and have some interesting properties. The most novel property is that they reflect the light that they do not transmit. Referring to the figure, imagine that fiber 1 is carrying two wavelengths, 850 nm and 1300 nm. Also imagine that the dichroic filter is designed so that it passes wavelengths longer than 1100 nm. This would be called a long-wave pass (LWP) filter. As the light exits fiber 1 it is first passed through the lens so that it will focus at a point. As the light hits the filter, the 1300 nm light passes through the filter and is collected by fiber 3. The 850 nm light exiting fiber 1 on the other hand reflects off of the filter and is collected by fiber 2. Thus the two wavelengths have been effectively separated, and the information carried on each wavelength can be independently decoded. The dichroic filter can offer a great deal of isolation in the transmission mode, but has poor isolation in the reflection mode. Usually these types of WDM's are built with both short-wave pass (SWP) and LWP filters and used in combinations to achieve the best system performance.

FIGURE 8.5

Bulk Optics WDM

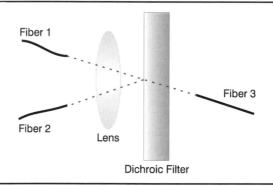

FUSED FIBER COUPLERS

By far the most popular type of coupler in use today is a fused fiber coupler. In this type of coupler, two or more fibers are twisted together and melted in a flame. The basic construction is shown in Figure 8.6 below.

FIGURE 8.6

Fused Fiber Coupler

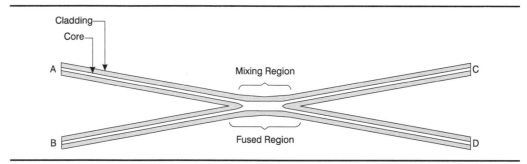

This simple construction technique can be used to make 50%-50% couplers, 99%-1% couplers and even WDM's. The length of the coupling region (the fused region) as well as the amount of twisting and pulling done on the fiber while it is melted determines the result. This type of coupler has become very popular because the basic materials are almost free; a few meters of fiber, a bit of potting compound, and a metal tube are all it takes for the most basic coupler. The magic is knowing how to melt, twist, and pull the fiber.

The most interesting type of fused fiber coupler is the WDM. It is only possible with single-mode fiber. Simple splitters can be made with the fused fiber coupler method for both single-mode and multimode fiber types. A WDM is formed by an interferometric action within the fused mixing region. Like an interferometer, this causes a sinusoidal response as the length is increased. WDM's are designed to operate at two specific wavelengths. By adjusting the minimum of the sinusoid to correspond to the first wavelength of interest and the maximum of the sinusoid to correspond to the second wavelength of interest, a WDM is formed.

The figure below shows the optical characteristics of a fused fiber coupler WDM that has been designed for wavelengths of 1300 nm and 1550 nm. The response of port 1 is the port that would be designated for use with 1300 nm while port 2 is the port that would be designated for use with 1550 nm. While the shapes of the curves aren't very familiar, it should be noted that they are basically formed by taking the logarithm of a sine function.

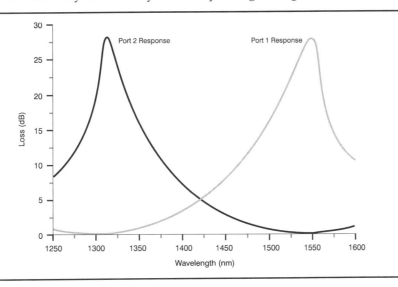

FIGURE 8.7

Optical Characteristics of a Fused Fiber WDM

The WDM shown in the figure has a maximum isolation of about 28 dB. That may be adequate for some applications, but many will require even greater isolation. This is usually accomplished by cascading two or more WDM's to increase the isolation. Usually this type of WDM can only be used with laser sources because of the relatively narrow isolation peak at each wavelength. The passband peak is quite broad on the other hand.

FIBER OPTIC SWITCHES

It is sometimes necessary to switch light between two or more fibers. Fiber optic switches are components that selectively route optical signals among different fiber paths without optical-to-electrical conversion. Switches can be grouped into three categories: opto-mechanical, electro-optical, and photonic switches.

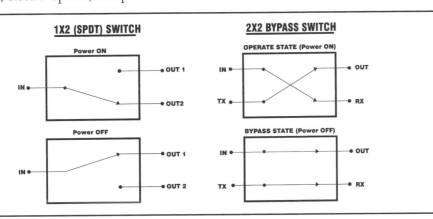

FIGURE 8.8

Typical Switch Configurations

(Illustration courtesy of AMP, Inc.)

Opto-mechanical switches involve a moving optical element such as a fiber, a lens assembly, or a prism, that routes the light signal from one fiber to another. These switches represent the majority of fiber optic switches. They find application in high-resolution stepper motors or piezoelectric motors and can expand to 1 x M and N x M configurations with up to 100 channels. Performance parameters include low insertion loss (0.5 dB), low backreflections (-55 dB), and high reliability (ten million cycle life). Opto-mechanical switches are physically limited to millisecond switching speeds, so they are not suitable for optical computing or central office switching. Despite this limitation, opto-mechanical switches are making important contributions in telecommunications, networks, and sensing systems, and several international standards bodies have completed specifications and test procedures including Bellcore TA-NWT-001073, U.S. Navy MIL-S-24725, IEC SC86B, EIA/ TIA PN-2984.

Electro-optical switches select lightwave paths inside active devices using a biasing electrical current. These switches do not have the physical speed limitations of opto-mechanical switches. Most consist of titanium diffused waveguides in lithium niobate substrates and offer nanosecond switching speeds; however, these devices suffer from high (20 dB) crosstalk and high insertion loss.

Photonic switches are true, all-optical switches that switch light channels in response to optical control signals. As a competing technology, they may eventually replace electro-optical switches. The most promising approach for photonic switching involves using self-excited electro-optic devices (SEEDs) that act as optical logic gates that would allow the switches to be used to perform logical operations for digital computing.

Chapter Summary

- Passive devices include splitters, tap ports, switches and WDM's.
- Common applications for couplers and splitters include local monitoring of a light source, distributing a common signal to several locations simultaneously, and making a linear, tapped fiber optic bus.
- Bus topology utilizes a tee coupler and is constructed so that a series of stations can listen to a single backbone of cable.
- A WDM is a passive device that allows two or more different wavelengths of light to be split into multiple fibers or combined onto one fiber.
- Fiber optic switches are components that selectively route optical signals among different fiber paths without optical-to-electrical conversion.
- Opto-mechanical switches involve a moving element such as a fiber, a lens assembly, or a prism.
- Electro-optical switches select lightwave paths inside active devices using a biasing electrical current.
- Photonic switches are true, all-optical switches that switch light channels in response to optical control signals.

Selected References and Additional Reading

Palladino, John R. 1990. *Fiber Optics: Communicating By Light*. Piscataway, NJ: Bellcore.

Pesavento, Gerry, and Dr. David Polinksy. 1993. "Fiberoptic Switches." *Fiber Optic Product News* Feb. 1993.

Sterling, Donald J. 1993. *Amp Technician's Guide to Fiber Optics*, 2nd Edition. New York: Delmar Publishers.

—. 1991. *Universal Transport System Design Guide*. Hickory, NC: Siecor Corporation.

SYSTEM DESIGN CONSIDERATIONS

The main system-level components of a fiber optic data link are the fiber optic transmitter, the fiber optic receiver, and the fiber itself. Each component plays a vital role in the quality of transmission. Careful decisions, based on operating parameters, must be made for each component of the system if high-quality transmission is to be achieved. The main questions involve data rates and bit error rates in digital systems, bandwidth, linearity, and signal-to-noise ratios in analog systems, and in all systems, transmission distances. These questions of how far, how good, and how fast define the basic application constraints. Once these basic factors are decided, it is time to evaluate the other factors involved.

System Factor	Consideration/Choices
Type of Fiber	Single-mode or Multimode
Operating Wavelength	780, 850, 1300, and 1550 nm typical.
Transmitter Power	Typically expressed in dBm.
Source Type	LED or Laser
Receiver Sensitivity and Overload Characteristics	Typically expressed in dBm.
Detector Type	PIN Diode, APD, or IDP
Modulation Code	AM, FM, PCM, or Digital
Bit Error Rate (BER) (Digital Systems Only)	10^{-9}, 10^{-12} Typical
Signal-to-Noise Ratio	Specified in decibels (dB).
Number of Connectors	Loss increases with the number of connectors.
Number of Splices	Loss increases with the number of splices.
Environmental Requirements	Humidity, Temperature, Exposure to Sunlight
Mechanical Requirements	Flammability, Indoor/Outdoor Application

TABLE 9.1

Factors for Evaluating Fiber Optic System Design

Many of these considerations are directly related to other considerations. For example, detector choice will impact the receiver sensitivity which will affect the necessary transmitter output power. Output power then impacts the transmitter light emitter type which will affect the fiber type and connector type that may be used. Obviously, designing a fiber optic link is not trivial; however, there are logical ways to proceed. One way is to first analyze the fiber optic link power budget, also called the optical link loss budget.

OPTICAL LINK LOSS BUDGET

Figure 9.1 shows a simple means of visualizing the key optical calculations that must be performed in designing a fiber optic link. In order for a link to be practical, it must tolerate some range of optical loss. Ideally, but not always, it should work back-to-back (i.e., with the shortest possible fiber). And of course, it should work with some longer length of fiber. Many factors affect how far one can transmit over fiber. The key factors are:

- Transmitter optical output power
- Operating wavelength
- Fiber attenuation
- Fiber bandwidth
- Receiver optical sensitivity

The designer can often adjust any or all of these variables to create a product that meets the needs of a given application.

FIGURE 9.1

Optical Link Loss
Budget

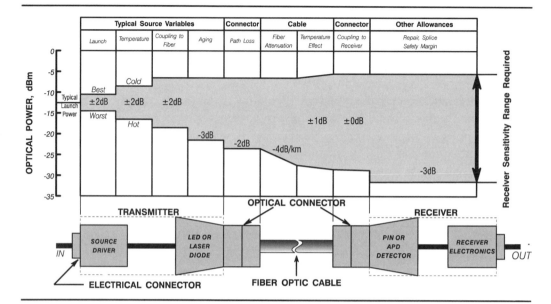

The graphic shows a hypothetical link and its corresponding link budget. Start with the transmitter output power on the left side of the chart. The typical launch power is -12.5 dBm. However, the transmitter LED output power can vary by ±2 dB due to manufacturing variability of the LED itself. So the output power can be as high as -10.5 dBm or as low as -14.5 dBm. The block is shaded between these two values.

Further transmitter variations of ±2 dB result from the effects of temperature on the electronics and the electro-optics (e.g., LED or laser). Another potential ±2 dB of loss is due to variations in the optical coupling to the transmitter output. The effects of aging should be included in the system's design. Typically 1-3 dB is assigned to this factor. The next factor involves the losses due to optical connectors that may be in the optical path. The graphic allows 2 dB for this factor. For this system, the loss due to the optical fiber itself amounts to 4 dB/km of length. Multiply this value times the actual length to determine the loss due to the fiber. There are also temperature effects associated with most fibers. Another ± 1 dB is allowed for temperature effects on the fiber and cable.

The variation in loss at the receiver is the next factor. This is a large-area detector since there is no variation in this parameter. The last factor is the safety margin that should be built into all systems. Typically 3 dB is allowed for this safety margin. At each step, any variation causes the shaded band to enlarge. On the right side of the chart the receiver has to cope with optical inputs as high as -5.5 dBm and as low as -31.5 dBm. Or stated differently, the receiver would need to have an optical loss range or optical dynamic range of 26 dB.

A discussion of the decibel is necessary to understand these link loss values. The decibel (dB) is a convenient means of comparing two powers. The loss a given link can tolerate is often rated in dB. For example, a given AM video link may tolerate a maximum of 9 dB of optical loss. How much light actually reaches the receiver? Table 9.2 describes the decibel to power conversion. According to the table, 12% of the optical power actually reaches the receiver, so 88% of the light output by the transmitter was lost somewhere along the way.

If the link could tolerate 20 dB of optical loss, then only 1% of the transmitter's optical output would reach the receiver. To determine the amount of light reaching the receiver, take any two values that total the dB of optical loss in question. For example, 15 dB is the total of 10 dB and 5 dB. The corresponding power out for 15 dB is 3.2% according to Table 9.2. This value is also attainable by multiplying the corresponding percent values for the two dB readings, 10 dB and 5 dB, to get the desired result, e.g. 10% times 32% is 3.2%. Thus, 3.2% of the light actually reaches the receiver.

dB	Power Out as a % of Power In	% of Power Lost	Remarks
1	79%	21%	...
2	63%	37%	...
3	50%	50%	1/2 the power
4	40%	60%	...
5	32%	68%	...
6	25%	75%	1/4 the power
7	20%	80%	1/5 the power
8	16%	84%	1/6 the power
9	12%	88%	1/8 the power
10	10%	90%	1/10 the power
11	8.0%	92%	1/12 the power
12	6.3%	93.7%	1/16 the power
13	5.0%	95%	1/20 the power
14	4.0%	96.0%	1/25 the power
15	3.2%	96.8%	1/30 the power
16	2.5%	97.5%	1/40 the power
17	2.0%	98.0%	1/50 the power
18	1.6%	98.4%	1/60 the power
19	1.3%	98.7%	1/80 the power
20	1.0%	99.0%	1/100 the power
25	0.3%	99.7%	1/300 the power
30	0.1%	99.9%	1/1000 the power
40	0.01%	99.99%	1/10,000 the power
50	0.001%	99.999%	1/100,000 the power

TABLE 9.2

Decibel to Power Conversion

A decibel is always a ratio between two numbers. Equation 9.1 gives the calculation of a decibel.

Eq. 9.1
$$dB = 10 \cdot \log_{10}\left(\frac{P_1}{P_2}\right)$$

For fiber optics, the ratio is generally the transmitter output power compared to the receiver input power as in the examples above. The decibel describes all loss mechanisms in the optical path of a fiber optic link. This includes fiber loss (usually described as decibels per kilometer or dB/km), connector loss, and splice loss. In all cases, Table 9.2 provides easy reference for converting the decibel value to a percentage. There is also a unit of power called dBm which indicates the actual power level referred to 1 milliwatt. The conversion table is useful here as well. In the case of the dBm, the ratio is between some power level to be described and 1 milliwatt. Typically the number is negative because power levels are typically less than 1 milliwatt in fiber optics. Table 9.3 shows some typical dBm values, converts them to power, and relates them to actual applications.

TABLE 9.3

Typical dBm Values

dBm Value	% of 1 Milliwatt	Power	Application
0.0	100%	1.0 milliwatt	Typical Laser Peak Output
-13.0	5%	50.0 microwatts	Typical LED Peak Output
-30.0	0.1%	1.0 microwatt	Typical PIN Receiver Sensitivity
-40.0	0.01%	100.0 nanowatts	Typical APD Receiver Sensitivity

Newcomers to purchasing fiber optics often mistakenly feel they must specify transmitter power, receiver sensitivity, *and* optical link loss budget. This is an overly constrained, over-specified system. In most cases, the customer needs to specify optical link loss budget and nothing else. A customer needing a 10 dB maximum optical link loss budget will never know the difference between a system with a transmitter with a 0 dB output and a receiver with -10 dBm sensitivity and a system with a transmitter with a -10 dB output and a receiver with -20 dBm sensitivity. As long as all other requirements such as bit error rate (BER) are met, the application's loss requirements have been met. By specifying only the required maximum optical loss, the most economical transmitter/receiver pair can be utilized. The only time that transmitter optical output power and receiver optical sensitivity need to be specified is when the transmitter and receiver are bought separately. In that case, the maximum optical link loss budget need not be specified.

RISE TIME BUDGET

The optical link loss budget analyzes the link to ensure that sufficient power is available throughout the link to meet the demands of a given application, but this power is only one part of the link requirement. The other part is bandwidth or *rise time*. The components in the link must be turned on and off fast enough and fiber dispersion must be low enough to meet the bandwidth requirements of the application. It is easy to overlook bandwidth in many fiber optic applications; the current media hype constantly emphasizes fiber's unlimited bandwidth, but this isn't always the case. Adequate bandwidth for a system can be assured by developing a *rise time budget*. As noted in the previous chapters on light emitters and light detectors, active devices have a finite response time to inputs. The devices do not turn on or turn off instantaneously. Rise and fall times determine the overall response time and the resulting bandwidth. In Chapter 3 we learned that dispersion also limits bandwidth. When the bandwidth of a component is given, the 10% to 90% rise time can be derived from:

Eq. 9.2 $$t_r = \frac{0.35}{BW}$$

where:

t_r = Rise time.
BW = Bandwidth.

This equation accounts for multimode dispersion in the fiber. The rise time budget must also include the rise times of the transmitter and the receiver. For the receiver, rise time/bandwidth may be limited by either the rise time of the components or the bandwidth of the RC time constant. Because connectors, couplers, and splices do not affect system speed, they need not be accounted for in the rise time budget as they were in the optical link loss budget. Once all the necessary component rise times have been determined, the system rise time can be derived from:

Eq. 9.3
$$t_{sys} = 1.1\sqrt{t_{r1}^2 + t_{r2}^2 + \ldots t_{rn}^2}$$

The 1.1 factor at the beginning of the equation allows for a 10% degradation factor in the system rise time. The system rise time budget can set any component rise time simply by rearranging the equation to solve for the unknown rise time. The table below gives an example of a rise time analysis for a simple fiber optic system.

Element	Bandwidth	Rise Time (10%-90%)
Tx Drive Electronics	200 MHz	1.75 ns
LED (850 nm)	100 MHz	3.50 ns
Fiber (1 km)	90 MHz	3.89 ns
PIN Detector	350 MHz	1.00 ns
Rx Electronics	180 MHz	1.94 ns

TABLE 9.4

Rise Time Analysis

The system rise time would be found by applying Equation 9.3 to the individual element rise times listed above. Equation 9.3 yields a system rise time of 6.53 ns. Equation 9.2 can then be used to determine that the system bandwidth is 53.6 MHz. The slowest element in the system dominates in setting the overall system bandwidth. For example, using the system given in Table 9.4, changing the LED to a more expensive 200 MHz type would only improve the system by 16% to 62.4 MHz because of the effect of the low fiber bandwidth.

SENSITIVITY ANALYSIS

A sensitivity analysis determines the minimum optical power that must be present at the receiver in order to achieve the performance levels required for a given system. Several factors will affect this analysis:

Source Intensity Noise: This refers to noise generated by the LED or laser; there are two main types. *Phase noise* describes the difference in the phases of two optical wavetrains separated by time, cut out of the same optical wave. When a comparison of the two trains shows a stable phase difference, the wave is coherent. An unstable phase difference becomes a source of noise. *Amplitude noise* is caused by the laser emission process. Noise power increases with rising optical power, reaches a maximum at the threshold level, then decays at higher power levels.

Fiber Noise: Fiber noise specifically relates to modal partition noise. In a multimode laser, all the longitudinal modes are in competition, causing the spectral distribution to be time-dependent rather than the normal time-average distribution. Because each of these competing modes corresponds to a color, chromatic dispersion of the fiber separates or partitions these modes, causing a reduction in system bandwidth and increased noise. This effect is not present in single-mode fiber, although another phenomenon called modal noise can create similar difficulties when lasers are used with single-mode fibers.

Receiver Noise: The photodiode, the conversion resistor, and the amplifier all contribute to receiver noise. Shot noise is the dominant form of receiver noise. Shot noise relates to the statistical arrival of the photons at the detector and is caused by random fluctuations in the current that arise from the discrete nature of electrons. Receiver noise is directly proportional to the receiver's bandwidth.

Since the flow of electrons is a random process, a simple counting statistics equation can be used to estimate the noise level. The noise level will be approximately equal to the square root of the number of electrons flowing per unit time. Since a single electron has a charge of 1.6×10^{-19} Coulombs, then 1 Ampere of current is the flow of 6.25×10^{18} electrons per second. The noise associated with that flow is 2.5×10^9 electrons per second or 0.4 A. The same approach can be used to estimate error in random samplings such as public opinion polls. If 625 people are polled, the margin of error is 25 or 4% of the total.

Time Jitter and Intersymbol Interference: Time jitter is a short-term variation or instability in the duration of a specified interval. *Intersymbol interference* is the result of other bits interfering with the bit of interest. Where receiver noise is directly proportional to bandwidth, intersymbol interference is inversely proportional to the bandwidth. The eye diagram is a good way to see the effects of time jitter and intersymbol interference.

FIGURE 9.2

Eye Diagrams

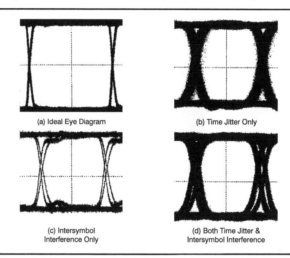

(a) Ideal Eye Diagram

(b) Time Jitter Only

(c) Intersymbol
Interference Only

(d) Both Time Jitter &
Intersymbol Interference

Figure 9.2a shows an ideal eye diagram that has no jitter or intersymbol interference. The lines are sharp. In Figure 9.2b, noise has been added which causes time jitter. This causes the lines to broaden. Only intersymbol interference has been added in Figure 9.2c, causing the sharp lines; however, multiple lines can be seen. This is due to amplitude distortion of some of the frequency elements. Figure 9.2d shows the combined effects of time jitter and intersymbol interference.

Bit Error Rate: Bit error rate (BER) is the main quality criterion for a digital transmission system. Any of the factors listed in the sensitivity analysis may cause bit errors. The bit error rate of a system can be estimated as follows:

Eq. 9.4
$$BER = Q \left[\sqrt{\frac{I_{MIN}^2}{4 \cdot N_0 \cdot B}} \right]$$

where:

N_0=Noise power spectral density (A^2/Hz).
I_{MIN}=Minimum effective signal amplitude (Amps).
B=Bandwidth (Hz).
Q(x)=Cumulative distribution function (Gaussian distribution).

SIGNAL-TO-NOISE RATIO

The signal-to-noise ratio is a common way to express the quality of the signal in a system. It is a ratio of average signal power and total noise and can be written as:

Eq. 9.5
$$SNR = \frac{S}{N}$$

In this ratio, signal (S) represents the information to be transmitted while noise (N) is the integration of all noise factors over the full system bandwidth. Expressed in decibels, the equation can be written as:

Eq. 9.6 $$\text{SNR (dB)} = 10 \cdot \log_{10}\left(\frac{S}{N}\right)$$

BER and SNR are interrelated; a better SNR yields a better BER. The optical power necessary to achieve a given SNR, and thus a given BER, is called the receiver's sensitivity. For instance, a signal-to-noise ratio of 11 dB would yield a BER of about 10^{-9}. Improving the SNR by less than 1 dB to 11.75 dB improves the BER by a factor of 1,000 times to 10^{-12}. The exact relationship between SNR and BER does vary depending on the data encoding method. This interdependent relationship, shown in Figure 9.3, is typical for *NRZ* (non-return to zero) data.

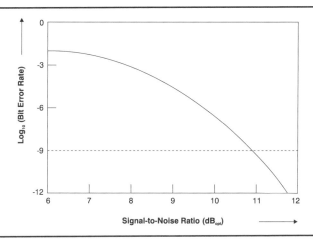

FIGURE 9.3

BER Dependence on Signal-to-Noise Ratio

MODULATION SCHEMES

The process of passing information over the communication link typically involves three steps: encoding the information, transmitting the information, and decoding the information. Encoding is the process that modifies the information so that it can be carried by the communication medium. Most communication systems incorporate some sort of encoding

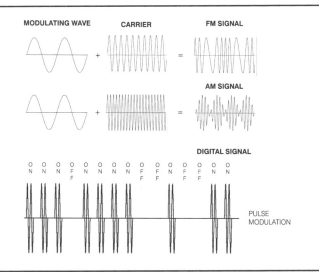

FIGURE 9.4

Types of Modulation Used for Encoding

for a variety of reasons. In some cases it improves the integrity of the transmission, in others it allows more information to be sent per unit time, and in some cases the encoding scheme takes advantage of some strength of the communication medium or overcomes some inherent weakness. This encoding process is called the modulation scheme. There are literally dozens of popular modulation schemes. Some are very fundamental techniques while others are minor or major variations on these basic themes. The four most common modulation schemes are frequency modulation (FM), amplitude modulation (AM), pulse-code modulation (PCM), and digital encoding.

Amplitude Modulation (AM)

Amplitude modulation or AM is the same scheme used in AM radio. It is a simple technique that often results in very low-cost hardware. There are basically two types of AM techniques; baseband and RF carrier. In a baseband system, the input signal directly modulates the strength of the transmitter output, in this case light. In the RF carrier AM technique, a carrier with a frequency much higher than the information to be encoded is used as the heart of the transmitter. The amplitude of the carrier wave varies according to the amplitude of the information being encoded. In a fiber optic system, the magnitude of the voltage input signal is directly translated into a corresponding light intensity. Figure 9.5 shows the function of a typical AM system.

FIGURE 9.5

Typical AM System

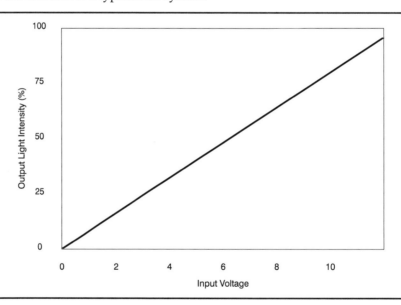

This simple technique has two main drawbacks however. First, the system requires the use of highly linear components throughout so that the signal is not distorted as it travels through the communication link. Second, since the information is encoded by varying light intensity, it becomes difficult to separate real signal level variations from changes in the optical loss in the fiber itself. For instance, if one was feeding a 100% maximum signal into the optical transmitter and if there was 10 dB of optical loss between the transmitter and receiver, the receiver would indicate that the signal level is 10%. Without some other means, the receiver cannot distinguish signal level changes from optical loss changes. One challenge of AM transmission over optical fiber is devising a scheme to compensate for the loss associated with the fiber. This problem may be solved two ways: by taking advantage of some special property of the input waveform (for video, the sync pulse is an invariant that can be used to distinguish optical loss from signal level variation) or using a technique that allows the signal level to be interpreted independently from optical loss. One means of accomplishing this is to send a pilot tone at a high frequency that is above or below the frequency of the information being encoded. The need for highly linear components can erase much of AM's advantage over other techniques because of the expense associated with obtaining highly linear LED's. In spite of the difficulties mentioned above, AM is generally the simplest and least expensive approach to encoding information for transmission over fiber.

Frequency Modulation (FM)

Frequency modulation, called FM, is a more sophisticated modulation scheme. It is well suited to the inherent properties of optical fiber since proper recovery of the encoded signals only requires measurement of timing information, one of fiber's strengths. FM is also immune to amplitude variations caused by optical loss, one of fiber's weaknesses. The heart of the FM modulator revolves around a high-frequency carrier. Now, instead of changing the amplitude of the carrier, the frequency of the carrier is changed according to differences in the signal amplitude. Part of the advantage of FM systems is buried in mathematical analyses that show that the signal-to-noise ratio at the receiver can be improved by increasing the deviation of the carrier. For instance, let us assume that a video signal is to be transmitted, and it has a 5 MHz bandwidth. If we use a 70 MHz carrier frequency and cause it to deviate 5 MHz by applying the video signal then we get about a 5 dB enhancement in the receiver's signal-to-noise compared to an AM system. If we increase the deviation of the carrier frequency to 10 MHz, then the improvement increases to 15.6 dB. This is one of FM's strong features. Another important advantage of FM is that it eliminates the need for highly linear optical components that are required for AM systems. Often optical systems employing FM encoding refer to the technique as pulse-frequency modulation (PFM). This simply means that the FM signal is limited (converted to digital 0's and 1's) before it is transmitted over the fiber. The result is the same. Generally the modulator is designed so that the frequency of pulses increases as the input voltage increases, but there is nothing magic about this convention.

FM optical systems almost always require more complex electronic circuitry than AM optical systems, but often the total cost is comparable since lower-cost optical components can be used in the FM system. Figure 9.6 demonstrates the operation of a typical FM system.

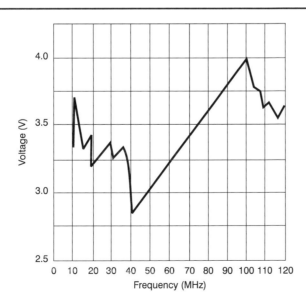

FIGURE 9.6

Typical FM System

Pulse-Code Modulation (PCM)

Pulse-code modulation or PCM converts an *analog* signal, the human voice for example, into *digital* format with a number of possible twists. The human voice can be represented by a series of numbers, the value of each number corresponding to the amplitude of the voice at a given instant of speech. Some variants of PCM are listed below. The first three describe analog-to-digital modulation, and the last two are strictly digital modulation schemes.

Pulse-Amplitude Modulation (PAM): Information is encoded by a stream of pulses with discrete amplitudes.

Delta Modulation (DM): Pulses are sent at a constant rate with duration determined by the first derivative of the input signal.

Adaptive Delta Modulation (ADM): Similar to DM with the ability to adjust the slope of the tracking signal.

Phase Shift Keying (PSK): Information is sent over a constant carrier frequency. The phase of the carrier is shifted between two levels as determined by the digital bit to be sent.

Differential Phase Shift Keying (DPSK): A variant of PSK that allows for more straight-forward decoding.

One example of this method of digital modulation is outlined in Table 9.5.

TABLE 9.5

Typical Digital System Operation

Input Signal	Output Code
0 Volts	0000
1 Volt	0001
...	...
9 Volts	1001
10 Volts	1010

ANALOG VERSUS DIGITAL

The world around us is an analog world. Analog means that most variables (e.g., temperature, sound, color and brightness) exhibit a continuous range of difference. Sound is analog in that it continuously varies in both amplitude and frequency within a given range. Another example could be a 60 W light on a dimmer switch. The dimmer adjusts the light within a certain range. The light levels are continuously variable in that there is no one discrete level. The brightness can be adjusted anywhere from completely off to completely on. However, no human sense can distinguish a truly infinite range. There is a point at which two amplitudes appear identical. In hearing, the smallest discernible unit is one decibel.

Digital implies numbers—distinct units. In a digital system all information exists in numerical values of digital pulses. A three-way lamp is an example of a digital system. Each setting brings the bulb up to a specific level of brightness. No levels exist between these three settings. That analog information can be converted to digital and that digital information can be converted to analog is important to electronic communication.

FIGURE 9.7

Analog and Digital Signals

(Illustration courtesy of AMP, Inc.)

Digital Basics

The *bit* (short for binary digit) is the basic unit of digital information. This unit has only two values: 1 (one) or 0 (zero). Electronically, the bit is equivalent to the circuit being on or off where 0=Off and 1=On. One-bit information is limited to these two values, however, if we use two bits of information, more information can be communicated. Look again at a three-way lamp. Using two bits we have 00=Off, 01=On (Dim), 10=Brighter, 11=Brightest. The more bits used in a unit, the more information can be sent. An 8-bit group is called a *byte*. A byte gives 256 different meanings to a pattern of 1's and 0's. Each time a bit is added, the number of possible combinations doubles.

Digital systems can be illustrated by pulse trains such as the one seen in Figure 9.8. A pulse train represents the 1's and 0's of digital information. The pulse train can depict high-voltage and low-voltage levels or the presence and absence of a voltage. In dealing with pulses, engineers must consider the shape of a pulse.

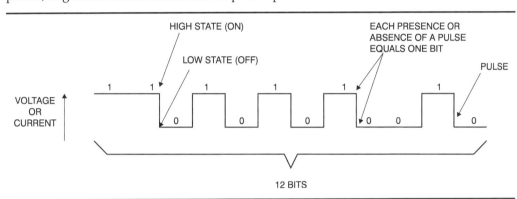

FIGURE 9.8

Typical Pulse Train

A *pulse* has five main characteristics, illustrated in Figure 9.9. Amplitude is the height of the pulse. Rise time indicates the amount of time required to turn the pulse on and is typically the length of time required to go from 10% to 90% of amplitude. Rise time is critical because it sets the upper speed limit of the system. The speed at which pulses turn off and on determines the fastest rate at which the pulses can occur. *Fall time* is the opposite of rise time, representing the time taken to turn the pulse off. Pulse width is the width, expressed in time, of the pulse at 50% of its amplitude. Bit period is the time required for the pulse to go through a complete cycle. Most digital systems are clocked; pulses must occur in the time allotted by the system for a bit period. A clock is an unchanging pulse train that provides timing by defining the bit periods.

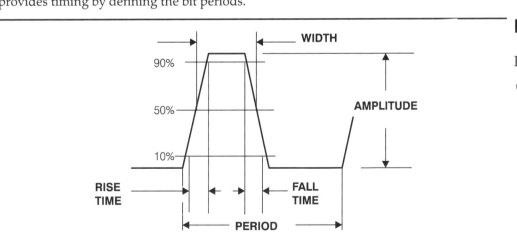

FIGURE 9.9

Parts of a Pulse

(Illustration courtesy of AMP, Inc.)

COST/PERFORMANCE CONSIDERATIONS

The cost of a fiber optic transmission system can be a critical consideration. While minimizing the cost is often a primary goal, there are performance trade-offs to consider. Component considerations such as light emitter type, emitter wavelength, connector type, fiber type, and detector type will have an impact on both the cost and performance of a system. Common sense goes a long way in designing the most cost-effective system to meet an applications requirements. A properly engineered system is one that meets the required performance limits and margins with little extra. Excess performance capability often means the system costs too much for the specific application.

With an understanding of how a fiber optic system is designed, it is time to turn the discussion to what applications these systems might be designed for.

Chapter Summary

- The key factors that determine how far one can transmit over fiber are transmitter optical output power, operating wavelength, fiber attenuation, fiber bandwidth, and receiver optical sensitivity.
- The decibel (dB) is a convenient means of comparing two power levels.
- The optical link loss budget analyzes a link to ensure that sufficient power is available to meet the demands of a given application.
- Rise and fall times determine the overall response time and the resulting bandwidth.
- A sensitivity analysis determines the amount of optical power that must be received for a system to perform properly.
- Bit errors may be caused by source intensity noise, fiber noise, receiver noise, time jitter, and intersymbol interference.
- Amplitude modulation (AM), the same scheme used in AM radio, includes baseband and RF carriers and is a simple technique that often results in very low-cost hardware.
- Frequency modulation (FM) is well suited to the inherent properties of optical fiber since proper recovery of the encoded signals only requires measurement of timing information.
- Analog means continuously changing in value.
- In a digital system all information exists in discrete numerical values.
- The bit is the basic unit of digital information and has only two values: one or zero.
- The five characteristics of a pulse are rise time, period, fall time, width, and amplitude.

Selected References and Additional Reading

Baack, Clemens. 1986. *Optical Wideband Transmission Systems.* Florida: CRC Press, Inc.

Hentschel, Christian. 1988. *Fiber Optics Handbook.* 2nd edition. Germany: Hewlett-Packard Company.

Palladino, John R. 1990. *Fiber Optics: Communicating By Light.* Piscataway, NJ: Bellcore.

Sterling, Donald J. 1993. *Amp Technician's Guide to Fiber Optics*, 2nd Edition. New York: Delmar Publishers.

—. 1991. *Universal Transport System Design Guide.* Hickory, NC: Siecor Corporation.

Yeh, Chai. 1990. *Handbook of Fiber Optics: Theory and Applications.* New York: Academic Press, Inc.

FIBER OPTIC APPLICATIONS

10

Advantages of fiber optic systems such as light weight, small size, large bandwidth, and EMI immunity make these systems applicable to a wide range of fields and uses.

BROADCAST

It has been observed that the broadcast industry is at the threshold of two major technological revolutions. The first is the rapid move to digitized video. The second is a move to some form of enhanced definition television such as *HDTV* (High-Definition Television) or enhanced *NTSC* (National Television Standards Committee). These two revolutions in combination make a move to fiber optic technology all but inevitable in the broadcast industry. Fiber optic links can support both video and audio broadcast transmissions as well as data transmission. Video transport signal types include multichannel (4, 12, 40, 60, 80 channels are common), point-to-point RS-250, and digitized video (NTSC, CCIR 656, EU95, SMPTE 259). Audio transport signal types include the multichannel audio snake, point-to-point CD quality (stereo), and digitized audio.

Digital Video

As CD technology revolutionized the audio industry, digitized video will revolutionize the broadcast industry. Because of the bandwidth required to transmit digital video, fiber is the clear choice in this application, especially in the studio environment. However, high levels of compression using standards such as *MPEG* will undoubtedly be used for mass distribution of digital video. In MPEG compressed applications, fiber is not essential because bandwidth is low.

The first all-digital video broadcast occurred at the 1994 Winter Olympics. Fiber optic links were selected to connect distant outside events such as the downhill ski events and the cross-country ski events to the production studio. The topography of the venues located outside Lillehammer, Norway made microwave links unusable. Digital Video/Serial Data Fiber Optic Video Links, built by Force, Incorporated of Christiansburg, VA were used to connect these distant venues to the production studio offering a field test that successfully demonstrated fiber's advantages in this type of application.

FIGURE 10.1

Digital Video/Serial Data Fiber Optic Link

(Photo by Force, Inc.)

The critical fiber parameters for broadcast are light weight, lightning immunity, high bandwidth, long distance, and better signal quality. Actual applications include: intra-studio broadcasting, inter-studio broadcasting, electronic news gathering (ENG), signals to TV camera pan/tilt/zoom pedestals, multimedia distribution systems, and campus video distribution systems.

CATV

Once dominated by such transmission media as twisted pair, copper coaxial cable, satellite, and microwave transmission, broadband networks are now looking to fiber for the transmission of radio frequency (RF) signals. This transition is the result of an increased consumer demand for new services, speed, bandwidth, and cost-containment. While all-digital systems may ultimately prevail, they are still prohibitively expensive to install and operate. Recently, hybrid fiber/coaxial cable networks have gained wide acceptance as an alternative to copper-only systems, allowing for a more cost-effective transition. CATV transmission is discussed in further detail in Chapter 11.

HIGH-RESOLUTION IMAGING

Fiber optics carry high-resolution video images typically encoded as RGB (red-green-blue) signals. Bandwidth can exceed 100 MHz per color. The higher bandwidth and distance capability make this application useful in medical imaging and computer workstations. Other applications include CAD/CAM/CAE computer modeling, air traffic control remote imaging, flight simulation imaging and signal distribution for flight training, telemetry, geophysical computer imaging, and map making, high-end graphic design, and publishing. A summary of screen resolution and RGB bandwidth is given in Table 10.1.

TABLE 10.1

RGB Bandwidth and Screen Resolution

Resolution (Non-Interlaced)	Scan Rate (Hz)	Analog RGB Bandwidth (MHz)
640 x 480	60	18.4
	72	22.1
800 x 600	60	28.8
	72	34.6
1024 x 768	60	47.2
	72	56.6
1280 x 1024	60	78.6
	72	94.4
1600 x 1280	60	122.9
	72	147.5
2560 x 2048	60	314.6
	72	377.5

TELECOMMUNICATIONS

The telecommunications industry is the heaviest user of fiber optic technology. Fiber optic systems are being employed in a wide range of communication devices. Fiber optics offer greater bandwidth and smaller attenuation. In addition, fiber optic cable costs less than metallic cable and is less susceptible to crosstalk. Fiber was first used in trunk lines that connected central offices with long distance toll centers. Another application gaining wide use is the subscriber loop, a circuit that connects a central office to subscriber telephones. With the high bandwidth of fiber, the telephone company could offer other services such as video and information services. Two approaches are used in subscriber loops. They are *fiber-to-the-curb (FTTC)* and *fiber-to-the-home (FTTH)*. As each name suggests, in FTTH systems the transceiver is located inside the subscriber's home while transceivers in a FTTC system remain outside the home.

The advantages of fiber to the telecommunications industry go beyond supplying phone services. Other advantages are:

- Telephone companies can increase system capacity without digging additional underground cable lines.
- Transatlantic fiber optic cable offers continuous communication between North America and Europe.
- Fiber optic cable holds an advantage over satellite communications; satellites are subject to bandwidth limitations, and produce echo and delay effects due to the much longer transmission path. These effects are not created by optical fiber.
- Fiber plays a role in improving the operation of such communications applications as public address (PA) systems, microwave communications, and emergency phone systems.

As video communications and fiber-to-the-curb and fiber-to-the-home systems increase, the range of applications for this industry will only increase and diversify.

DISTANCE LEARNING (TELE-CLASSROOMS)

Distance learning revolutionizes education, linking classrooms across a campus or a nation. The application involves video/audio links that connect a teacher in one classroom to students in other classrooms. Communication is two-way by video cameras and audio systems connected via fiber optics. The advantages of fiber include no EM radiation, immunity to EMI and EMP, distance capability, and better signal quality. As the cost of a college education skyrockets, the ability to teach more people over greater distance will benefit two ways. Education will become more accessible to those who desire it, and the educational facility can expand its student base to cover a wider area.

TELECONFERENCING

Similar to distance learning, teleconferencing employs fiber optic systems to connect municipalities and other government units by both video and audio. A specific example of this application, known as electronic magistrate, allows law enforcement officials to arraign suspects via video camera with audio. In rural communities where the arrest site and the courtroom are some distance from each other, this allows officers to process suspects from remotes rather than making a trip to the county seat. The money saved in officer man-hours alone makes this application cost-effective. In addition, teleconferencing connects remote municipalities to the state seat of government and allows business conferences to be conducted face-to-face from a variety of remote locations.

DATA COMMUNICATIONS

With bandwidths and data rates out of the reach of copper coax cable, the uses for fiber in datacom are numerous. Low-speed data links such as RS-422, RS-485, and RS-232 have found use in security and surveillance industries, distance learning, monitoring systems, and LAN's. High-speed data links capable of data rates up to 2.5 Gb/s and above will find a niche in the computer industry as well as military applications. These important applications deserve closer examination.

Computers

High-speed fiber optic links allow quick transmission of data from mainframe to remote. The bandwidth capabilities of fiber allow for a local area network (LAN) operating at high data rates. Applications in computers fall into two major categories: peripheral interconnection and local area network setup. In both areas, the critical fiber parameters include immunity to EMI and EMP, high bandwidth, and long distance. The low cost of installation and ease of upgrade are factors as well. In terms of computer peripherals, fiber connects one computer to other computers and computers to smart sensors. In local area networks, the use of fiber is more extensive.

FIGURE 10.2

Various Configurations of High-Speed Data Links

(Photo by Force, Inc.)

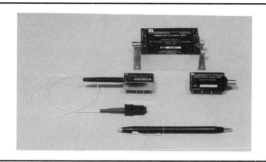

Local Area Networks

A local area network or LAN is an electronic communications network that interconnects equipment such as computers, printers, fax servers, modems, and plotters (to name a few) in a limited geographical area like an office building or a campus. All equipment can communicate with other equipment in the LAN. Each point of attachment is a node; each node is an addressable point, capable of sending and receiving information from the other nodes. The use of fiber optics is an attractive approach to LAN's for two reasons. It extends transmission distances, and it offers EMI immunity.

Star couplers and tee coupler bus configurations are often used in LAN's. The star network and the ring network are also used. A star topology connects all nodes at a central point through which all messages pass. The ring structure connects all nodes serially with one another; messages flow from one node to the next in one direction. Hybrid topologies combine more than one of these three basic topologies. For example, several stars could be arranged in a ring or bus configuration.

Fiber Optics In LAN's

The two most popular LAN's currently in use are IEEE 802.3 *ethernet* and IEEE 802.5 *token ring*. Both systems incorporate fiber optic interconnecting devices but transmit the signals over copper coax twisted pairs. The fiber distributed data interface (FDDI) is the first local area network designed in all aspects to use fiber optics. Its performance has been impressive, offering a data rate of 100 Mb/s over 100 km with up to 1000 attached workstations. Compare this to the 16 Mb/s rate for an 802.5 token ring or the 10 Mb/s rate for ethernet.

FIGURE 10.3

FDDI Switch

(Photo courtesy of AMP, Inc.)

Erbium-Doped Fiber Amplifiers

The use of erbium-doped fiber amplifiers (EDFA's) in high-speed networks is gaining momentum. EDFA's allow information to be transmitted over longer distances without the need for repeaters. The fiber is doped with erbium, a rare earth element that happens to have the appropriate energy levels in its atomic structure for amplifying light at 1550 nm. When a weak signal at 1550 nm enters the fiber, the light stimulates the erbium atoms to release their stored energy as additional 1550 nm light. This process continues as the signal passes down the fiber, growing stronger and stronger as it goes. Figure 10.4 illustrates a typical erbium-doped fiber amplifier.

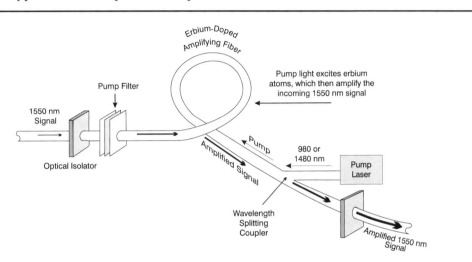

FIGURE 10.4

Erbium-Doped Fiber Amplifier

CONTROL SYSTEMS AND INSTRUMENTATION

Control Systems

Fiber optics allows control systems to be integrated through multiplexing signals and connecting modems, and high-speed links. The applications for control systems are finding use in a variety of industries. For example, in chemical plants, optical fiber senses the physical and chemical parameter changes during a chemical process and activates the control mechanisms. Because the optical sensors are small and impervious to many adverse environments, the sensors can be placed in more than one location and wired in series to provide constant monitoring during a chemical process. In aerospace applications, light weight and EMI/RFI immunity of optical fiber have replaced heavy, EMI sensitive copper cables. Future applications to monitor critical airplane controls during flight are anticipated. Optical fiber is already used in automobile instrument panel illumination, but the use of sensors to monitor controls is in development. These sensors could monitor lights, exhaust emissions, spark plug firing performance, or the firing mixture of the automobile. Radar control systems and traffic control systems that interconnect traffic signals and monitor traffic flows are already in use, and more areas are installing such highway monitoring systems.

Instrumentation

Fiber optics can be utilized in scientific instruments in many ways. Sensing capabilities are important, but size is also a factor, and fiber optic systems can be fitted to almost any system design. Developments such as the endoscope, a variation of the fiberscope, allow engineers to look into the cavity of an engine or into a reactor during operation. Fiber optic interferometric instruments allow a controlled environment and an environment under test to be simultaneously monitored and the differences to be calibrated and measured. Fiber optic holography has use in many industries; highly accurate three-dimensional images could check for flaws in aircraft structure, boiler welding and other objects. Other fiber optic instruments can measure velocity.

MILITARY

Fiber optics plays an important role in military applications in a variety of ways. A single optical fiber can replace miles and pounds of copper wire used in many types of control systems. On ships, fiber optic links carry video signals from surveillance cameras to a central control location, eliminating interference caused by powerful radar systems located near these cameras. There are fiber optic radar systems, fiber optic missile launchers, and fiber optic torpedo launchers that allow the missile to be monitored and course corrected for a direct hit. Tactical ground control systems in military LAN's let field posts monitor and gather environmental information and stay in contact with other field stations. Fiber offers a security advantage since optical fiber is much more difficult to tap than metallic wire.

FIGURE 10.5

Fiber Optic Data Converter Set Mounted on Hawk Missile System Hardware

(Photo by Northrop Grumman Corp. and Force, Inc.)

Figure 10.5 shows the data converter set developed to upgrade the Hawk missile system from copper cable to optical fiber. Developed in cooperation with Force, Inc. and Northrop Grumman Corp., the upgrade used an AT&T tactical fiber optical cable assembly (TFOCA) to replace the heavy copper cable used in the original design. The copper cable connected the control unit with the radar system and was limited to 500 feet. This endangered personnel since the Hawk uses an active radar illuminator and is therefore susceptible to radar-homing missiles. The copper-based system also weighed 300 lbs., making it difficult to transport and deploy. Using the data converter set as a transparent drop-in replacement with the tactical fiber optical cable assembly, the maximum distance was increased to 6,500 feet, while reducing the cable weight to only 140 lbs. Force, Inc. and Northrop Grumman Corp. applied mature fiber optic technology to replace the copper links without affecting system functions and without redesigning any existing Hawk Missile hardware.

SECURITY AND SURVEILLANCE

Fiber optics are ideal in security applications because they are very difficult (but not impossible) to tap without detection. The fact that these systems radiate no electromagnetic energy makes non-invasive eavesdropping techniques useless. Optical fiber carries both video and voice surveillance and incorporates data links such as RS-232, RS-485, and access control. Perimeter security, tunnel/highway monitoring, airport security, and access control use fiber optics with excellent results. In the early 1990's airport security requirements became more stringent, requiring electronic card-key access systems to screen the entry of all personnel. These systems implement video monitoring and data access at all gates. An employee's card would access a gate only if the video image being transmitted from that gate to a control center matched the image data stored in a computer. The distances involved, especially in larger airports, make fiber the best choice for these systems.

FIGURE 10.6

Fiber Optic Video, Audio, and Data Links for Security and Surveillance Applications

(Photo by Force, Inc.)

Chapter Summary

- The broadcast industry is at the threshold of two major technological revolutions, digitized video and high definition television or enhanced NTSC.
- Because of the bandwidth required to transmit digital video, fiber is the clear choice in this application.
- Because all-digital systems are prohibitively expensive to install and operate, hybrid fiber/coaxial cable networks have gained wide acceptance as an alternative improvement to copper-only systems.
- High-resolution imaging applications include the endoscope, CAD/CAM/CAE computer modeling, air traffic control remote imaging, flight simulation imaging, telemetry, geophysical computer imaging, map making, and high end graphic design.
- The telecommunications industry is the heaviest user of fiber optic technology.
- Two approaches of bringing fiber optic technology to the consumer are fiber-to-the-curb (FTTC) and fiber-to-the-home (FTTH).
- Distance learning links classrooms across a campus or a nation.
- Teleconferencing employs fiber optic systems to connect municipalities and other government units by both video and audio.
- In computers, high-speed fiber optic links allow quick transmission of data from mainframe to remote.

- A local area network (LAN) is an electronic communication network that interconnects equipment such as computers, printers, fax servers, modems, and plotters in a limited geographical area like an office building or a campus.
- The two most popular LAN's currently in use are IEEE 802.3 ethernet and IEEE 802.5 token Ring.
- Erbium-doped fiber amplifiers allow information to be transmitted over longer distances without the need for repeaters.
- Control systems can be integrated using fiber optics by multiplexing signals, modems, and high-speed links.
- Fiber optics can be utilized in scientific instruments such as sensors, interferometric devices, and holographic imagers.
- Military applications of fiber optics technology include fiber optic links on ships that carry video signals from surveillance cameras to a central control location, fiber optic radar systems, fiber optic missile launchers, and fiber optic torpedo launchers.
- Security applications benefit from the use of fiber optics because the links are very difficult to tap without detection.

Selected References and Additional Reading

—. 1991. *Basics of Fiber Optics: Applications for CATV.* State College, PA: C-Cor Electronics, Inc.

Bazaar, Charles. 1991. "Fibre Channel: The Future of High-Speed Connectivity." *Fiberoptic Product News.* May 1995: 34-35.

Cottingham, Charles F. 1990. "The Fiberoptic Challenge for the '90s and Beyond." *Fiberoptic Product News.* 10th Anniversary Issue: 31-32.

Engineering Staff, Codenoll Technology Corp. 1990. *The Fiber Optic LAN Handbook.* New York: Codenoll Technology Corporation.

Henkemeyer, Rich. 1995. "How to Future-Proof a Hybrid Fiber/Coaxial Cable Network." *Lightwave Magazine.* May 1995: 40-41.

Kopakowski, Edward T. 1995. "Metropolitan Area Networks for Interactive Distance Learning." *Fiberoptic Product News.* September 1995: 17+.

Paff, Andy. 1995. "Hybrid Fiber/Coaxial Cable Networks to Expand into Interactive Global Platforms." *Lightwave Magazine.* May 1995: 32-38.

Shoemake, Mike and Lynn Woods. 1995. "FDDI in the" [sic]. *Fiberoptic Product News.* August 1995: 3-5.

Yeh, Chai. 1990. *Handbook of Fiber Optics: Theory and Applications.* New York: Academic Press, Inc.

VIDEO OVER FIBER

11

Video signals are complicated. They encode continually changing pictures, and in many cases, sound. If one looks closely at a television set, one would see that the picture is actually made up of many horizontal lines drawn one after another. The video signal contains the information to draw these lines, detailing whether parts of the line should be dark or light and how the colors should be displayed. Sound is also encoded in the signal.

VIDEO SIGNAL QUALITY PARAMETERS

There are dozens of parameters used to describe the quality of a video signal. The ones encountered most often are signal-to-noise ratio (SNR), *differential gain* (DG), and *differential phase* (DP). These are defined as follows:

SNR: Signal-to-noise ratio measures how clear and crisp the picture is. A picture that is very sharp and clear has a large or high SNR. Conversely, a picture that shows lots of snow or other interference has a small or low SNR. An SNR of 67 dB is said to be "studio quality" video while the typical cable system delivers 45-50 dB to the average home.

DG: Differential gain measures the portion of the video signal that controls how bright a given dot will be. In other words, a perfect link that has zero DG distortion would show everything in the picture at exactly the right brightness. However, if the link has a large amount of DG distortion, the brightness shades would be incorrect. For instance an object that is supposed to be white might be gray. DG usually ranges from 1% to 5%.

DP: Differential Phase measures the portion of the video signal that controls the color, hue, or shade. A link that has zero DP distortion will show all colors exactly right. A link with large amounts of DP distortion will show colors incorrectly. For instance, a leaf that is supposed to be green might appear yellow. DP is usually less than a few degrees.

METHODS OF ENCODING VIDEO SIGNALS

There are three predominant methods of encoding a video signal: amplitude modulation (AM), frequency modulation (FM) and digital modulation. The difference between various modulation schemes can be understood by examining their corresponding frequency spectra. The very simple baseband AM occupies the region from DC to about 5 MHz and requires the least bandwidth. The RF carrier modulation spectrum is similar; it has been shifted to some nonzero frequency (F). This approach requires additional bandwidth and offers no advantage over baseband operation in a system where a single channel is carried on each fiber. However, it allows multiple channels to be combined onto a single fiber. With vestigial-sideband AM, the spectrum is again shifted to a nonzero frequency (F), and the lower *sideband* has been removed by filtering. It allows the spectrum to be used much more efficiently than straight RF carrier AM, requiring half the bandwidth per channel. Table 11.1 offers a comparison between AM, FM, and digital transmission of video signals over fiber.

TABLE 11.1

Comparison of
AM, FM and
Digital Fiber Optic
Links

Parameter	AM	FM	Digital
Signal-to-Noise Ratio	Low-Moderate	Moderate-High	High
Performance vs. Attenuation	Sensitive	Tolerant	Invariant
Transmitter Cost	Moderate-High	Moderate	High
Receiver Cost	Moderate	Moderate-High	High
Receiver Gain Adjustment	Often Required	Not Required	Not Required
Installation	Adjustments Required	No Adjustments Required	No Adjustments Required
Multichannel Capability	Good Capability Requires High Linearity Optics	Fewer Channels	Good
Performance Over Time	Moderate	Excellent	Excellent
Environmental Factors	Moderate	Excellent	Excellent

The notable difference between sine wave FM, square wave FM, and pulse-frequency modulation is the presence of harmonics. The square wave FM spectrum signal contains only odd-order harmonics. The pulse-frequency modulation spectrum contains all odd- and even-order harmonics. This yields a very cluttered spectrum that is poorly suited for multiple channel stacking. However, its value as a single-channel transmission scheme is not lessened.

RF carrier AM, vestigial-sideband AM, and sine wave FM are best suited for multiple channel transmission because they lack harmonics. Multiple channels can be combined for transmission over a single fiber by assigning different carrier frequencies to each video signal. The resulting modulated carriers are summed to yield a single composite electrical signal. For instance, a 4-channel sine wave FM system could occupy the frequencies of 70 MHz, 90 MHz, 110 MHz and 130 MHz.

ADVANTAGES OF FM

The inherent advantage offered by FM compared to AM is that the output signal amplitude is independent of link loss, whereas the AM received signal amplitude directly reflects the link loss. This dictates that the AM link must have some type of gain control at the receiver. Sometimes, this is a manual control because of the difficulty of implementing an automatic gain control (*AGC*) on recovered baseband video. This complicates the installation of some AM links because the user must inject a test signal and use an oscilloscope to set the correct receiver gain level. It also means that the link will not automatically adapt to changes in the link loss. Link loss could change over time due to fiber degradation or transmitter power variations. The received optical power will almost certainly change when optical connectors in the link are demated and remated. Ideally, an AM system requires an AGC to be practical.

A picture containing a range of levels from black to white will be 140 IRE units high, which also corresponds to 1.0 V_{p-p}. Thus, one IRE unit equals about 7.14 mV. A picture containing only black levels will only be 40 IRE units high, or about 0.286 V_{p-p}. If a simple AGC is used, one that always strives to output 1.0 V_{p-p}, the resulting output waveform for an all-black picture will be severely distorted. The sync-pulse, usually 40 IRE units high, will become 140 IRE high. This will not be properly interpreted by most monitors. Therefore, a simple peak-to-peak AGC cannot be effective in setting the gain of an AM receiver. A more sophisticated circuit that analyzes the input waveform and sets the sync pulse level to 40 IRE units is required.

The drawback to this approach is that the circuit has to know what picture format to expect. It has to assume that the picture format is NTSC, PAL, or another type. A circuit designed to work with NTSC may not be compatible with other formats. Long-term expandability of the link is limited. One possible scenario is that NTSC-compatible-only AM links today may be obsolete if a change over to high-definition television (HDTV) occurred.

FM transmission can be designed to be independent of the encoding format. Inherent in the FM modulation/demodulation technique is proper recovered signal level and complete independence of the input video format. The recovered signal level is independent of link loss and varies only by component tolerances in the transmitter and receiver. The output level typically equals the input level plus or minus a small percentage. This reduces long-term maintenance and ensures proper performance and consistent quality over the duration of the link's life.

The other factor to consider is that FM can yield a higher signal-to-noise ratio (SNR) than AM under similar link loss conditions. In a comparison of the estimated SNR for an AM versus FM scenario, each modulation scheme can be summarized by a single factor relating the video SNR to the fiber optic link carrier-to-noise. The FM β is unity for this comparison; a larger value of β would result in improved SNR performance. Additional improvements would be obtained for β values ranging from one to ten.

Of course the penalty to be paid for larger β values is a large increase in transmission bandwidth. With all other factors equal, an FM system with a β of two has about an 8 dB advantage over AM. This translates into either longer distance links, more loss margin, higher picture quality, or some combination of the three factors. This analysis overlooks nonlinearities in the link and their effect on picture quality. Nonlinearities in AM fiber optic links can cause cross-modulation products that results in lines and bars in the transmitted picture. While this interference is not noise, it can be quite annoying.

DISADVANTAGES OF FM

FM's key disadvantage is the requirement of more bandwidth. If one considers an NTSC video signal with a 5 MHz bandwidth allotment, then a baseband AM system only requires 5 MHz of bandwidth. However, an FM system requires at least seven times this bandwidth for a total of 70 MHz to accommodate the 70 MHz FM carrier. This requires higher bandwidth from the electro-optical components and the associated electrical components.

A good quality 62.5/125 μm fiber has a bandwidth-distance product of at least 400 MHz•km. Thus the carrier frequency of 70 MHz becomes a factor at a distance of 5 km or longer. It becomes a problem at 1.5 km distance when 850 nm LED's are used. Multimode fiber used with an 850 nm LED has very limited bandwidth because of high dispersion.

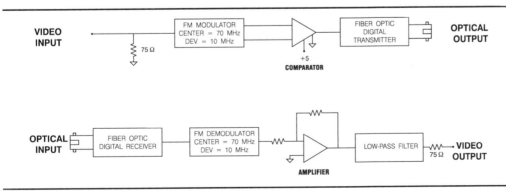

FIGURE 11.1

FM Video Link

Figure 11.1 shows the typical architecture of an FM video link. The top of the figure illustrates an FM transmitter. In this design, the FM modulator centers around 70 MHz with a deviation of 10 MHz. Once through the modulator, the analog signal is converted to a digital signal by the comparator and is then converted to light by the fiber optic digital transmitter. The FM video receiver operates in the reverse manner, converting the digital signal to an electronic signal. The signal is decoded by the FM demodulator and is then filtered through a low pass filter that removes the carrier frequency components.

FM VIDEO LINK PERFORMANCE

Figure 11.2 illustrates the performance typically achieved by a fiber optic video link showing signal-to-noise ratio of the video signal versus fiber length in meters. The performance is level to 9,000 meters of fiber and rolls off to 16,000 meters where it drops below the minimum specification of 48 dB signal-to-noise ratio.

FIGURE 11.2

SNR versus Fiber Length

(Data based on the performance of Force, Inc. fiber optic FM video links.)

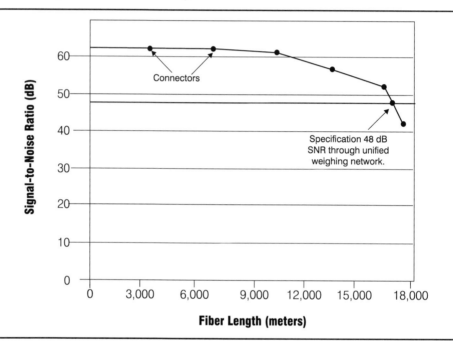

Figure 11.3 demonstrates FM video links as compared to AM video links. At 0 dB optical loss, AM has a higher 70 dB signal-to-noise ratio, but at 3 dB of optical loss, the AM signal-to-noise ratio drops below the FM SNR. This continues for the remainder of the optical loss range. At a practical (and typical) loss of 10 dB, the FM link has a 13 dB signal-to-noise advantage over the AM link. The slope of the AM video link indicates that the signal-to-noise ratio drops 2 dB every time the optical loss increases by 1 dB. This contrasts to the FM link that shows nearly flat signal-to-noise ratio performance over the usable loss range. At the loss limit, the signal-to-noise ratio drops by 2 dB per 1 dB of optical loss like the AM link.

DIGITAL VIDEO

The digital transmission of video signals over optical fiber is accomplished by converting the analog video inputs to digital values. At the transmitting end, the baseband video channels are routed to analog-to-digital (A/D) converters. Once the analog information has been put into a digital form, the digital channels are time-division multiplexed (TDM) and sent to the laser transmitter. The digital signal is converted into light pulses; the laser is on for a one and off for a zero. At the receiving end, the light pulses are converted back

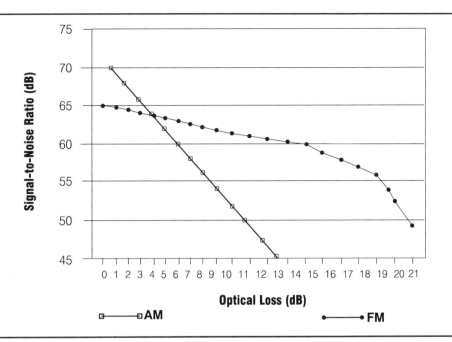

FIGURE 11.3

FM Video Link
versus AM Video
Link

*(FM data based on the performance of
Force, Inc. fiber optic FM video links.)*

into electrical pulses. The pulses are time-division demultiplexed (TDD) and sent through a digital-to-analog (D/A) converter. This converts the information back into a baseband video signal. Figure 11.4 shows a block diagram of this process.

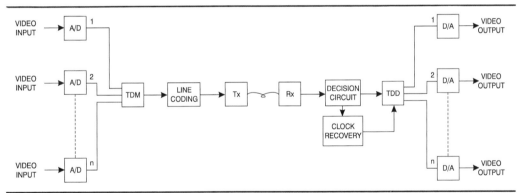

FIGURE 11.4

Digital Video
Transmission Block
Diagram

The analog to digital conversion is a three step process:

Sampling: This is the act of obtaining a sequence of instantaneous values of a particular signal, usually at regular time intervals. The sampling rate is the number of samples taken per unit of time. Typically, analog signals are sampled at a rate at least twice the highest frequency component of the signal.

Quantizing: This process divides the continuous range of values of a signal into nonoverlapping subranges that are not necessarily equal. Each subrange is given a discrete value. If a sampled signal falls within a given subrange, the sample is given the corresponding discrete value.

Coding: This is the process of assigning a binary code to designate each of the quantizing levels. A coding scheme can be non-return to zero (NRZ), return to zero (RZ) manchester, etc.

Digital video has several inherent advantages where video transmission is concerned. Digital systems do not require linear light sources, allowing the system to have a wide range of non-critical operating parameters. They can be used with both single-mode or multimode fiber, and have a high immunity to noise.

Several standards exist or have been proposed for digitized video transmission. All result in a serial digital data stream. The most important standards are shown in Table 11.2. Of the standards listed, the first four are currently the most active. EU95, a digitized HDTV standard, is the focus of several european community development programs. The SMPTE (Society of Motion Picture and Television Engineers) Committee has proposed a second digitized HDTV standard. One cannot help but marvel at how much bandwidth is required by these standards compared to standard broadcast analog video's bandwidth of 4.5 MHz. Digital compression will help some, but mostly on the distribution of these signals. Studios will use full bandwidth or minimally compressed (lossless) signals that will require the use of fiber as the prevalent means of signal distribution. Copper coax is nearly worthless at gigabit data rates.

TABLE 11.2

Digital Video Formats

Standard	Sample Rate	Word Rate	Line Serial Date Rate
NTSC	14.318 MHz	14.318 M/s	143.18 Mb/s
PAL (4 f_{sc})	17.7 MHz	17.7 M/s	177 Mb/s
SMPTE 259 (4:2:2)	13.5 MHz	27 M/s	270 Mb/s
CCIR 656	13.5 MHz	27 M/s	270 Mb/s
EU95	72.0 MHz	144 M/s	1440 Mb/s
Proposed SMPTE Digitized HDTV Standard	74.25 MHz	148.5 M/s	1485 Mb/s

SERIAL DATA TRANSMISSION FORMATS AND STANDARDS

All of the standards in Table 11.2 describe the conversion process from an analog video signal to a digitized signal. The standards describe the scale factors, word formats and overall data structures. These standards do not describe how to serialize data or send it over copper coax cable or fiber. The first attempt at this definition is in SMPTE standard T14.224, *Serial Digital Interface for 10-bit, 4:2:2 Component and 4 fsc NTSC Digital Signals*. The current draft only addresses copper coax transmission. Table 11.3 lists the key parameters of the serial digital data link associated with SMPTE T14.224. The broadcast industry already produces equipment that works with the SMPTE T14.224 standard, so producing a T14.224 compatible fiber optic link means that customers can now effortlessly switch between copper and fiber as needed.

TABLE 11.3

SMPTE T14.224 Signal Levels and Specifications

Parameter	Requirement
Input/Output Load	75 Ohm
Return Loss (Electrical)	15 dB Minimum
Data Amplitude (Into 75 Ohms)	800 mV ±10% peak-to-peak
DC Offset	±0.5 Volts
Rise/Fall Times (20-80%)	<1.50 ns
Jitter & Distortion	<±250 ps
Electrical Connector	BNC

COMPRESSION

Certainly one of today's hot technologies is digital compression. It is impacting computers and video in a big way. There are two basic types of compression, lossless and lossy. Lossless compression algorithms do what the name implies; they return the data or picture to its exact original state. Lossy algorithms throw away some information in the original picture. Lossless algorithms are the only type ever used on data, while lossy algorithms are used exclusively for pictures. Lossless algorithms can only achieve a factor of two or three compression, while lossy algorithms can achieve a factor of 100 or more compression. Various standards such as *JPEG*, MPEG-1, and MPEG-2 are often used in conjunction with images and moving pictures.

CATV TRANSMISSION

Cable television has its roots in community antenna television and retains the acronym of its predecessor. As consumer demands for services increase, bandwidth and signal quality requirements for cable television transmission also increase, and optical fiber is becoming a major player in this area of video transmission. While cable systems began as strictly copper systems, fiber backbones for CATV networks are commonplace. The hybrid fiber/coaxial cable (HFC) network has become widespread in the CATV industry. This network uses a fiber optic trunk line as the backbone and coaxial cable lines for the drops to individual sites.

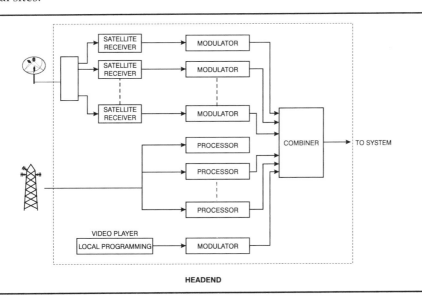

FIGURE 11.5

CATV Headend Block Diagram

A CATV system consists of several components. The first, obvious component is the cable. As mentioned, copper cable's dominance is waning as bandwidth demands make optical fiber the only choice. Passive components such as signal dividers, signal combiners, and taps create multiple paths in the network allowing greater coverage for more users. Amplifier stations along these paths recover the signal strength after it has been attenuated by the cable. These stations are placed in the transmission line at approximate intervals.

There are three layers of a CATV system. The trunks cascade one after another to cover greater distances. The feeder reaches directly from the trunk to uniformly cover smaller areas, and the drop is the final short-distance connection to the individual receiver, i.e., the cable customers's television. Signals are collected from many of sources including satellites and distributed from the *headend* of the system. The signals are amplified and converted to the desired television channel frequency before distribution to the cable system.

Chapter Summary

- Three common parameters used to describe video quality are signal-to-noise ratio (SNR), differential gain (DG), and differential phase (DP).

- The three predominant methods of encoding a video signal are amplitude modulation (AM), frequency modulation (FM), and digital modulation.

- RF carrier AM, vestigial-sideband AM, and sine wave FM are best suited for multichannel transmission.

- The inherent advantage offered by FM over AM is that the output signal amplitude is independent of link loss, whereas the uncompensated AM received signal amplitude directly reflects the link loss.

- FM's key disadvantage is the requirement for more bandwidth.

- The digital transmission of video signals over optical fiber is accomplished by converting the analog video inputs to digital values.

- Some inherent advantages of digital video systems are that they do not require linear light sources, can be used with both single-mode or multimode fiber, and have a high immunity to noise.

- Digital video formats include NTSC, PAL (4 f_{sc}), SMPTE 259 (4:2:2), CCIR 656, and EU95.

- A typical CATV system consists of signal dividers, signal combiners, taps, and amplifier stations.

- The three layers of a CATV system are the trunk, the feeder, and the drop.

Selected References and Additional Reading

—. 1991. *Basics of Fiber Optics: Applications for CATV.* State College, PA: C-Cor Electronics, Inc.

Engineering Staff, Codenoll Technology Corp. 1990. *The Fiber Optic LAN Handbook.* New York: Codenoll Technology Corporation.

Fibush, David K. 1995. "Video Compression Overview." *Video Systems.* December 1995: 32+.

Henkemeyer, Rich. 1995. "How to Future-Proof a Hybrid Fiber/Coaxial Cable Network." *Lightwave Magazine.* May 1995: 40-41.

Kopakowski, Edward T. 1995. "Metropolitan Area Networks for Interactive Distance Learning." *Fiberoptic Product News.* September 1995: 17+.

Paff, Andy. 1995. "Hybrid Fiber/Coaxial Cable Networks to Expand into Interactive Global Platforms." *Lightwave Magazine.* May 1995: 32-38.

DATA OVER FIBER

12

We have seen how video signals are transmitted over fiber; however, fiber excels at digital data transmission because of characteristics such as high bandwidth, low loss, and EMI/RFI/lightning immunity. These characteristics are more important in some applications than in others. In security applications, EMI/RFI/lightning immunity is the paramount consideration. For very long distance applications, low loss drives the choice to fiber. For high-speed applications, such as gigabit data rates between computers, the high bandwidth is the determining factor. After the initial decision has been made to use fiber optics for a data transmission requirement, other decisions must be made such as:

- Operating wavelength
- Optical loss budget
- Fiber type
- Fiber size
- Optical connector type

Whereas analog transmission over fiber often uses a carrier transmission, digital data transmission typically uses baseband transmission. This means that a logic *0* is sent as a low light level (or the light may be completely off) and a logic *1* is sent as a high light level. Often, some sort of data coding (e.g., 4B5B, 8B10B, etc.) is employed to guarantee some minimum transition density. This is required because it is difficult to design a fiber optic data link that includes true DC or steady-state data rates. DC can be accommodated at low data rates (typically <1 MHz), but becomes increasingly difficult at higher data rates. This limitation is due to the relatively poor DC performance of amplifiers capable of high frequencies.

True DC response is difficult because negative light does not exist. Light is an unbalanced transmission media. When transmitting data using an electrical signal on copper cable, a balanced signal provides optimum signal fidelity, especially at higher data rates. A logic *0* might be sent using a -1.0 Volt level, while a logic *1* might be sent using a +1.0 Volt level. This is a balanced, symmetrical transmission scheme. Recovering the data is a simple matter of comparing the received signal to a threshold of 0.0 Volts regardless of how much the signal is attenuated during transmission. For instance, if the signal was attenuated by a factor of ten, the received voltage levels would be +0.1 Volts and -0.1 Volts. A threshold of 0.0 Volts would still be perfectly centered allowing optimum data recovery.

DC data transmission with light is not quite so easy. Since negative light cannot be generated, the best one can do is use a zero light level for a logic *0* and a high light level for a logic *1*. This makes recovering the data very challenging at the receiver. As the light is attenuated through the fiber, the receiver threshold needs to be adjusted to 50% of the peak received light level. However, the receiver has no way of determining the peak level if a steady logic *0* is being sent. To correct this, the logic threshold is set at something like 5% of the largest peak signal. This can lead to varying duty cycle distortion as the input light level changes. DC coupled data links also cannot tolerate as much optical loss as an AC coupled data link.

AC coupling the receiver with a capacitor prevents these difficulties. This removes the average DC voltage from the signal allowing the threshold to again be set to a constant 0.0 Volts. The value of this capacitor sets the low-frequency data rate of the link. Typically the low-frequency limit is set to be 100 to 1,000 times lower than the lowest data rate. For instance, a 10-200 Mb/s data link would have its low frequency limit set to 1 kHz to 10 kHz to ensure proper operation. Generally, the transmitter is DC coupled. It is also necessary to tailor the upper frequency limit of the transmitter and receiver to match the data rate range. If the upper frequency limit is too low, the recovered data will exhibit intersymbol interference. If the upper frequency limit is too high, excessive noise will degrade sensitivity.

Data coding takes an unknown, unpredictable data stream and converts it into a data stream with a predictable transition density. An AC coupled receiver must have constant transitions at its input for proper operation. There are two types of coding schemes used, cyclic bit scrambling and block coding. Consider a fictitious block coding scheme 3B6B. Table 12.1 shows how this scheme might work. The input data stream is divided into 3-bit groups. These form the input code. The lookup table then converts the 3-bit input code into a 6-bit output code. This scheme has the disadvantage of generating twice as many bits, doubling the data rate, but it also guarantees that the transition density is always 50%. Even if the input data stream is always zeroes, the scrambled output code will have an average duty cycle of 50%. Real scrambling codes are not as inefficient than the example given. Block coding schemes such as 4B5B or 8B10B increase the bit rate by 25%. The alternative data scrambling technique, cyclic bit scrambling does not increase the bit rate, but it also does not provide as much confidence in the output duty cycle.

TABLE 12.1

3B6B Coding Scheme

Input Code	Output Code
000	100011
001	100101
010	100110
011	101001
100	101010
101	101100
110	110001
111	110010

COMMON DATA INTERFACE CHARACTERISTICS

There is an almost uncountable variety of interfaces used in data links; a few common interfaces are worth mentioning. These include TTL, CMOS, RS-485, ECL, and PECL. Table 12.2 lists these common interfaces and the key attributes associated with them.

TABLE 12.2

Common Data Interface Characteristics

Interface Type	Logic Low Voltage	Logic High Voltage	Receiver Threshold	Transmission Impedance	Frequency Range
CMOS	0.0 Volts	+5.0 Volts	+2.5 Volts	N/A	0-200 MHz
TTL	0.4 Volts	+3.4 Volts	+1.4 Volts	N/A	0-200 MHz
RS-485	-0.2 Volts	+0.2 Volts	+0.0 Volts	120 Ω	0-2.5 MHz
ECL	-1.8 Volts	-0.8 Volts	-1.3 Volts	50 Ω	0-2,000 MHz
PECL	+3.2 Volts	+4.2 Volts	+3.7 Volts	50 Ω	0-2,000 MHz

The values shown in Table 12.2 are typical values. Refer to manufacturers' data sheets and published specifications for full details.

KEY DATA LINK ATTRIBUTES

The key parameters associated with fiber optic data transmission are:

- Data rate
- Bit error rate (BER)
- Eye diagram parameters
- Jitter
- Protocol

A link's data rate limitation is obvious. It is designed in by the fiber optic data link manufacturer. BER is determined by signal-to-noise limitations in the receiver. Noise is usually accurately modeled as a *Gaussian* function. If the signal-to-noise ratio is sufficiently high, the link will be essentially error free. However, when the signal-to-noise ratio drops, errors will increase. Table 12.3 shows the relationship between signal-to-noise ratio and BER. (Note that this relationship depends on the exact coding scheme being used and several modulation parameters.) The relationship becomes very steep at optical signal-to-noise ratios above 10 dB. At an optical SNR of 11.0 dB, the BER is 8.87×10^{-11}. Increasing the optical SNR to 12.0 dB improves the BER to 8.64×10^{-14}, an advancement by a factor of 1,000 times.

The required BER depends on the application and, to some extent, the data rate. For instance, specifying a BER of 10^{-12} for a 2400 Baud RS-232 link would mean that one error occurs every 13.2 years, making an acceptance test impossible to perform. However, if the data rate increases to 2.5 Gb/s, then one error would occur every 6.7 minutes. Typically the information contained in Table 12.3 is used to predict very high BER values. For instance, the optical loss could be increased until a BER of 10^{-3} is observed. Then the optical power could be increased the appropriate number of dB per Table 12.3 until the desired BER is obtained. Typically low-speed data links (< 10 Mb/s) use a BER of 10^{-9}. A BER of 10^{-12} is becoming common for gigabit data links, and a few requirements are even pushing the BER requirement to 10^{-15}.

SNR$_{OPT}$	BER
6.0	2.13×10^{-2}
6.5	9.80×10^{-3}
7.0	3.49×10^{-3}
7.5	9.61×10^{-4}
8.0	2.05×10^{-4}
8.5	3.38×10^{-5}
9.0	4.32×10^{-6}
9.5	4.27×10^{-7}
10.0	3.27×10^{-8}
10.5	1.94×10^{-9}
11.0	8.87×10^{-11}
11.5	3.15×10^{-12}
12.0	8.64×10^{-14}
12.5	1.84×10^{-15}
13.0	3.02×10^{-17}

TABLE 12.3

Relationship Between SNR$_{Optical}$ and BER

The eye diagram is a powerful tool for analyzing the overall quality of a fiber optic data link. Key eye parameters are eye width and eye height. Figure 12.1 shows a typical eye diagram.

FIGURE 12.1

Typical Eye
Diagram

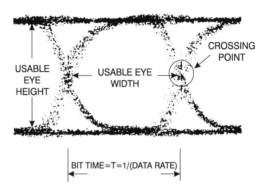

An eye diagram is observed on an oscilloscope by monitoring the receiver data output and triggering on the data clock. Usually the data clock is available from the bit error pattern generator. A pseudorandom data pattern is used to generate the worst case eye diagram. The blurry outline seen in Figure 12.1 is due to noise (jitter) and intersymbol interference. Noise in the receiver is converted to jitter. Noise and jitter increase as the optical input to the receiver decreases. Intersymbol interference is due to non-flat gain somewhere in the fiber optic link. A pseudorandom sequence will contain a wide range of frequency components up to the actual data rate. If the link has flat gain across this entire range, no intersymbol interference will be seen. Intersymbol interference can also be caused by distortion or overload of some components.

The key features of the eye diagram are the crossing points and the open area in the eye. Ideally, the crossing points should be symmetrical and centered, and the open area should be as large as possible. The distance between the crossing points is the reciprocal of the data rate. The other key eye diagram attributes are the usable eye opening width (time) and the usable eye opening height (voltage). Since noise is Gaussian in nature, the eye will gradually close the longer it is observed. The exact values of the eye width and height depend on the desired BER contour. Figure 12.2 shows the BER contours associated with the eye diagram shown in Figure 12.1. Contour A corresponds to a BER of 10^{-6}, contour B corresponds to a BER of 10^{-9}, and contour C corresponds to a BER of 10^{-12}. Usable eye width and usable eye height decrease as the BER requirement gets smaller and tougher to meet.

FIGURE 12.2

BER Contours of a
Typical Eye
Diagram

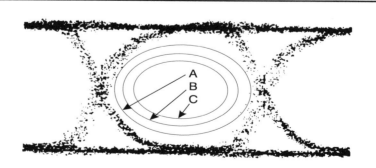

Table 12.4 shows the conversion factor to convert root mean square (RMS) jitter to peak-to-peak jitter. RMS jitter is measured and then converted to peak-to-peak jitter for a given BER. This allows the usable eye to be extrapolated to very small BER values. Jitter increases as the square root of sum of squares while data dependent jitter (intersymbol interference) increases linearly.

BER	SCALE FACTOR
10^{-4}	7.438
10^{-5}	8.530
10^{-6}	9.506
10^{-7}	10.398
10^{-8}	11.224
10^{-9}	11.994
10^{-10}	12.722
10^{-11}	13.411
10^{-12}	14.069

TABLE 12.4

Conversion for Root Mean Square Jitter to Peak-to-Peak Jitter

Figure 12.3 shows an example of how jitter adds across three circuit elements.

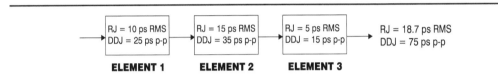

RJ = 10 ps RMS DDJ = 25 ps p-p **ELEMENT 1** — RJ = 15 ps RMS DDJ = 35 ps p-p **ELEMENT 2** — RJ = 5 ps RMS DDJ = 15 ps p-p **ELEMENT 3** — RJ = 18.7 ps RMS DDJ = 75 ps p-p

FIGURE 12.3

Jitter Buildup

The total jitter for the system shown in Figure 12.3 is as follows:

Assume the BER is 10^{-9}. The total system random jitter is:

$$\sqrt{10^2 + 15^2 + 5^2} = 18.7 \text{ ps RMS}$$

The total system data dependent jitter is:

$$25 + 35 + 15 = 75 \text{ ps}$$

Now if the data rate is 500 Mb/s, then the bit time is 2,000 ps. Use the information in Table 12.4 to convert RMS jitter to peak-to-peak jitter:

$$\text{Jitter}_{p\text{-}p} = 11.994 \cdot 18.7 = 224 \text{ ps}$$

So now the total jitter is:

$$224 + 75 = 299 \text{ ps}$$

The usable eye width will be:

$$2000 - 299 = 1701 \text{ ps}$$

DISTORTION IN FIBER OPTIC DATA LINKS

The ideal data link, fiber optic or otherwise, introduces no distortion into the data being transmitted. Why does distortion matter in a digital data link? Isn't it just a matter of ones and zeroes? Yes and no. If the data link introduces excessive distortion, the ability to recover the ones and zeroes will be hampered. Figure 12.4 shows the effect of one type of distortion, noise, and how it hampers the ability to recover the data. The noise closes the eye diagram which makes data recovery more difficult. The key types of distortion are noise (jitter), data dependent distortion, and duty cycle or pulse width distortion. Noise can be measured in a digital link as random jitter (RJ). Random jitter is simply the time-domain manifestation of voltage noise. The topology of a fiber optic receiver was described in Chapter 6. Basically a receiver consists of a photodiode, an amplifier, such as a transimpedance amplifier, and a digitizing element, often a decision circuit. Figure 12.4 shows a typical receiver topology and the relationship between voltage noise at the transimpedance amplifier output and the random jitter at the output of the decision circuit. As the amplitude of the voltage noise increases, the amount of time-domain random jitter will increase as well. Random jitter is usually measured as a RMS value since it is basically Gaussian in nature. The information in Table 12.4 can be used to convert the RMS values to *peak-to-peak (p-p)* values. Note that the exact scaling factor depends on the BER. Random jitter can be further broken down as phase noise. Phase noise is the spectral density of the jitter versus offset frequency. It is often an important parameter when designing phase-locked loops and clock recovery circuits. Random jitter is caused by electronics noise in the fiber optic link.

FIGURE 12.4

Fiber Optic Receiver—Noise to Jitter Conversion

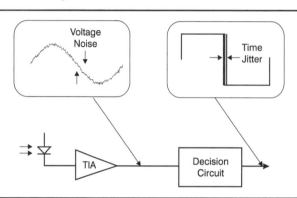

There are several sources of this noise. The primary contributors in most modern fiber optic links are the transmitter light emitter (LED or laser diode) and the receiver photo-diode preamplifier. Other elements contribute but are usually minor factors in modern fiber optic links. At low optical losses (i.e., high optical power at the receiver input) the transmitter is generally the limiting factor. At high optical losses (i.e., low optical power at the receiver input) the receiver noise usually dominates. Figure 12.5 shows the typical random jitter performance of a high-speed fiber optic link.

FIGURE 12.5

Random Jitter versus Optical Loss

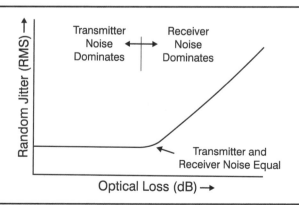

The next type of distortion is data dependent distortion or data dependent jitter. Data dependent jitter (DDJ) is also known as intersymbol interference. DDJ is caused by a frequency dependent amplitude distortion of the data stream. A pseudorandom NRZ data stream has a frequency spectrum that is a sin(x)/x function. The sin(x)/x function is shown is Figure 12.6.

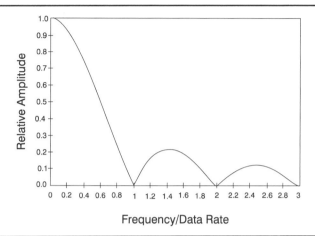

FIGURE 12.6

Frequency Spectrum of a Pseudorandom NRZ Data Stream

For the data to be faithfully transmitted, the data link needs to transmit the portion of the spectrum that contains most of the energy with minimal distortion. If the data link has non-flat frequency response in this critical region, then intersymbol interference will result.

It is possible to test random jitter and data dependent jitter relatively independently. Random jitter can be tested using a '101010...' test pattern. Data dependent jitter can be tested by constructing a pattern that contains two distinct frequency components. For instance a pattern such as '111110000010' exercises the link at its maximum toggle rate (the last two bits) and at a rate that is 20% of the maximum rate (the first 10 bits). This will show any DDJ distortion very clearly.

The last type of distortion is duty cycle or pulse width distortion. This is a systematic distortion caused by asymmetrical rise and fall times in some elements in the system. For instance, if a component has faster rise times than fall times, it will consistently make pulses too wide and vice versa.

COMMON TRANSMISSION STANDARDS

There are many types of fiber optic data links. The variety of links reflects a wide range of applications, data rates, and formats. Figure 12.7 shows many of the transmission standards now in use. The TV remote control reference point at 100 b/s is a free-space optical link that connects the television to the remote control. The other extreme, labeled future telecom, indicates the direction that telecommunication links are moving. Standards such as 14,400 baud usually connect computer modems to the phone line. The transmission standard labeled digitized audio refers to compact disk or CD audio that has a data rate of 512 kb/s. Data rates from RS-232 to RS-422/485 are short-haul data rates, while European CEPT/CCITT, Japan NTT, CCITT, BELL/CCITT, DS1-DS4 and SONET, OC1-OC48 relate to current telecommunications links. MIL-STD-1553 is the military standard used in military craft. The transmission standard ESCON is gaining popularity as a means to connect computers to peripherals and computers to computers.

FIGURE 12.7

Worldwide
Communication
Standards

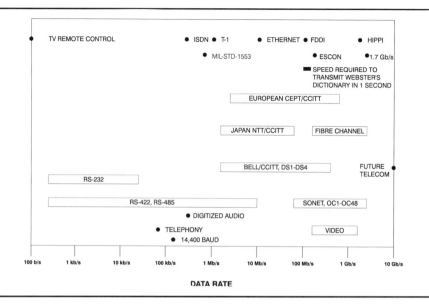

All transmission standards are arranged from lowest (100 b/s) to highest (10 Gb/s). A position on the chart around 100 Mb/s indicates the transmission speed required to send Webster's dictionary in one second. Another point of reference not shown could be placed at 3.2 Gb/s. This would indicate the speed required to transmit the Encyclopedia Britannica in one second.

HIGH-SPEED DATA TRANSMISSION

Data rate separates high-speed data links from other data links. This value indicates the number of data bits transmitted per second (bps or b/s). Generally data rates over 100 Mb/s are considered high-speed. In many ways, it is true to say that fiber can achieve data rates that are out of the reach of copper coax cable. Fiber's key attributes are very advantageous for high-speed data transmission: large bandwidth, low attenuation, long distance capability, EMI/RFI, and lightning immunity, are all critical parameters for high-speed data transmission. It is true, however, that high bit data rates place demands on system components that low bit data rates do not; therefore, special consideration must be given in selecting the components to develop a high-speed circuit. Frequency range, frequency-dependent components, parasitic components (such as capacitors), delay times: these variables and more must be given additional consideration when designing a high-speed data link.

The rapid growth of high-speed data links in network applications has lead to the development of many protocols such as Fibre Channel, FDDI, SONET, and ATM (see Figure 12.7). Local area networks (LAN), wide area networks (WAN), and point-to-point communications may each have protocols uniquely suited to their requirements. To further complicate matters, LAN's, WAN's, and point-to-point links are being interconnected and internetworked.

FIGURE 12.8

An Internetworked
System

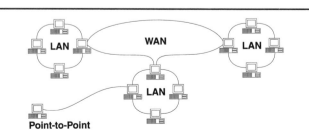

Most available fiber optic data links are designed to handle a single data protocol. Recently, fiber optic data links capable of meeting the requirements of several data protocols simultaneously have begun to appear. These are known as multi-protocol links.

In designing a multi-protocol link the designer must meet both physical and performance requirements for all of the protocols that the link is expected to handle. At the physical layer, the requirements for many of the protocols are very similar. The optical transmitter must take an incoming electrical signal and convert it to an optical signal suitable for transmission over fiber. Conversely, the receiver must take an incoming optical signal and convert it to an electrical signal. Most protocols also require some way to disable the transmitter and require a loss-of-signal flag on the receiver.

On the other hand, there are factors to challenge the development of a multi-protocol high-speed data link. The optical dynamic range or optical loss range and *receiver sensitivity* vary from protocol to protocol. Optical dynamic range describes the range of optical input powers for which the receiver will deliver acceptable performance. Receiver sensitivity is the minimum average optical input power to the receiver for which it will deliver acceptable performance. Bandwidth requirements and fiber compatibility are also protocol-specific issues. The link bandwidth defines the maximum rate at which the link can transfer data. Data links are specified in terms of maximum data rate in bits/second (b/s or bps). When designing a data link, one must know how much bandwidth in Hertz (Hz) is required to transfer a given data rate. Generally, the required bandwidth in Hz is 56% of the maximum data rate. This formula is given in Equation 12.6.

Eq. 12.6 $$BW\ (Hz) = 0.56 \bullet Data\ Rate\ (b/s)$$

For example, a data rate of 1 Gb/s would require about 560 MHz of bandwidth. Considering all these factors, to handle more than one data protocol, a multi-protocol high-speed data link must meet the following criteria:

- The receiver's sensitivity must be low enough to handle the lowest required sensitivity of the protocols in use.
- The receiver's optical dynamic range must be wide enough to handle the widest requirement.
- The transmitter's optical output power must be set such that the link can handle all loss budget requirements.
- The link bandwidth must be such that the highest data rates can be transferred while maintaining all specified levels of performance.

Such a link was developed in 1994 by Force, Incorporated for use in both FDDI and fibre channel networks. These two important high-speed protocols bear further discussion.

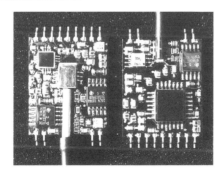

FIGURE 12.9

Multi-Protocol High-Speed Data Link

(Photo by Force, Inc.)

FDDI

The fiber distributed data interface standard was first introduced in the 1980's as the next step above ethernet and token-ring networks. The topology is a token-passing ring topology that uses two counter-rotating rings. The primary ring carries the information while the secondary ring is used as a backup in case of a component or other failure in the primary ring. Figure 12.10 illustrates this topology.

FIGURE 12.10

FDDI Dual Ring
Topology

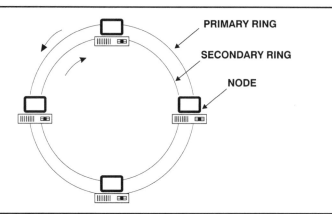

Data transmission is 100 Mb/s, but because FDDI uses a 4B5B transmission code, one extra bit is added for every four bits of data, making the actual data rate 125 Mb/s. The data is transferred between nodes which include a transmitter, a receiver, an electronic interface to the terminal, and an optional optical bypass switch. The transmitter/receiver pair act as a repeater. The receiver detects and amplifies the signal which is then transmitted to the next node.

Because of the dual ring topology, FDDI networks are highly reliable. In the event of failure such as a cut cable or inoperable node, the network bypasses the problem area by converting the network to a single ring until the problem is fixed.

FIBRE CHANNEL

This high-speed network protocol operates at an average data transfer rate of 132-1062.5 Mb/s. Fibre channel was initially developed to augment the small computer system interface (SCSI) protocol which transfers data at 20 Mb/s. The advance of such applications as multimedia, image processing, supercomputer modeling, and transaction processing to and from mass storage has led to the need for a much faster transfer protocol. With a minimum data rate of 132 Mb/s, a maximum data rate of 1062.5 Mb/s, and enormous bandwidth, fibre channel is meeting these high-speed transfer requirements. In fact, work is being done to extend the data rates to 2125 Mb/s and 4250 Mb/s.

Fibre channel topologies may be point-to-point or switched. As a point-to-point topology, the protocol has many attributes of a channel. It establishes dedicated, point-to-point connections between devices rather than using a shared medium. This allows the full bandwidth to be provided to each connection. It is also hardware-intensive, relying on a frame header to trigger the routing of arriving data to the proper *buffer*. The switch topology is used to connect a number of different devices where the switch acts as a central box that manages multiple point-to-point connections. In this topology, fibre channel has many of the attributes of a network. It is bidirectional, and it decouples the data encoding from the physical media, enabling a high degree of scalability. All physical control signals are replaced by logical constructs, eliminating the need for a channel control bus.

One advantage of fibre channel is its ability to transport other network protocols such as asynchronous transfer mode (ATM), FDDI, ethernet, SCSI, and high performance parallel interface (HIPPI) over a single medium with the same hardware connection. Fibre channel achieves this by taking advantage of the fact that all network and channel protocols use buffers to hold the data being sent or received. Fibre channel provides the means to transfer this data between the sending buffer at the source and the receiving buffer at the destination without concern for how the data is formatted. Fibre channel uses 8B/10B encoding developed by IBM in 1983.

With a better understanding of fiber optic components and systems, it is time to examine the methods used for testing and troubleshooting fiber optic installations.

Chapter Summary

- When using fiber for data transmission, factors such as operating wavelength, optical loss budget, fiber type, fiber size, and optical connector type must be decided.
- Data transmission over fiber typically uses baseband transmission.
- Data scrambling takes an unknown, unpredictable data stream and converts it into a data stream with a predictable transition density.
- A few common data interfaces are TTL, CMOS, RS-485, ECL, and PECL.
- The key parameters associated with fiber optic data are data rate, bit error rate (BER), eye diagram parameters, jitter, and protocol.
- The required BER depends on the application and, to some extent, the data rate.
- The eye diagram is a powerful tool for analyzing the overall quality of a fiber optic data link.
- Although digital signals transmit data as a series of ones and zeros, distortion can still occur.
- Types of digital distortion include random jitter, data dependent jitter, and duty cycle or pulse width distortion.
- Sources of noise in digital data systems may be the transmitter light emitter or the receiver photodiode preamplifier.
- Fiber optic data links have transmission standards that range from DC to 10 Gb/s.
- High-speed data links are an integral component in the rapidly growing areas of high-speed communications and high-speed networks, with standards such as fibre channel, FDDI, SONET, and ATM.
- Optical dynamic range describes the range of optical input powers for which the receiver will deliver acceptable performance.
- Sensitivity is the minimum average optical power to the receiver for which it will be able to deliver acceptable performance.
- Because of their dual ring topology, fiber distributed data interface (FDDI) networks are highly reliable.

- Fibre channel operates at an average data transfer rate of 132-1062.5 Mb/s.
- One advantage of fibre channel is its ability to transport other network protocols.

Selected References and Additional Reading

Bazaar, Charles. 1995. "Fibre Channel: The Future of High-Speed Connectivity." *Fiberoptic Product News*. May 1995: 34-35.

Hentschel, Christian. 1988. *Fiber Optics Handbook*. 2nd edition. Germany: Hewlett-Packard Company.

Kuecken, John A. 1987. *Fiberoptics: A Revolution in Communications*. Blue Ridge Summit, PA: Tab Professional and Reference Books.

Newell, Wade S. 1995. "Multi-Protocol High-Performance Serial Digital Fiber Optic Data Links." from *Digital Communications Design Conference, Day 3*, Joseph F. Havel, ed. Proceedings of Design SuperCon '95, Santa Clara, CA, Feb. 28-Mar. 2, 1995.

Shoemake, Mike and Lynn Woods. 1995. "FDDI in the" [sic]. *Fiberoptic Product News*. August 1995: 3-5.

Sterling, Donald J. 1993. *Amp Technician's Guide to Fiber Optics*, 2nd Edition. New York: Delmar Publishers.

TESTING & MEASUREMENT TECHNIQUES

13

Standardized measurement and testing techniques are important to precisely define the parameters under which a fiber optic system or any of its components will function. In evaluating the performance of a system, all the components that comprise the system must be considered. Techniques for measuring fiber optic systems fall into two major categories: functional testing and performance testing. Functional testing involves determining whether a component of the fiber optic system is operating. For example, testing fiber continuity with the use of an *optical time-domain reflectometer* represents functional testing. Performance testing involves how the system performs relative to the expected performance determined by the optical link loss budget.

FIBER OPTIC TEST EQUIPMENT

An essential part of any fiber optic system is the test equipment required to install and troubleshoot it. Several important types of test equipment are now in wide use. These are as follows:

Optical Power Meter: This instrument measures the amount of optical power in a fiber. Most models handle several wavelengths and provide relative (dB) as well as absolute (dB or Watts) measurements. Multiple adapters are usually required to deal with different optical connector types.

Optical Light Source: An optical light source injects a stable test light signal into a fiber. Most models offer continuous wave (CW) modes. Some offer modulation modes as well. Typical test modulation frequencies are 270 Hz, 1 kHz and 2 kHz.

Optical Loss Meter: This instrument combines an optical power meter and an optical light source into a single instrument. It is also called an *optical loss test set* (OLTS).

Fiber Identifier: A fiber identifier traces (locates) the path of a fiber. This is accomplished by injecting a tracer signal into the fiber and using some sort of tapping device on the fiber. The tapping device causes a small amount of light to be detected as it leaks out of the fiber. The light that leaks out is captured and processed to determine if it matches the tracer signal.

Talk Set: Talk sets are used to coordinate maintenance activities over an optical fiber. Often, optical fibers are underground or in shielded rooms where walkie-talkies are unusable. In these cases, the fiber itself is the best means of communicating. Talk sets are available for operation on one fiber or two fibers and offer half-duplex or full-duplex communications.

Optical Time Domain Reflectometer (OTDR): This instrument diagnoses installed fibers. The OTDR is unique because it only requires access to one end of the fiber. (Note: While the OTDR only requires access to a single end of the fiber, accurate measurements typically require that measurements be taken from both ends and then averaged.) Measurements are obtained by analyzing light backscattered from the fiber. An OTDR can measure and locate fiber loss, splice loss, connector loss and breaks in the fiber.

Optical Spectrum Analyzer: Optical spectrum analyzers determine the wavelength of light. This device is usually used to determine the emission spectrum of a light source.

Fiber Optic Attenuator: A fiber optic attenuator, also called an optical attenuator, simulates the loss that would be caused by a long length of fiber. This device is typically used to perform receiver testing. While an optical attenuator can simulate the optical loss of a long length of fiber, the dispersion that would be caused by a long length of fiber is not accurately simulated.

Backreflection Meter: Modern high bit rate digital fiber optic links and most laser-based analog links require very low backreflection to operate properly. This instrument quantifies the amount of backreflection in the fiber path.

Bandwidth Sets: Very popular in the early days of fiber, these sets were used to determine the bandwidth of a length of fiber. Dramatic improvements in the consistency of fiber and the level of testing provided by fiber manufacturers has made this all but obsolete.

Local Injector Detectors: These sets are used to measure and/or assist in the tuning of fiber splices. Some are simple optimizers, while others provide an absolute indication of splice loss. They do so by injecting and detecting light through the sides of the fiber, usually using sophisticated microbending techniques. For rotary splice tuning, the light is detected through the splice itself rather than the fiber.

MEASURING SYSTEM COMPONENTS

Light Source: Light sources include lasers, LED's, broadband sources, or monochromators. Lasers and LED's are widely used as sources. Important characteristics include output power, speed, output pattern (numerical aperture), spectral width, fiber-type compatibility, ease of use, lifetime, and cost. Broadband sources were once popular but are now seldom used. Typical broadband sources include quartz, halogen, or xenon arc lamps with narrow bandpass interference filters. The filter typically has a bandpass that approximates the output of the source to be used in the proposed system to better account for wavelength-dependent fiber characteristics such as NA, attenuation, and dispersion. Monochromators isolate narrow portions of light by dispersing light into its component wavelengths. Most commercial monochromators exhibit very low energy on the output side, and they select a very narrow bandwidth.

Mode Scramblers: Mode scramblers mix light to excite every possible mode of transmission within the fiber. The easiest *mode scrambler* to make is a 15 cm long tube at least 7 cm in diameter filled with 1-mm lead shot through which the bare fiber is passed. Another type uses a row of one-eighth inch diameter brass pins through which the fiber is zigzagged. The resulting bends in either type cause mode coupling that fills the fiber. A more complex scrambler is a butt-welded (fusion spliced) length of alternating graded-index, step-index, and graded-index fibers. The step-index fiber generally has a length of one meter. The discontinuities that result mix the light.

FIGURE 13.1

Three Types of
Mode Scramblers

(Illustrations courtesy of AMP, Inc.)

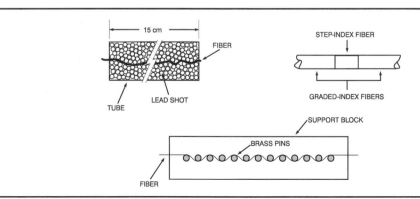

Core Mode Filter: Mandrel wrap core mode filters allow high-order mode signals from the core to be removed. High-order modes traveling through several hundred meters of fiber leak into the cladding and are lost. This results in an exit numerical aperture less than the material NA of the fiber. A fiber that has reached modal equilibrium, along with the reduced NA, is said to exhibit long-launch conditions. Rather than testing connector loss over several hundred meters of fiber, core mode filters simulate this distance. The standard recommended core mode filter for smaller fibers is 12.5-mm diameter mandrel around which the fiber is wrapped five times under zero tension. The mandrel wrap reduces the exit NA to about 50% of the fiber's material NA. The mandrel wrap also reduces the light-emitting area of the core of a graded-index fiber by about 50%. This reduction in the emitting area affects the performance of a connector or a splice during loss measurements.

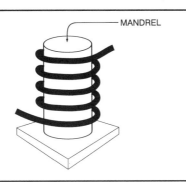

FIGURE 13.2

Mandrel Wrap

(Illustration courtesy of AMP, Inc.)

Cladding Mode Stripper: The use of the mandrel wrap described above scrambles the modes or strips the high-order modes. This stripped light has nowhere to go except the cladding. In short fiber runs or in setups where the mandrel wrap occurs at the end of the fiber, this light deflected to the cladding can be substantial. It is necessary to strip the *cladding modes* with a device known as a cladding mode stripper. This device incorporates a fiber, stripped of its cladding buffer, and covered in Corona Dope (available from TV repair suppliers) or some other liquid with a refractive index higher than the cladding. Corona dope has advantages: it is low-cost, it has a high refractive index, and the coating is black. Figure 13.3 illustrates a cladding mode stripper.

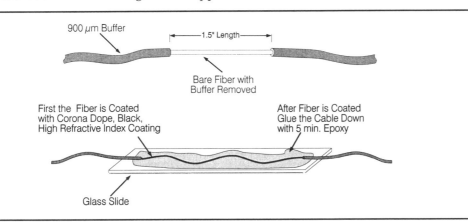

FIGURE 13.3

Cladding Mode Stripper

Detector System: Optical multimeters, also called optical power meters, measure optical power levels. The meter is completely electronic with sensors that plug into the unit. Different sensors are available for use at different power levels and operating wavelengths. Adapters permit bare fibers or a variety of popular connectors to be used with the sensor. Some multimeters include the Photodyne, Inc. Model 22XLA, the SOAR

Corporation Model 1800 series, and Hewlett-Packard's 3478A multimeter. A drawback of the multimeter is that in many applications both ends of the fiber must be available. An optical time-domain reflectometer allows testing when only one end of the fiber is available. This device relies on the backscattering of light that occurs in an optical fiber for detection.

MEASUREMENT TECHNIQUES FOR COMPONENTS

Fiber

The measurements critical to evaluating an optical fiber vary depending on whether the fiber is multimode or single-mode. Tests for multimode fiber measure attenuation, multimode dispersion, chromatic dispersion, numerical aperture, and core diameter. Important single-mode fiber tests measure attenuation, chromatic dispersion, cutoff wavelength, and spot size.

Attenuation is a critical fiber parameter. Testing is complicated by the propagation of a number of modes, each of which propagates differently from the other modes. In order to account for these differences, attenuation is tested by exciting the fiber to *equilibrium mode distribution (EMD)*. EMD represents the modal distribution through a long length of fiber. An optical source and a power meter provide the means for testing attenuation. The insertion loss for a short reference fiber is compared to the insertion loss of the longer length test fiber. This test is only useful if care is taken to accomplish the same coupling efficiency in both cases. By analyzing the fiber's backscatter signal with an optical time-domain reflectometer, the uniformity of the attenuation along the fiber length can be determined.

FIGURE 13.4

EMD—Excitation of Graded-Index Fiber

(Illustration courtesy of Hewlett-Packard.)

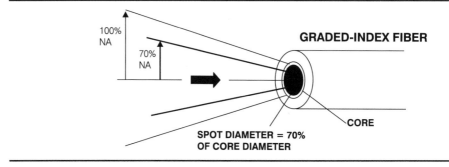

Multimode dispersion is the term used to describe the pulse broadening caused by the different velocities of the individual modes. By exciting the fiber with a short pulse in which the modal distribution is EMD, and measuring the pulse width at the end of the fiber, dispersion can be measured. It is important to note that a narrow spectral width such as given by a laser diode should be used. Another form of dispersion, chromatic dispersion, is a measure of pulse broadening due to the different velocities of different colors contained in the spectrum of the source. This broadening is directly related to the spectral width of the source. Chromatic dispersion is a material property, and cannot be measured directly. Instead, multimode dispersion adds quadratically to the measurement result.

Numerical aperture and core diameter determine how much power can be launched into multimode fiber. The NA is defined by the maximum angle of the guided rays in the fibers. It is measured at the output (in the far field) because the maximum angle at output is equal to the maximum angle at input. Measurements should be taken when all modes are fully excited. The core diameter is measured in the near field of the fiber output. Power density is measured, again, while all modes are fully excited. Figure 13.5 shows the output from a multimode fiber. Numerical aperture is not universally defined; some manufacturers define it at the 50% power points, while others define it at the 5% power points. This specmanship can confuse potential users. If the manufacturer wants to show a large NA he will pick 5% power points and vice versa. Be sure to read the fine print.

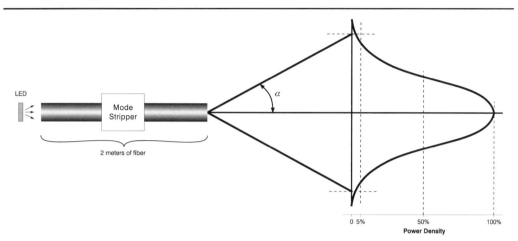

FIGURE 13.5

Numerical Aperture Measurement

(Illustration courtesy of Hewlett-Packard.)

When the measurement wavelength is longer than the cutoff wavelength, only one mode will propagate. Under this condition, measuring the attenuation of a single-mode fiber is less complicated than the attenuation measurements of multimode fiber. To maintain a constant input excitation of the fiber, the measurement should be taken in two steps. First measure the output power at the far end. Then cut the fiber near the input end and measure the output again. The difference in power levels (in units of optical dB) is the attenuation. This is known as the cutback method, and as with multimode fiber, the use of an OTDR to analyze backscatter will provide information on the uniformity of the attenuation.

Single-mode fiber is affected mainly by chromatic dispersion. Chromatic dispersion can be attributed to three factors: material dispersion, waveguide dispersion, and profile dispersion. Material dispersion is caused by the wavelength-dependance of the fiber's refractive index and the associated differences in the speed of light. It is generally described in units of ps/(nm•km).Waveguide dispersion is caused by the wavelength-dependence of the modal characteristics of a fiber. This is a result of the geometry of the fiber which causes the propagation constant of each mode to change if the wavelength changes. In multimode fiber this effect is washed out by the presence of so many modes, but in single-mode fiber the result is a refractive index somewhere between the core and the cladding. A fiber's profile is the contrast of refractive indices of the core and cladding of the fiber. Profile dispersion is the result of different wavelength-dependent refractive indices of the core and the cladding due to the different materials involved. In multimode fiber this is a critical parameter because the profile can only be optimized for one wavelength. In single-mode fibers profile dispersion is treated as part of waveguide dispersion. In order to test chromatic dispersion, narrow pulses of different wavelengths (colors) are launched, and the difference in arrival times is measured.

In single-mode fiber, cutoff wavelength defines the lowest wavelength which should be used if high bandwidth is important. Below the cutoff wavelength, more than one mode will propagate. The cutoff wavelength is measured by launching a wide spectrum into a short fiber and measuring the attenuation of each spectral component.

Mode field diameter is more important than core diameter in single-mode fiber. This is because the fundamental mode in a single-mode fiber can be approximated by a Gaussian beam. The Gaussian beam is fully defined by two numbers: the spot radius and the wavelength. The radiation characteristics of the end of a fiber can be directly calculated by these numbers as can waveguide dispersion.

Automated switching systems for testing components such as cable, connectors, or couplers are beneficial in keeping down the cost of test equipment, because several devices can be tested simultaneously. Figures 13.6 and 13.7 demonstrate a setup to test cable using an automated switching system.

FIGURE 13.6

Cable Testing

(Illustration courtesy of DiCon, Inc.)

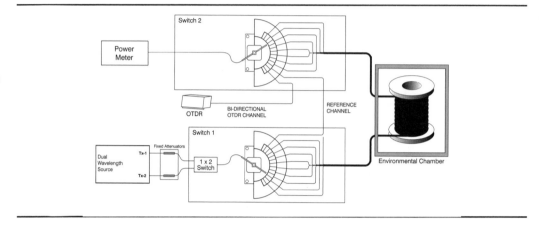

FIGURE 13.7

Backscatter Test for Fiber

(Illustration courtesy of DiCon, Inc.)

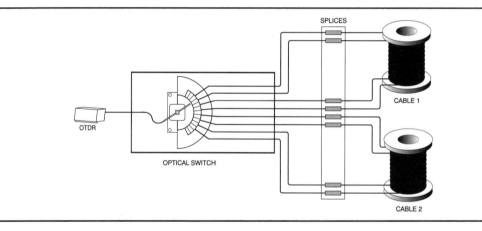

Source Measurements

Two types of sources, LED's and laser diodes, must be considered when measuring fiber optic systems. LED measurement tests include output power, modulation bandwidth, center wavelength, spectral width, source size, and far field pattern. Tests for laser diodes include optical parameters, output power, modulation bandwidth, center wavelength/ number of modes, chirp/line width, and mode field of the Gaussian beam as well as electrical parameters such as threshold current, slope efficiency, forward voltage, monitor current, etc.

LED's exhibit a nearly linear dependence of the output power on the drive current. Power from the fiber pigtail can be measured by a power meter. Because of the wide spectral width of the emitted radiation, the accuracy of the measurement depends on the wavelength-dependence of the power meter's detector.

In measuring the modulation bandwidth, the LED is intensity-modulated with a sweep-generator, and a PIN diode reconverts the optical signal back to the electrical domain. The frequency response can be observed on an oscilloscope. Another approach uses a network analyzer with an optical-to-electrical conversion at the receiver end.

Center wavelength as well as spectral width can be measured with an optical spectrum analyzer. The result of this measurement will influence the pulse broadening of the fiber. The size of the radiating area as well as far field should be measured directly at the LED chip (without the fiber). The size can be determined by analyzing the microscopic image (near field), whereas the measurement of the angular power-distribution in some distance from the source will yield the far-field. Narrow near-fields and far-fields are necessary for high coupling efficiency to the fiber.

Output power can be measured with a variable current source and a power meter. For lasers, threshold current defines the onset of the stimulated emission.

Modulation bandwidth measurement techniques for lasers are identical to those described for LED's. However, because lasers operate at a much higher bandwidth, the measurement equipment must be able to read higher bandwidth. Center wavelength as well as the number of modes should be measured with an optical spectrum analyzer.

Chirp is the undesired wavelength shift caused by intensity modulation. Laser diodes that emit a single linewidth are affected by chirp. Interferometric methods are capable of measuring both chirp and linewidth.

The radiation characteristics of a laser diode can be approximated by an elliptic version of the Gaussian beam. A Gaussian beam has a finite beam width which smoothly transits into a light cone of fixed numerical aperture. It is rotationally symmetric to the direction of propagation. The ellipticity is caused by the fact that the emitting area is a stripe rather than a circle. A far-field measurement (analysis for the power density at some distance from the radiating area) will deliver the parameters of the Gaussian beam.

Detector Measurements

Tests for PIN photodiodes and APD's include diameter, spectral responsivity, bandwidth, and dark current/NEP. APD's are additionally tested for multiplication factors and excess noise. Detectors perform two functions: signal detection in receivers and optical power measurement. When the detector's function is signal detection, the smallest possible diameter is the desirable measurement because the *noise equivalent power (NEP)* is proportional to the active diameter, and the bandwidth is inversely proportional to the active area. When the detector is acting as an optical power detector, the desirable diameter is the largest possible measurement because it increases power measurement accuracy.

For both PIN photodiodes and APD's, spectral responsivity is strongly dependent on wavelength. The measurement is made with a wavelength-calibrated combination of a tungsten lamp and a tunable monochromator. Ideal responsivity would be proportional to the wavelength; however, actual readings may vary considerably. The multiplication factor measured in APD's is a result of high voltage which leads to the multiplication of the number of generated carriers. This multiplication factor is measured the same as spectral responsivity.

Detector bandwidth can be measured by exciting the detector with a sinewave-modulated laser source. A network analyzer with an electrical-to-optical conversion of the generator signal can be used to perform the measurement. Noise equivalent power is critical because of its influence on the achievable sensitivity of the receiver. Ideally, the NEP is proportional to the square root of the dark current. A more accurate measurement can be taken using a spectrum analyzer. In APD's, the excess noise factor is an additional noise contribution caused by the multiplication process mentioned above. It too is measured with a spectrum analyzer.

Interconnection Loss Measurements

The ideal interconnection of one fiber to another would have two fibers that are optically and physically identical held by a connector or splice that squarely aligns them on their center axes. However, in the real world, system loss due to fiber interconnection is a factor. Insertion loss is the primary consideration for connector performance. There are three types of insertion loss: fiber-related loss, connector-related loss, and system factors that contribute to loss. Because of the discrepancy between insertion-loss testing and connector performance, users must understand the test methods used to measure insertion loss. The best test results are obtained when lengths of fibers are attached to the source and detector as permanent parts of the test setup. This avoids variations in results that are caused by source and detector interconnection losses from test to test.

Insertion loss tests will reduce the influence of fiber-related losses. A general test should be reproducible and provide applicable results. Most tests measure the output power (P_1) of a length of fiber. The fiber is then cut in the middle and terminated with a connector or splice. The output power (P_2) is measured again. Insertion loss is given by:

$$\text{Eq. 13.1} \qquad \text{Loss (dB)} = 10 \cdot \log_{10}\left(\frac{P_1}{P_2}\right)$$

The length of fiber must be broken perfectly in the middle to produce an identical fiber on each side of the splice. This method purposely eliminates fiber-induced losses in order to evaluate connector performance independently of fiber-related variations. Three sets of launch conditions are of interest.

1. Short-launch, short-receive: Represented by short fibers with no mandrel wrap on the transmitting or receiving ends. Short-launch, short-receive conditions exhibit losses that increase with the slightest mechanical offset of the connection. Lateral misalignment is a critical parameter under these conditions.

2. Long-launch, short-receive. A mandrel wrap is on the transmitting end but not on the receiving end. This condition reduces the exit NA of the transmitting fiber, and end-separation losses are smaller. Since all of the receiving core can be used, greater separation of the fibers can be tolerated.

3. Long-launch, long-receive. This condition is created by using a mandrel wrap at both the transmit and receive end and shows greater sensitivity to lateral misalignment than the other two conditions. Because the effective core area of both fibers is reduced, any offset increases loss more significantly.

The insertion loss test assumed that two pieces of identical fiber were used. However, if two different types of fibers are connected, then NA mismatch loss and diameter mismatch loss must be accounted for.

NA mismatch loss occurs when the numerical aperture of the transmitting fiber (t) is larger than that of the receiving fiber (r). NA mismatch loss is illustrated in Figure 13.8.

FIGURE 13.8

NA Mismatch Loss

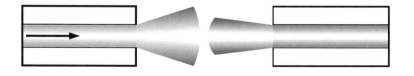

The calculated loss for numerical aperture mismatch is approximated by:

$$\text{Eq. 13.2} \qquad \text{Loss}_{NA} = 10 \cdot \log_{10}\left(\frac{NA_r}{NA_t}\right)$$

As illustrated in Figure 13.9, core diameter mismatch occurs when the core diameter of the transmitting fiber (t) is larger than the core diameter of the fiber at the receiving end (r). Cladding diameter mismatch is similar to core diameter mismatch loss except the cladding of the transmitting fiber differs in diameter from the cladding of the receiving fiber. Either mismatch prevents the cores from aligning.

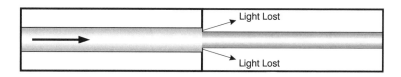

FIGURE 13.9

Core-Diameter Mismatch Loss

Both types of diameter mismatch loss are approximated by:

$$\text{Eq. 13.3} \qquad \text{Loss}_{dia} = 10 \cdot \log_{10}\left(\frac{dia_r}{dia_t}\right)$$

This equation is only accurate if all of the modes in the fiber are excited. When only low-order modes are excited, the loss is greatly reduced and may not be present at all.

Concentricity, also known as eccentricity, occurs because the core may not be perfectly centered in the cladding. Ellipticity or ovality describes the fact that the core or cladding may be elliptical rather than circular. The alignment of the two elliptical cores will vary depending on how the fibers are brought together. These forms of connector loss are illustrated in Figure 13.10.

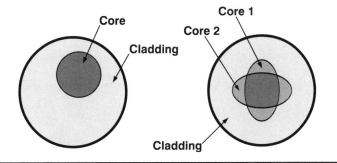

FIGURE 13.10

Concentricity and Ellipticity

Connector-related loss can also result from the mechanical misalignment of the optical fiber cores. There are several types of misalignment loss: lateral displacement, angular misalignment, and end separation.

A connector should align the fibers on their center axes, but when one fiber's axis does not coincide with the other fiber's axis, lateral displacement occurs. A displacement of only 10% of the core axis diameter results in a loss of about 0.5 dB. The ends of mated fibers should be perpendicular to the fibers' axes and to each other. Failure to be perpendicular is called angular misalignment.

Some connectors hold the two fibers slightly apart to prevent the fibers from rubbing against each other and damaging their end polishes. *Fresnel reflection loss* or end separation loss is caused by the difference in the refractive indices of the two fibers and the air that fills the gap between the two fibers. Some connector manufacturers believe the use of index-matching gel in the gap reduces Fresnel reflection loss, but others do not recommend using index-matching gel. This gap may collect small flecks of abrasive contaminates that will damage the end finishes, and the addition of index-matching gel could compound this contamination.

In a single-mode interconnection with a flat end finish, Fresnel reflection loss can be as much as -11 dB, a level sufficient to disrupt the operation of most lasers. This loss can be reduced by rounding the fiber end of one fiber during polishing (called a PC or physical contact finish). While it would seem practical to use a flat finish and butt the ends, getting two perfectly smooth, flat finishes is nearly impossible. With a rounded finish, fibers always touch on a high point near the light-carrying fiber core. (See Figure 13.11.)

FIGURE 13.11

Fiber End Face
Finishes

(Illustration courtesy of AMP, Inc.)

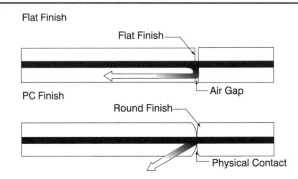

System-related factors in connector loss involve the launch and receive conditions. These conditions result from the mode distribution in the fibers. The performance of the connector depends on modal conditions and the connector's position in the system. Four different conditions exist:

1. Short launch, short receive.
2. Short launch, long receive.
3. Long launch, short receive.
4. Long launch, long receive.

These launch conditions must be controlled in order to provide repeatable measurements. Long-launch conditions are generally preferred. Long-launch or receive conditions mean that equilibrium mode distribution (EMD) exists in the fiber. The Electronic Industry Association (EIA) recommends a 70/70 launch: 70% of the fiber core diameter and 70% of the fiber NA should be filled. This recommendation corresponds to the EMD in a graded-index fiber. EMD can be reached three ways: by the optical approach (illustrated in Figure 13.4), filtering, or long fiber length. In general, connector losses under long-launch conditions range from 0.4-0.5 dB. Under short-launch conditions, losses are in the 1.3-1.4 dB range.

Splice loss testing for preterminated cable involves first measuring the power (P_1) transmitted through two separate fibers. Then the cable is joined in a splice bushing, and a new transmission power reading is made (P_2) through the combined cable. The equation for calculating splice loss is written:

Eq. 13.4 $$\text{Splice Loss} = 10 \cdot \log_{10}\left(\frac{P_2}{P_1}\right)$$

where:

P₁=Power reading through the first cable.

P₂=Power through the combined cables.

Systems Measurements

System measurements include fiber continuity, bit error rate (BER), sensitivity, and eye pattern. These are the functional tests. Fiber continuity is tested using an OTDR which locates breaks in the optical fiber. Bit error rate is defined as the number of false bits of information divided by the total number of bits that have been transmitted. BER is tested by modulating the source with a well-defined, long, bit-sequence and comparing the received sequence to the transmitted one. Sensitivity is the lowest power level that leads to the desired BER. It can be tested with the BER test method; in this case, an optical attenuator is used to reduce the transmitted power until the error rate exceeds the predefined value. Eye patterns are viewed with an oscilloscope.

TROUBLESHOOTING

The following troubleshooting guide describes some common problems that may be encountered in fiber optic systems.

PROBLEM	CHECK
No optical power out of transmitter or transceiver.	Check power connections. If there is less than the specified voltage between power pins, a higher current power supply is required. Be sure the power supply polarity is correct.
	Be sure the data input is present. Many data links put out no light if a logic 0 is input. Be sure that the input is toggling between 0's and 1's, otherwise no output light may be present.
No optical power out of the fiber at the receiver optical input port.	Check power at the optical output port of the transmitter or transceiver. If power is present at this port, ensure that the proper fiber is connected at the optical input port. Verify the integrity of the fiber.
Receiver output electrical signal is noisy or intermittent.	Check that the optical loss does not exceed the rated value between transceivers or between the transmitter and the receiver. High loss may be caused by bad connectors, improperly seated connectors, or bad splices.
	Check the wavelength of transceivers or transmitter and receiver. Detectors are typically optimized for one wavelength. Mixing 850 nm and 1300 nm units, for instance, may result in poor or no performance.
	Be sure that the transmitter and receiver enclosures are grounded.
	FM video links at 850 nm are generally bandwidth limited at distances over 1.5 km. When this occurs, the receiver output will not be usable even when sufficient optical power is received.
No signal out of the receiver.	Verify signal input at transmitter. Be sure an electrical input is present at the transmitter, and verify the signal has the proper amplitude, frequency and impedance. Be sure that power supply voltages and connections are correct.
	Check receiver power connection. If there is less than the specified voltage between power pins, a higher current power supply may be required. Be sure the power supply polarity is correct.
Signal amplitude out of the receiver is too large.	Verify that the receiver is terminated into the proper impedance. Many data links and most audio links require this terminating resistor; if it is omitted, the output will be two times too large. If the value is incorrect, the output may be too large or too small, depending on the value of the resistor.
Signal out of the receiver is distorted.	Verify the input signal at the transmitter.
	Verify the fiber size, and verify that the receiver power is within specifications. Receiver optical input may be too high.

PROBLEM	CHECK
Data errors occur.	Check that power supply voltage is correct and clean for transmitter and receiver or transceiver.
	Check that enclosures are properly grounded.
	Check that data inputs and outputs are properly terminated.
	Check that input data levels and data rates are correct.
	Check that the optical input level to the receiver is within valid limits.
Signal out of diplexer demultiplexer is noisy.	Check the copper or fiber optic link between the diplexer mux/demux pair. Check that losses in the optical path do not exceed the loss budget of the transmitter/receiver pair.
Audio signal amplitude out of diplexer de-multiplexer is too large or distorted.	Verify the signal input at the multiplexer.
	Verify that the diplexer demultiplexer audio output is terminated into the specified impedance.
	Typically the audio input to the diplexer multiplexer must be 1-Volt RMS maximum.

Chapter Summary

- Fiber optic systems may undergo functional testing or performance testing.
- Test equipment for fiber optic systems include an optical power meter, optical light source, optical loss meter, fiber identifier, talk set, optical time domain reflectometer, optical spectrum analyzer, optical attenuator, backreflection meter, bandwidth sets, and local injector detectors.
- Mode scramblers mix light to excite every possible mode within the fiber.
- Mandrel wrap core mode filters allow high-order modes in the core to be removed.
- Cladding mode strippers remove modes from the fiber cladding.
- Automated switching systems for testing components such as cable, connectors, or couplers are beneficial in keeping down the cost of test equipment.
- Center wavelength, as well as spectral width can be measured with an optical spectrum analyzer.
- Chirp is the undesired wavelength shift in lasers caused by intensity modulation.
- Tests for PIN photodiodes and APD's include diameter, spectral responsivity, bandwidth, and dark current/NEP.
- Detector bandwidth can be measured by exciting the detector with a sinewave-modulated laser source.
- Concentricity, or eccentricity, occurs because the core may not be perfectly centered in the cladding.
- Ellipticity, or ovality, describes the fact that the core or cladding may be elliptical rather than circular.

Selected References and Additional Reading

—. 1982. *Designers Guide to Fiber Optics*. Harrisburg, PA: AMP, Incorporated.

Hecht, Jeff. 1993. *Understanding Fiber Optics*. 2nd edition. Indianapolis, IN: Sams Publishing.

Hentschel, Christian. 1988. *Fiber Optics Handbook*. 2nd edition. Germany: Hewlett-Packard Company.

—. 1992. *Just the Facts*. New Jersey: Corning, Incorporated.

Sterling, Donald J. 1993. *Amp Technician's Guide to Fiber Optics*, 2nd Edition. New York: Delmar Publishers.

FUTURE TRENDS

14

Predicting the future has always been a challenging and usually unsuccessful undertaking. In spite of the mountain of evidence that it can't be done, we will attempt it anyway. First let us look at major trends that affect the fiber optics industry.

1. Expansion into mass markets
2. Cost reduction
3. Better education of potential users
4. Miniaturization
5. Competing technologies
6. Mass-produced devices
7. Standards
8. New materials
9. Technology refinements
10. New technology

The order of the items is intentional. The main thrusts in the fiber optics industry today are covered in the first four to seven items depending on the specific product and market that is being considered. Most prognosticators of the future focus on number ten, new technology. The truth is that the marvels in today's research labs may not materially affect the fiber optics industry for at least a decade. Items one and two go hand-in-hand. Fiber optics has always suffered from the chicken and egg syndrome. Prices would be low when high volumes were realized, but high volumes would not be realized until prices were low. For example, many large, well-financed fiber optics companies went bankrupt in the early 1990's gambling that fiber-to-the-home (FTTH) was going to take off at any moment. It didn't, and so far it hasn't. It probably will, but exactly how and when is the multi-million dollar question. Let's look at each of the ten points listed in some detail to get a better picture of what is happening and what is likely to happen.

EXPANSION INTO MASS MARKETS

So why hasn't fiber-to-the home (FTTH) or fiber-to-the curb (FTTC) happened yet? The technology to implement such a system has existed for years, but it is not economically feasible. The chicken and egg syndrome is the main problem. The right combination of need or market demand for services that could be provided, technology and financing has not yet been found. The final resolution of the problem may or may not involve fiber in a large way. Fiber optics has found mass market applications in many other areas however. The long-distance telephone industry is the one that has most thoroughly embraced fiber optics. Who can imagine a long-distance carrier bragging that his company uses state-of-the-art copper cables? Fiber optics in that industry is no longer a marketing advantage, it is a necessity. The CATV industry has used fiber optics extensively in headend and hub distribution of cable TV. In both of these examples, the move to fiber optics was a win-win situation. Fiber offered the cheapest solution, and it offered the best solution. Note that we didn't simply say that fiber was cheaper and better, we said it was the cheapest and the best. The distinction is a fine one, but it's another point that fools many futurists.

FIGURE 14.1

Fiber-to-the-Curb and Fiber-to-the-Home System

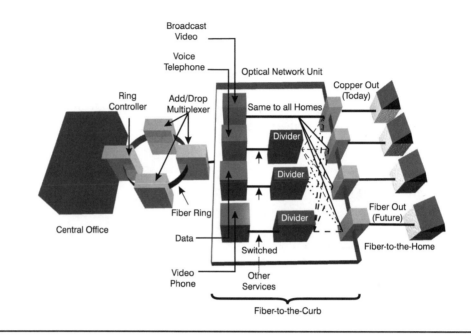

Other industries such as the broadcast industry have just now started flirting with fiber optics, but for broadcast, the advantages are not as clear cut. Sending studio-quality analog video signals a few hundred feet is an easy task for copper. Currently, fiber cannot compete with the cost of a 10-foot piece of copper coax cable. Sure, fiber optics offers better signal quality, lower weight, EMI immunity and so on, but if copper does the job at 1/10th the cost of fiber optics, copper will be the choice. The broadcast industry does have some changes on the near horizon that could alter the balance. The first change is the move to HDTV which will require much more bandwidth. The second is the move to digitized video which curiously is more difficult to transmit over long lengths of copper coax cable than analog video. Both of these trends in the broadcast industry mean that the acceptance of fiber optics will increase as the change to the new technologies is embraced. Fiber optics will play a big part in the broadcast studio, but its role in the mass distribution of video to the home is not so certain. Digital compression of video signals is the main factor that allows other technologies, even copper cable, to effectively compete for this market.

Other applications such as process control were predicted to be ideal for fiber optics, yet fiber has barely penetrated this market. Why? Again, the need and cost factors drove the choice to copper cable. On the other hand, new markets such as distance learning, that didn't even exist a few years ago, appear to be well suited for fiber optics.

On the digital side, fiber is playing an increasing role in computer communications. Standards such as fibre channel and HIPPI have spurred this deployment. The main uses to date have been inter-computer communication at high speeds. But computers are becoming so fast that the first major application of fiber *inside* a computer has appeared on the market. Cray Research's new Triton supercomputer uses high-performance fiber optic links developed by Force, Incorporated to distribute high-speed clock signals to multiple processors. Here the distances are modest, but the speeds are so high and the signal must be so clean, that fiber is the only viable alternative.

FIGURE 14.2

High-Speed Link for Cray Research's Triton Super-computer

(Photo by Force, Inc.)

Fiber optics is also finding broad application in transportation and highway monitoring. Smart highways often employ a large of number of sensors, communications stations and cameras that are well served by fiber. The increased public concern over safety has led many cities to install hundreds of cameras in and around the city to allow the police to constantly monitor these areas. The distances involved again make fiber a logical choice.

FIGURE 14.3

Smart Highway Systems use Fiber Optics to Transmit Signals to Variable Message Signs

(Photo by Force, Inc.)

Predicting which mass markets will be key in the future of fiber optics is partially impossible since many of those markets may not even exist today. Five years ago terms like *information superhighway, smart highways,* and *distance learning* were largely unknown and were not even considered in projections. Today, these are seen as key markets for fiber optics. The same will happen in the next five years.

FIGURE 14.4

Distance Learning Allows Teachers to Reach to Many Classrooms Over a Great Distance

COST REDUCTION

As mentioned earlier, the chicken and egg syndrome has been a key inhibitor to fiber's mass deployment. (No doubt the same could be said for most new technologies.) In spite of the difficulties, prices for all types of fiber optic components and systems have dropped dramatically during the last decade, and so far, there is no sign that this trend has levelled off. The cost of key fiber optic components such as lasers, LED's and detectors has dropped at a yearly rate of 5-20%. Fiber itself has levelled off due to a shortage of capacity, but it is often a small part of a full system cost. Fiber optic connectors have continued to make big strides in cost reduction due to innovative and simplified designs and the introduction of new materials. This trend is likely to continue for years to come. Cost reduction is the catalyst that will allow fiber to move into expanded markets.

BETTER EDUCATION OF POTENTIAL USERS

As is the case with all new technologies, many users fear change. There is always strong back pressure to keep the status quo. Fiber optics has been feared for two main reasons. Most users had a hard time accepting that a glass fiber wasn't fragile, and fiber optic connectors were very finicky and difficult to deal with. While some of these concerns were rooted in truth in the early days of fiber, they are no longer concerns. Still, many potential users of fiber are largely unfamiliar with fiber and its advantages. Books such as this one were written to combat this lack of user knowledge, and the fiber optics industry must continue to educate new users and new markets if it is to flourish.

MINIATURIZATION

One common theme in many fiber optic applications is to make all of the components smaller. This trend will continue for several years, although for many products the limit of diminishing returns has already been reached. The limiting factor is the size of fiber optic connectors. Unfortunately, most are huge compared to their electrical counterparts. Recent revivals of ribbon cable and mass termination connectors may be an answer, but the relatively negligible impact of similar products almost a decade ago makes us adopt a wait-and-see attitude. Much of the miniaturization has been achieved through the availability

FIGURE 14.5

On the left, the 20 Mb/s data link 15 years ago weighed 10 lbs. By comparison, the modern 1.5 Gb/s link on the right is 1.5" long and weighs 3 oz.

(Photos by Force, Inc.).

of more highly integrated electronics circuits. This will continue a few more years, but the point of diminishing returns has nearly been reached. The fiber optics industry is so diverse with so many different products that it may not be economical to develop a "complete link on a chip" because the sales volumes would not be sufficient to pay for the development costs. Because of this, the level of integration may stay at today's building block level for all but the highest-volume products.

COMPETING TECHNOLOGIES

Twenty years ago, the futurists predicted that the use of copper cable would be all but gone for all communication applications, a bold, but wrong, prediction. Copper is still here today and will be twenty years from now. In general, copper will be the first choice for short distances, and fiber will be the first choice for long distances. For most applications, there is a crossover distance at which copper and fiber prices are equal. This crossover distance is bandwidth-dependant: the higher the bandwidth, the shorter the crossover distance. At distances shorter than the crossover distance, copper is cheaper and vice versa. Ten years ago, this crossover distance was several kilometers. Today it may be as short as 100 meters for some applications.

Fiber's competition today will not be its only competition in the future. Five years ago, no one would have considered sending a NTSC video signal over 1 kilometer of twisted pair cable. Today, it can be done, and that has taken away many of fiber's opportunities. Other improvements in competing technologies will continue to be made in the future, but the inherent advantages of fiber will continue to make it an essential technology.

MASS-PRODUCED DEVICES

Much of the cost reduction that has been achieved in fiber optic components and systems is due to the mass production of components. Just 15 years ago, a good quality laser for fiber optics could cost $10,000. Today, fiber-ready short wavelength lasers such as those used in CD players cost about 1/200th this amount. The difference isn't so much technology as it is volume. The $10,000 lasers were made at volumes of tens of pieces per month. The $50 CD laser is made at volumes of about 1 million pieces per month. The basic laser chip in fact can be bought for under $10. The additional cost is due to the lensing and precision alignment required to make the device fiber-ready. The telephone industry was responsible for making the early fiber optic components affordable. As more fiber optic components are mass produced, prices will tumble dramatically. Earlier in this document it was pointed out that 1300 nm components cost more than 850 nm components and 1550 nm components cost even more. Much of this disparity is due to the volumes associated with each of these components. If 1550 nm components were produced at the same volumes as 850 nm components, the price gap would close considerably.

FIGURE 14.6

Mass Production of Components will Reduce the Costs of Fiber Optics

STANDARDS

Standards have always played an important role in the fiber optics industry. Ten years ago, the industry suffered from too few standards. Today, some of the industry suffers from too many standards or too many outdated and incomplete standards. The conventional wisdom states that a technology will only be accepted on a mass basis once standards become widespread. Fiber, so far, has grown significantly without a widespread unified set of standards. This is partially because fiber optics plays a key role in so many diverse markets that no one standard could ever hope to be applicable. Many standards that are now widespread, like 62.5/125 μm multimode fiber are *de facto* standards that resulted from years of open and often bitter competition in the marketplace. Not many people today will remember that years ago one camp pushed for 85/125 μm as the standard for multimode while the other camp pushed for 62.5/125 μm fiber size. The latter won and 85/125 μm fiber is but a dim memory. Standards are also difficult to implement in an industry that moves as quickly as the fiber optics industry. The first FDDI standard was virtually obsolete by the time it was introduced. At the start of the process to develop the FDDI standard, 125 Mb/s seemed like a good high-speed data rate. By the time the standard was ready, several years later, 125 Mb/s was laughably slow. Standards are important to fiber optic's future, but many of the largest mass applications will probably happen with or without them.

NEW MATERIALS

In the early days of fiber optics, components tended to use expensive and exotic materials. Fiber optic connectors initially used stainless steel ferrules or even precision drilled jewels to achieve alignment. Later, most connectors changed to ceramics. Today, fiber optic connectors with plastic ferrules are appearing, an unthinkable feat only a decade ago. Other fiber optic components have evolved from complex laser-welded, hermetically sealed enclosures to low-cost, unsealed enclosures. Improvements in the environmental tolerance of the basic components have enabled them to brave the elements with a minimum of packaging protection.

FIGURE 14.7

New Materials are Being Developed Every Day

All of these changes have enabled dramatic strides in lowering costs. These trends are likely to continue. Widespread use of plastics in fiber optic components and connectors will continue. Epoxies were once frowned upon in many fiber optic assemblies. Today, improvements in epoxies make them acceptable alternatives to laser welding in some applications.

TECHNOLOGY REFINEMENTS

When a technology reaches the level of maturity that the fiber optics industry has reached, improvements in the technology tend to be evolutionary rather than revolutionary. This means that small incremental changes and improvements will be made to products each generation, much like the automobile industry.

Look at what has happened with laser diodes in the fiber optics industry. Over the years we have seen dozens of breakthrough laser designs come and go. Along the way we have had Fabry-Perot designs, cleaved-coupled designs, DFB designs and so on. Some were simply bad ideas that quickly died from starvation in the open market. Some have thrived and formed the basis for improved designs. Today is the age of the MQW (multi-quantum well) laser. It is one of the more dramatic improvements in laser technology that have been seen in the last decade. So strong are its advantages that nearly every manufacturer in the world has switched to it in less than one year. Not only is its performance better, it handles high temperatures better, produces higher yields and is thus cheaper. MQW will undoubtedly be only a stopping point on the road to even better designs.

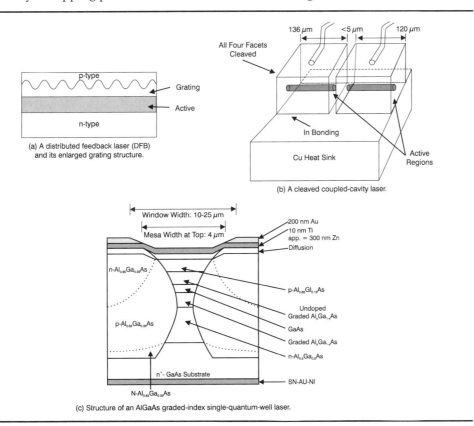

FIGURE 14.8

Evolution of Laser Structures

(a) A distributed feedback laser (DFB) and its enlarged grating structure.

(b) A cleaved coupled-cavity laser.

(c) Structure of an AlGaAs graded-index single-quantum-well laser.

All things considered, this evolutionary mode is good for the fiber optics industry as well as its customers. Things will not change so fast that manufacturers cannot recoup development costs or customers cannot keep up with the technology. Over 90% of the fiber optic systems that will be deployed over the next decade will be based on technology that is already in manufacturing today. Much of the real innovation will be at the system level, not the underlying component technology. Look at the CD player. Lasers have existed for years as have techniques for digitizing audio, but putting all of these elements together to create the CD player was revolutionary. The same will happen in the fiber optics industry.

NEW TECHNOLOGY

Well here's the promise that excites everyone. The new glimmering bits of technology are the ones that get the press and get investors and potential customers inspired. In reality, probably only 10% or less of the "breakthroughs" ever get out of the lab. It's hard to know which to bet on, but the marketplace usually decides quickly. If a breakthrough doesn't yield better performance *and* lower cost, it will die. And while you're tallying up the score of a new piece of technology, don't forget to consider the competing technologies that are ever present and ever improving. Fiber optic sensors are a great example of this.

In the early 1980's fiber optic sensors were touted as the solution to the world's measurement problems. Here finally, was a lightweight, small, EMI resistant sensor. The problem was, most of the people developing such sensors had no idea what existing technologies could already do. While the fiber optic sensor industry struggled to make a pressure sensor that could achieve 1% accuracy over a 0°C to +40°C temperature range for less than $1,000, you could buy a state-of-the-art capacitive pressure sensor for $200 that guaranteed 0.01% accuracy from -20°C to +70°C. Despite this, there is still an interest in fiber optic sensors in specialty applications.

So what looks promising today? Certainly erbium-doped fiber amplifiers (EDFA) have already made an impact and will be around for years; however, their present applicability is rather limited. We have heard that *soliton pulses* are the answer to the future. Soliton pulses have a special shape that allows them to regenerate at various points along the fiber rather than degrade. In this way, soliton pulses compensate for dispersion. While soliton technology is still in the labs, intense interest will move it into real world applications. When combined with EDFA technology, soliton technology may bring the promise of gigabit data rates transmitted over thousands of kilometers.

What about long wavelength optics? Theory states that if you could operate a fiber at a wavelength of 3 μm to 4 μm, rather than the 1.3 μm or 1.55 μm wavelengths widely used today, fiber losses could drop to 0.001 dB/km. That means that a single continuous fiber stretched around the equator would have a loss of only 40 dB, achievable without the use of repeaters. There are serious practical problems however. The glass for such fibers is fluorine based rather than silica based. Unfortunately water attacks this fiber like acid, so making connections is challenging. After a decade of trying, researchers have not been able to get close to the theoretical loss of 0.001 dB/km raising serious doubts about the feasibility. The other problem is that the lasers, LED's or detectors would likely have to operate at liquid nitrogen temperatures. Still, researchers persevere.

Another new technology with potential is dense-wavelength multiplexed systems. From time to time there are amazing reports of sending six, twelve or even twenty simultaneous wavelengths over a single fiber, increasing its capacity by this same factor. The theory is tantalizing, but they are not practical systems. The main limitation is buying lasers at six, twelve or twenty closely spaced wavelengths. The process of growing lasers just isn't that precise. Sure, you can sort through a hundred lasers and find six with the wavelengths that you need for one system, but who could afford to throw away the other 94 lasers? In addition, keeping the wavelength tolerance and stability of these wavelengths to 1 nm over time temperature is difficult without having very deep pockets. Perhaps as the process of growing lasers becomes more precise, these systems may become feasible.

Coherent communications has also been publicized as the future of fibers optics. In this system, the laser transmitter emits light at one frequency which is modulated using amplitude, phase or frequency modulation, by the signal. At the receive end, the light is mixed with light from a second laser, and the intermediate frequency is then detected and converted to an output signal. This scheme avoids noise encountered in direct detection, making the receiver sensitive to faint signals which in turn increases the allowable optical loss between the transmitter and the receiver. This greater sensitivity in coherent communications should allow much higher data rates than the non-coherent communications used today. Increased sensitivity means the receiver requires fewer photons of light to detect the optical signal. One problem to this transmission scheme is that today's electronics can't run fast enough to take full advantage of the technique. In the meantime, don't write off non-coherent communications. It will continue to evolve year after year taking incremental advantage of the improvements in electronics components and other pieces of technology. As a result, the edge offered by coherent communications will decline.

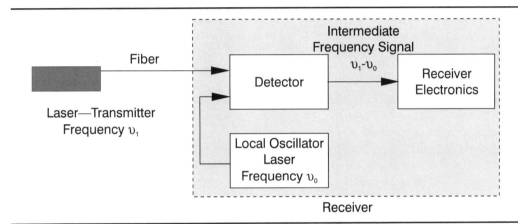

FIGURE 14.9

Coherent Transmission Scheme

Much is made of the fact that today's fiber optic technology uses less than 1% of fiber optics terahertz communication capability. Why the limitation? As with coherent communications, the electronics can't keep up. Even if the electronics could keep up, how smart would it be to put every phone conversation in the world onto a single fiber? If that fiber broke, chaos would result. We have already seen the results of having so many conversations on a single fiber in recent years as major phone companies have lost service for multi-state regions for hours because *the* cable for that region was cut. The need for capacity per fiber has to be balanced against reliability of the system.

A technique that has been extolled for years as the next great advancement is *external modulation*. An external modulator is like an electronic shutter that can be used to modulate light passing through it. Most of today's gigabit systems use direct modulation of lasers. External modulators are supposedly superior because they eliminate nasty laser problems like laser chirp and ringing. Still, a fiber optic transmitter using external modulation requires a laser, and thus it will always cost more to build than a direct modulation system. When external modulators were first introduced, direct modulation transmitters were rather limited. Rather than taking the market by storm, external modulators created competition that forced direct modulation systems to improve. Currently, external modulators are only used in very high power laser transmitters, and research continues.

Integrated optics, which can broadly refer to two technologies, will certainly play a part in fiber's future. The first technology is the integration of light emitters and/or detectors onto a substrate that holds some or all of the signal processing circuitry. This has been promoted as the way of the future for nearly a decade but has yet to capture even 1% of the market. The reason is that the performance is not up to the level that can be achieved with existing components. The process also has yield problems that drive up cost, and development cost is high. Perhaps these obstacles will be overcome in the future, but this technology must jump ahead of the declining price curve now being following by discrete optic and electronic designs.

The second type of integrated optics is the creation of optical splitters and WDM's in single substrates. This differs significantly from the bulk optical techniques used in most existing devices. Again, this technology is plagued by performance and yield problems; however, as production difficulties are overcome, these IOCs will become essential in the high-speed networks that make up the information superhighway.

The future of fiber optics will certainly be exciting. How closely it will resemble the viewpoints given here remains to be seen. Important new technology will most likely be refinements of existing technology, but those technologies still in the labs today may yet emerge as new real-world applications. Those that do will certainly change the industry but only if they offer better performance and lower cost than existing and competing technologies. As is the case with most things, time will tell.

Chapter Summary

- Major trends for the future of fiber optics include expansion into mass markets, cost reduction, education of potential users, miniaturization, mass-produced devices, new materials, and new technology.

- The development of standards such as fibre channel, FDDI, and HIPPI have spurred the deployment of fiber optics in computer communications.

- The improvement of competing technologies such as digital video transmission over copper cable will reduce fiber's opportunities.

- Technology refinements for the fiber optics industry have reached the level of maturity where they are evolutionary rather than revolutionary.

- New technology being developed in today's test labs will only be embraced if it offers better performance and lower cost than today's technology.

Selected References and Additional Reading

Higgins, Thomas V. 1995. "Optoelectronics: The Next Technological Revolution." *Laser Focus World.* November 1995: 93-102.

Kopakowski, Edward T. 1995. "Metropolitan Area Networks for Interactive Distance Learning." *Fiberoptic Product News.* September 1995: 17+.

Lewis, Kirk. 1995. "Integrated Optics Building Better Products." *Photonics Spectra.* June 1995: 119-124.

Newell, Wade S. 1995. "Multi-Protocol High-Performance Serial Digital Fiber Optic Data Links." from *Digital Communications Design Conference, Day 3*, Joseph F. Havel, ed. Proceedings of Design SuperCon '95, Santa Clara, CA, Feb. 28-Mar. 2, 1995.

Yeh, Chai. 1990. *Handbook of Fiber Optics: Theory and Applications.* New York: Academic Press, Inc.

APPENDICES

GENERAL REFERENCE MATERIAL

CONCERNING NUMBERS

One encounters an extreme range of numbers in the fiber optics business. For example, a high-speed data link may respond in the picosecond range while a low-speed RS-232 link may respond in the millisecond range. What do these prefixes mean?

Large and small numbers can be difficult to express and to imagine. The International System of Units, also called SI or the metric system, commonly uses prefixes to express large and small numbers. To understand fiber optics and much of electronics, one must be familiar with these prefixes to get some idea of the size of the number being expressed. For example, deep red light at one end of the visible spectrum has a wavelength of 700 nanometers (nm) which is 700 billionths of a meter or .00002756 inches. Deep violet visible light has a wavelength roughly half of red's: 350 nm or .00001378 inches. Wavelengths of light are given in small numbers. If, however, we express light as a frequency, we get a large number. Deep red has a frequency of 430 THz or 430,000,000,000,000 Hz. Violet's frequency is some 850 THz or 850,000,000,000,000 Hz. The following SI prefixes are encountered most.

Tera	(T)	1,000,000,000,000	10^{12}	trillions
Giga	(G)	1,000,000,000	10^9	billions
Mega	(M)	1,000,000	10^6	millions
Kilo	(k)	1,000	10^3	thousands

Milli	(m)	.001	10^{-3}	thousandths
Micro	(μ)	.000 001	10^{-6}	millionths
Nano	(n)	.000 000 001	10^{-9}	billionths
Pico	(p)	.000 000 000 0001	10^{-12}	trillionths

Mil	1/1000 of an inch	1 mil = 25 microns
Micron	1/1,000,000 of a meter	1 micron = 1 micrometer (μm)

CONSTANTS

The following is a partial list of constants that are pertinent to the use of fiber optics and electronics:

1. Speed of Light in a Vacuum

$C = 2.998 \times 10^8$ m/s = 29.98 cm/ns = 2,998 nm/ps = 0.9836 ft./ns

2. Speed of Light in Fiber (Refractive index = 1.45)

$C = 2.068 \times 10^8$ m/s = 20.68 cm/ns = 2,068 nm/ps = 0.6783 ft./ns

3 .Electronic Charge

$q = 1.602 \times 10^{-19}$ Coulombs

4. Planck's Constant

$h = 6.625 \times 10^{-34}$ J•s = 4.135×10^{-15} eV•s = 6.625×10^{-27} erg•s

5. Boltzman's Constant

$k = 1.3804 \times 10^{-23}$ J/°K = 8.616×10^{-5} eV/°K = 1.3804×10^{-16} erg/°K

6. Energy-Wavelength Conversion Factor

$\text{hc/e} = 1239.8 \text{ eV nm} = 1.2398 \text{ eV} \bullet \mu m$

7. Wien's Displacement Law Constant

$\text{max} \bullet T = 2.9878 \times 10^6 \text{ nm} \bullet {}^\circ K = 2897.8 \ \mu m \bullet {}^\circ K$

8. kT at +300°K

0.0259

9. Characteristic Impedance of Vacuum

$Z_0 = 376.7 \ \Omega$

10. Pi (π)

3.1415926535

CONVERSION FACTORS

The following is a list of common conversions that pertain to fiber optics and electronics:

To Convert From:	To:	Multiply by:
μm (micrometers)	mils	.03937
	inches	.0000397
	Angstroms	10,000
	nanometers	1,000
mm (millimeters)	inches	.03937
	feet	.00328
cm (centimeters)	inches	.39370
	feet	.03281
m (meters)	feet	3.280
	inches	39.37
	cm	100
	mm	1,000
km (kilometers)	miles	.62137
	feet	3280.8
	mm	1,000,000
kg (kilograms)	pounds	2.2046
g (grams)	oz. (ounces)	0.03527
	pounds	.0022045
°C (Centigrade)	°F (Fahrenheit)	1.8 then add 32
°F	°C	subtract 32 then divide by 1.8
°K (Kelvin)	°C	= °C + 273.16
N (Newton)	pounds	.2247
kg/km	pounds/mile	3.5480
kPa	PSI	.14511

GLOSSARY OF TERMS

Absorption That portion of optical attenuation in optical fiber resulting from the conversion of optical power to heat. Caused by impurities in the fiber such as hydroxyl ions.

AC Alternating current.

Acceptance Angle The half-angle of the cone within which incident light is totally internally reflected by the fiber core. It is equal to $\sin^{-1}(NA)$.

AD or ADC Analog-to-digital converter. A device used to convert analog signals to digital signals.

AGC Automatic gain control. A process or means by which gain is automatically adjusted in a specified manner as a function of input level or another specified parameter.

AM See amplitude modulation.

Amplitude Modulation (AM) A transmission technique in which the amplitude of the carrier is varied in accordance with the signal.

Analog A continuously variable signal. A mercury thermometer, which gives a variable range of temperature readings, is an example of an analog instrument. Opposite of digital.

Angstrom (Å) A unit of length in optical measurements where:

$$1\text{Å} = 10^{-10} \text{ meters}$$
$$= 10^{-4} \text{ micrometers}$$
$$= 10^{-1} \text{ nanometers}$$

The angstrom has been used historically in the field of optics, but it is not an SI (International System) unit. Rarely used in fiber optics, nanometers is preferred.

Angular Misalignment Loss at a connector due to fiber end face angles being misaligned.

ANSI American National Standards Institute.

APC Angle polished connector. An 5°-15° angle on the connector tip for the minimum possible backreflection.

APD See avalanche photodiode.

APL Average picture level. Video parameter.

AR Coating Antireflection coating. A thin, dielectric or metallic film applied to an optical surface to reduce its reflectance and thereby increase its transmittance.

Armor A protective layer, usually metal, wrapped around a cable.

ASCII American standard code for information interchange. A means of encoding information.

ASIC Application-specific integrated circuit. A custom-designed integrated circuit.

ASTM American Society for Testing and Materials.

Asynchronous Data that is transmitted without an associated clock signal.

Asynchronous Transfer Mode (ATM) A digital transmission switching format, with cells containing 5 bytes of header information followed by 48 data bytes. Part of the B-ISDN standard.

ATE Automatic test equipment.

Attenuation The decrease in signal strength along a fiber optic waveguide caused by absorption and scattering. Attenuation is usually expressed in dB/km.

Attenuation Constant For a particular propagation mode in an optical fiber, the real part of the axial propagation constant.

Attenuation-Limited Operation The condition in a fiber optic link when operation is limited by the power of the received signal (rather than by bandwidth or distortion).

Attenuator *1.* In electrical systems, a usually passive network for reducing the amplitude of a signal without appreciably distorting the waveform. *2.* In optical systems, a passive device for reducing the amplitude of a signal without appreciably distorting the waveform.

Avalanche Photodiode (APD) A photodiode that exhibits internal amplification of photocurrent through avalanche multiplication of carriers in the junction region.

Average Power The average level of power in a signal that varies with time.

Axial Propagation Constant For an optical fiber, the propagation constant evaluated along the axis of a fiber in the direction of transmission.

Axis The center of an optical fiber.

B See bel.

Backscattering The return of a portion of scattered light to the input end of a fiber; the scattering of light in the direction opposite to its original propagation.

Bandwidth The range of frequencies within which a fiber optic waveguide or terminal device can transmit data or information.

Bandwidth-Limited Operation The condition in a fiber optic link when bandwidth, rather than received optical power, limits performance. This condition is reached when the signal becomes distorted, principally by dispersion, beyond specified limits.

Baseband A method of communication in which a signal is transmitted at its original frequency without being impressed on a carrier.

Baud A unit of signaling speed equal to the number of signal symbols per second, which may or may not be equal to the data rate in bits per second.

Beamsplitter An optical device, such as a partially reflecting mirror, that splits a beam of light into two or more beams. Used in fiber optics for directional couplers.

Bel (B) The logarithm to the base 10 of a power ratio, expressed as $B = \log_{10}(P_1/P_2)$, where P_1 and P_2 are distinct powers. The decibel, equal to one-tenth bel, is a more commonly used unit.

Bending Loss Attenuation caused by high-order modes radiating from the outside of a fiber optic waveguide which occur when the fiber is bent around a small radius. *See also* macrobending, microbending.

Bend Radius The smallest radius an optical fiber or fiber cable can bend before increased attenuation or breakage occurs.

BER See bit error rate.

Bidirectional Operating in both directions. Bidirectional couplers operate the same way regardless of the direction light passes through them. Bidirectional transmission sends signals in both directions, sometimes through the same fiber.

Birefringent Having a refractive index that differs for light of different polarizations.

Bit The smallest unit of information upon which digital communications are based; also an electrical or optical pulse that carries this information.

BITE Built-in test equipment. Features designed into a piece of equipment that allow on-line diagnosis of failures and operating status.

Bit Error Rate (BER) The fraction of bits transmitted that are received incorrectly.

BNC Popular coax bayonet style connector. Often used for baseband video.

Broadband A method of communication where the signal is transmitted by being impressed on a high-frequency carrier.

Buffer *1*. In optical fiber, a protective coating applied directly to the fiber. *2*. A routine or storage used to compensate for a difference in rate of flow of data, or time of occurrence of events, when transferring data from one device to another.

Bus Network A network topology in which all terminals are attached to a transmission medium serving as a bus.

Butt Splice A joining of two fibers without optical connectors arranged end-to-end by means of a coupling. Fusion splicing is an example.

BW See bandwidth.

Bypass The ability of a station to isolate itself optically from a network while maintaining the continuity of the cable plant.

Byte A unit of eight bits.

C Celsius. Measure of temperature where water freezes at 0° and boils at 100°.

Cable One or more optical fibers enclosed within protective covering(s) and strength members.

Cable Assembly A cable that is connector terminated and ready for installation.

Cable Plant The cable plant consists of all the optical elements including fiber connectors, splices, etc. between a transmitter and a receiver.

Carrier-to-Noise Ratio The ratio, in decibels, of the level of the carrier to that of the noise in a receiver's IF bandwidth before any nonlinear process such as amplitude limiting and detection takes place.

CATV Community antenna television. A television distribution method whereby signals from distant stations are received, amplified, and then transmitted by coaxial or fiber cable or microwave links to subscribers. This term is now typically used to refer to cable TV.

CCIR Consultative Committee on Radio.

CCITT Consultative Committee on Telephony and Telegraphy.

CCTV Closed-circuit television.

CD Compact disk. Often used to describe high-quality audio, CD-quality audio, or short-wavelength lasers; CD Laser.

Center Wavelength In a laser, the nominal value central operating wavelength. It is the wavelength defined by a peak mode measurement where the effective optical power resides. In an LED, the average of the two wavelengths measured at the half amplitude points of the power spectrum.

Central Office A common carrier switching office in which users' lines terminate. The nerve center of a telephone system.

CGA Color graphics adapter. A low-resolution color standard for computer monitors.

Channel A communications path or the signal sent over that path. Through multiplexing several channels, voice channels can be transmitted over an optical channel.

Chirp In laser diodes, the shift of the laser's central wavelength during single pulse durations due to laser instability.

Chromatic Dispersion All fiber has the property that the speed an optical pulse travels depends on its wavelength. This is caused by several factors including material dispersion, waveguide dispersion and profile dispersion. The net effect is that if an optical pulse contains multiple wavelengths (colors), then the different colors will travel at different speeds and arrive at different times, smearing the received optical signal.

Cladding Material that surrounds the core of an optical fiber. Its lower index of refraction, compared to that of the core, causes the transmitted light to travel down the core.

Cladding Mode A mode confined to the cladding; a light ray that propagates in the cladding.

Cleave The process of separating an optical fiber by a controlled fracture of the glass, for the purpose of obtaining a fiber end, which is flat, smooth, and perpendicular to the fiber axis.

cm Centimeter. Approximately 0.4 inches.

CMOS Complementary metal oxide semiconductor. A family of IC's. Particularly useful for low-speed or low-power applications.

CNR See carrier-to-noise ratio.

CO See central office.

Coating The material surrounding the cladding of a fiber. Generally a soft plastic material that protects the fiber from damage.

Coherent Communications In fiber optics, a communication system where the output of a local laser oscillator is mixed optically with a received signal, and the difference frequency is detected and amplified.

Color Subcarrier The 3.58 MHz signal which carries color information in a TV signal.

Composite Sync A signal consisting of horizontal sync pulses, vertical sync pulses, and equalizing pulses only, with a no-signal reference level.

Composite Video A signal which consists of the luminance (black and white), chrominance (color), blanking pulses, sync pulses, and color burst.

Compression A process in which the dynamic range or data rate of a signal is reduced by controlling it as a function of the inverse relationship of its instantaneous value relative to a specified reference level. Compression is usually accomplished by separate devices called compressors and is used for many purposes such as: improving signal-to-noise ratios, preventing overload of succeeding elements of a system, or matching the dynamic ranges of two devices. Compression can introduce distortion, but it is usually not objectionable.

Concatenation The process of connecting pieces of fiber together.

Concentrator A multiport repeater.

Concentricity The measurement of how well-centered the core is within the cladding.

Connector A mechanical or optical device that provides a demountable connection between two fibers or a fiber and a source or detector.

Connector Plug A device used to terminate an optical conductor cable.

Connector Receptacle The fixed or stationary half of a connection that is mounted on a panel/bulkhead. Receptacles mate with plugs.

Connector Variation The maximum value in dB of the difference in insertion loss between mating optical connectors (e.g., with remating, temperature cycling, etc.). Also called optical connector variation.

Core The light-conducting central portion of an optical fiber, composed of material with a higher index of refraction than the cladding. The portion of the fiber that transmits light.

Counter-Rotating An arrangement whereby two signal paths, one in each direction, exist in a ring topology.

Coupler An optical device that combines or splits power from optical fibers.

Coupling Ratio/Loss (C_R,C_L) The ratio/loss of optical power from one output port to the total output power, expressed as a percent. For a 1 x 2 WDM or coupler with output powers O_1 and O_2, and O_i representing both output powers:

$$C_R(\%) = \left(\frac{O_i}{O_1 + O_2}\right) \times 100\%$$

$$C_R(\%) = -10 \cdot \log_{10}\left(\frac{O_i}{O_1 + O_2}\right)$$

Critical Angle In geometric optics, at a refractive boundary, the smallest angle of incidence at which total internal reflection occurs.

Crosstalk (XT) *1.* Undesired coupling from one circuit, part of a circuit, or channel to another. *2.* Any phenomenon by which a signal transmitted on one circuit or channel of a transmission system creates and undesired effect in another circuit or channel.

CRT Cathode ray tube.

CSMA/CD Carrier sense multiple access with collision detection.

CTS Clear to send.

Cutback Method A technique of measuring optical fiber attenuation by measuring the optical power at two points at different distances from the test source.

Cutoff Wavelength In single-mode fiber, the wavelength below which the fiber ceases to be single-mode.

CW Abbreviation for continuous wave. Usually refers to the constant optical output from an optical source when it is biased (i.e., turned on) but not modulated with a signal.

DA or DAC Digital-to-analog converter. A device used to convert digital signals to analog signals.

Dark Current The induced current that exists in a reversed biased photodiode in the absence of incident optical power. It is better understood to be caused by the shunt resistance of the photodiode. A bias voltage across the diode (and the shunt resistance) causes current to flow in the absence of light.

Data Rate The number of bits of information in a transmission system, expressed in bits per second (b/s or bps), and which may or may not be equal to the signal or baud rate.

dB Decibel.

dBc Decibel relative to a carrier level.

dBµ Decibels relative to microwatt.

dBm Decibels relative to milliwatt.

DC Direct current.

DCE Data circuit-terminating equipment.

Decibel (dB) A unit of measurement indicating relative optic power on a logarithmic scale. Often expressed in reference to a fixed value, such as dBm (1 milliwatt) or dBµ (1 microwatt).

$$dB = 10 \bullet \log_{10}\left(\frac{P_1}{P_2}\right)$$

Demultiplexer A module that separates two or more signals previously combined by compatible multiplexing equipment.

DESC Defense electronics supply center.

Detector An opto-electric transducer used in fiber optics to convert optical power to electrical current. Usually referred to as a photodiode.

DFB See distributed feedback laser.

DG See differential gain.

Diameter-Mismatch Loss The loss of power at a joint that occurs when the transmitting fiber has a diameter greater than the diameter of the receiving fiber. The loss occurs when coupling light from a source to fiber, from fiber to fiber, or from fiber to detector.

Dichroic Filter An optical filter that transmits light according to wavelength. Dichroic filters reflect light that they do not transmit.

Dielectric Any substance in which an electric field may be maintained with zero or near-zero power dissipation. This term usually refers to non-metallic materials.

Differential Gain A type of distortion in a video signal that causes the brightness information to be distorted.

Differential Phase A type of distortion in a video signal that causes the color information to be distorted.

Diffraction Grating An array of fine, parallel, equally spaced reflecting or transmitting lines that mutually enhance the effects of diffraction to concentrate the diffracted light in a few directions determined by the spacing of the lines and by the wavelength of the light.

Digital A signal that consists of discrete states. A binary signal has only two states, 0 and 1.

Diode An electronic device that lets current flow in only one direction. Semiconductor diodes used in fiber optics contain a junction between regions of different doping. They include light emitters (LED's and laser diodes) and detectors (photodiodes).

Diode Laser Synonymous with injection laser diode.

DIP Dual in-line package.

Diplexer A device that combines two or more types of signals into a single output.

Directional Coupler A coupling device for separately sampling (through a known coupling loss) either the forward (incident) or the backward (reflected) wave in a transmission line.

Directivity See near-end crosstalk.

Dispersion The temporal spreading of a light signal in an optical waveguide caused by light signals traveling at different speeds through a fiber either due to modal or chromatic effects.

Dispersion-Shifted Fiber Standard single-mode fibers exhibit optimum attenuation performance at 1550 nm and optimum bandwidth at 1300 nm. Dispersion-shifted fibers are made so that both attenuation and bandwidth are optimum at 1550 nm.

Distortion Nonlinearities in a unit that cause harmonics and beat products to be generated.

Distortion-Limited Operation Generally synonymous with bandwidth-limited operation.

Distributed Feedback Laser (DFB) An injection laser diode which has a Bragg reflection grating in the active region in order to suppress multiple longitudinal modes and enhance a single longitudinal mode.

Dominant Mode The mode in an optical device spectrum with the most power.

Dope Thick liquid or paste used to prepare a surface or a varnish-like substance used for waterproofing or strengthening a material.

Double-Window Fiber This term is used two ways. For multimode fibers, the term means that the fiber is optimized for 850 nm and 1300 nm operation. For single-mode fibers, the term means that the fiber is optimized for 1300 nm and 1550 nm operation.

DP See differential phase.

DSR Data set ready.

DSx A transmission rate in the North American digital telephone hierarchy. Also called T-carrier.

DTE Data terminal equipment.

DTR Data terminal ready.

Dual Attachment Concentrator A concentrator that offers two attachments to the FDDI network which are capable of accommodating a dual (counter-rotating) ring.

Dual Attachment Station A station that offers two attachments to the FDDI network which are capable of accommodating a dual (counter- rotating) ring.

Dual Ring (FDDI Dual Ring) A pair of counter- rotating logical rings.

Duplex Cable A two-fiber cable suitable for duplex transmission.

Duplex Transmission Transmission in both directions, either one direction at a time (half-duplex) or both directions simultaneously (full-duplex).

Duty Cycle In a digital transmission, the fraction of time a signal is at the high level.

ECL Emitter-coupled logic. A high-speed logic family capable of GHz rates.

EDFA See Erbium-doped fiber amplifier.

Edge-Emitting Diode An LED that emits light from its edge, producing more directional output than surface-emitting LED's that emit from their top surface.

EGA Enhanced graphics adapter. A medium-resolution color standard for computer monitors.

EIA Electronic Industries Association. An organization that sets video and audio standards.

8B10B Encoding A signal modulation scheme in which eight bits are encoded in a 10-bit word to ensure that too many consecutive zeroes do not occur; used in ESCON and fibre channel.

802.3 Network A 10 Mb/s CSMA/CD bus-based network; commonly called ethernet.

802.5 Network A token-passing ring network operating at 4 Mb/s or 16 Mb/s.

Electromagnetic Interference (EMI) Any electrical or electromagnetic interference that causes undesirable response, degradation, or failure in electronic equipment. Optical fibers neither emit nor receive EMI.

Electromagnetic Radiation (EMR) Radiation made up of oscillating electric and magnetic fields and propagated with the speed of light. Includes gamma radiation, X-rays, ultraviolet, visible, and infrared radiation, and radar and radio waves.

Electromagnetic Spectrum The range of frequencies of electromagnetic radiation from zero to infinity.

ELED See edge-emitting diode.

Ellipticity Describes the fact that the core or cladding may be elliptical rather than circular.

EM Abbreviation for electromagnetic.

EMD See equilibrium mode distribution.

EMI Electromagnetic interference.

EMP Electromagnetic pulse.

EMR Electromagnetic radiation.

Endoscope A fiber optic bundle used for imaging and viewing inside the human body.

ENG Electronic news gathering.

E/O Abbreviation for electrical-to-optical converter.

Equilibrium Mode Distribution (EMD) The steady modal state of a multimode fiber in which the relative power distribution among modes is independent of fiber length.

Erbium-Doped Fiber Amplifier Optical fibers doped with the rare earth element erbium, which can amplify light in the 1550 nm region when pumped by an external light source.

ESCON Enterprise systems connection. A duplex optical connector used for computer-to-computer data exchange.

Ethernet A baseband local area network marketed by Xerox and developed jointly by Xerox, Digital Equipment Corporation, and Intel.

Evanescent Wave Light guided in the inner part of an optical fiber's cladding rather than in the core.

Excess Loss In a fiber optic coupler, the optical loss from that portion of light that does not emerge from the nominal operation ports of the device.

External Modulation Modulation of a light source by an external device that acts like an electronic shutter.

Extinction Ratio The ratio of the low, or OFF optical power level (P_L) to the high, or ON optical power level (P_H).

$$\text{Extinction Ratio (\%)} = \left(\frac{P_L}{P_H}\right) \times 100$$

Extrinsic Loss In a fiber interconnection, that portion of loss not intrinsic to the fiber but related to imperfect joining of a connector or splice.

Eye Pattern Also called eye diagram. The proper function of a digital system can be quantitatively described by its BER, or qualitatively by its eye pattern. The "openness" of the eye relates to the BER that can be achieved.

F Fahrenheit. Measure of temperature where water freezes at 32° and boils at 212°.

Failure Rate The number of failures of a device per unit of time.

Fall Time Also called turn-off time. The time required for the trailing edge of a pulse to fall from 90% to 10% of its amplitude; the time required for a component to produce such a result. Typically measured between the 80% and 20% points or alternately the 90% and 10% points.

FAR Federal acquisition regulation.

Faraday Effect A phenomenon that causes some materials to rotate the polarization of light in the presence of a magnetic field parallel to the direction of propagation. Also called magneto-optic effect.

Far-End Crosstalk See wavelength isolation.

FC A threaded optical connector that originated in Japan. Good for single-mode or multimode fiber and applications requiring low backreflection.

FCC Federal Communications Commission.

FC/PC See FC. A special curved polish on the connector for very low backreflection.

FDA Food and Drug Administration. Organization responsible for laser safety.

FDDI Fiber distributed data interface. 1. A dual counter-rotating ring local area network. 2. A connector used in a dual counter-rotating ring local area network

FDM See frequency-division multiplexing.

Ferrule A rigid tube that confines or holds a fiber as part of a connector assembly.

FET Field-effect transistor.

Fiber Grating An optical fiber in which the refractive index of the core varies periodically along its length, scattering light in a way similar to a diffraction grating, and transmitting or reflecting certain wavelengths selectively.

Fiber Optic Attenuator A component installed in a fiber optic transmission system that reduces the power in the optical signal. It is often used to limit the optical power received by the photodetector to within the limits of the optical receiver.

Fiber Optic Cable A cable containing one or more optical fibers.

Fiber Optic Communication System The transfer of modulated or unmodulated optical energy through optical fiber media which terminates in the same or different media.

Fiber Optic Gyroscope A coil of optical fiber that can detect rotation about its axis.

Fiber Optic Link A transmitter, receiver, and cable assembly that can transmit information between two points.

Fiber Optic Span An optical fiber/cable terminated at both ends which may include devices that add, subtract, or attenuate optical signals.

Fiber Optic Subsystem A functional entity with defined bounds and interfaces which is part of a system. It contains solid state and/or other components and is specified as a subsystem for the purpose of trade and commerce.

Fiber Optic Test Procedure (FOTP) Standards developed and published by the Electronic Industries Association (EIA) under the EIA-RS-455 series of standards. See Appendix E.

Fiber-to-the-Curb (FTTC) Fiber optic service to a node connected by wires to several nearby homes, typically on a block.

Fiber-to-the-Home (FTTH) Fiber optic service to a node located inside an individual home.

Fiber-to-the-Loop (FTTL) Fiber optic service to a node that is located in a neighborhood.

Fibre Channel An industry-standard specification that originated in Great Britain which details computer channel communications over fiber optics at transmission speeds from 132 Mb/s to 1062.5 Mb/s at distances of up to 10 kilometers.

Filter A device which transmits only part of the incident energy and may thereby change the spectral distribution of energy.

FIT Rate Number of device failures in one billion device hours.

Fluoride Glasses Materials that have the amorphous structure of glass but are made of fluoride compounds (e.g., zirconium fluoride) rather than oxide compounds (e.g., silica). Suitable for very long wavelength transmission.

FM See frequency modulation.

FOG-M Fiber optic guided missile.

FOTP See fiber optic test procedure.

4B5B Encoding A signal modulation scheme in which groups of four bits are encoded and transmitted in five bits in order to guarantee that no more than three consecutive zeroes ever occur; used in FDDI.

FP Fabry-Perot. Generally refers to a type of laser.

Frequency-Division Multiplexing (FDM) A method of deriving two or more simultaneous, continuous channels from a transmission medium by assigning separate portions of the available frequency spectrum to each of the individual channels. In optical communications, one also encounters wavelength-division multiplexing (WDM) involving the use of several distinct optical sources (lasers), each having a distinct center wavelength.

Frequency Modulation (FM) A method of transmission in which the carrier frequency varies in accordance with the signal.

Fresnel Reflection Loss Reflection losses at the ends of fibers caused by differences in the refractive index between glass and air. The maximum reflection caused by a perpendicular air-glass interface is about 4% or about -14 dB.

FSK Frequency shift keying. A method of encoding data by means of two or more tones.

FTTC See fiber-to-the-curb.

FTTH See fiber-to-the-home.

FTTL See fiber-to-the-loop.

Full-Duplex Simultaneous bidirectional transfer of data.

Fused Coupler A method of making a multimode or single-mode coupler by wrapping fibers together, heating them, and pulling them to form a central unified mass so that light on any input fiber is coupled to all output fibers.

Fused Fiber A bundle of fibers fused together so they maintain a fixed alignment with respect to each other in a rigid rod.

Fusion Splicer An instrument that permanently bonds two fibers together by heating and fusing them.

FUT Fiber under test.

FWHM Full width half maximum. Used to describe the width of a spectral emission at the 50% amplitude points.

FWHP Full width half power. Also known as FWHM.

G Giga. One billion or 10^9.

GaAlAs Gallium aluminum arsenide. Generally used for short wavelength light emitters.

GaAs Gallium arsenide. Used in light emitters.

GaInAsP Gallium indium arsenide phosphide. Generally used for long wavelength light emitters.

Gap Loss Loss resulting from the end separation of two axially aligned fibers.

Gate 1. A device having one output channel and one or more input channels, such that the output channel state is completely determined by the input channel states, except during switching transients. 2. One of the many types of combinational logic elements having at least two inputs.

Gaussian Beam A beam pattern used to approximate the distribution of energy in a fiber core. It can also be used to describe emission patterns from surface-emitting LED's. Most people would recognize it as the bell curve. The equation that defines a Gaussian beam is:

$$E(x) = E(0)e^{-x^2/w_0^2}$$

GBaud One billion bits of data per second or 10^9 bits.

Gb/s See GBaud.

Ge Germanium. Generally used in detectors. Good for most wavelengths (e.g., 800-1600 nm).

Genlock A process of sync generator locking. This is usually performed by introducing a composite video signal from a master source to the subject sync generator. The generator to be locked has circuits to isolate vertical drive, horizontal drive and subcarrier. The process then involves locking the subject generator to the master subcarrier, horizontal, and vertical drives so that the result is that both sync generators are running at the same frequency and phase.

GHz Gigahertz. One billion Hertz (cycles per second) or 10^9 Hertz.

Graded-Index Fiber Optical fiber in which the refractive index of the core is in the form of a parabolic curve, decreasing toward the cladding.

GRIN Gradient index. Generally refers to the SELFOC lens often used in fiber optics.

Ground Loop Noise Noise that results when equipment is grounded at points having different potentials thereby creating an unintended current path. The dielectric properties of optical fiber provide electrical isolation that eliminates ground loops.

Group Index Also called group refractive index. In fiber optics, for a given mode propagating in a medium of refractive index (n), the group index (N), is the velocity of light in a vacuum (c), divided by the group velocity of the mode.

Group Velocity 1. The velocity of propagation of an envelope produced when an electromagnetic wave is modulated by, or mixed with, other waves of different frequencies. 2. For a particular mode, the reciprocal of the rate of change of the phase constant with respect to angular frequency. 3. The velocity of the modulated optical power.

Half-Duplex A bidirectional link that is limited to one-way transfer of data, i.e., data can't be sent both ways at the same time.

Hard-Clad Silica Fiber An optical fiber having a silica core and a hard polymeric plastic cladding intimately bounded to the core.

HDTV High-definition television. Television that has approximately twice the horizontal and twice the vertical emitted resolution specified by the NTSC standard.

Headend *1*. A central control device required within some LAN/MAN systems to provide such centralized functions as remodulation, retiming, message accountability, contention control, diagnostic control, and access to a gateway. *2*. A central control device within CATV systems to provide such centralized functions as remodulation. *See also* local area network (LAN).

Hertz One cycle per second.

HIPPI High performance parallel interface as defined by ANSI X3T9.3 document.

Hydrogen Losses Increases in fiber attenuation that occur when hydrogen diffuses into the glass matrix and absorbs some light.

Hz See Hertz.

IDP See integrated detector/preamplifier.

IEEE Institute of Electrical and Electronic Engineers.

Index-Matching Fluid A fluid whose index of refraction nearly equals that of the fiber's core. Used to reduce Fresnel reflection at fiber ends. *See also* index-matching gel.

Index-Matching Gel A gel whose index of refraction nearly equals that of the fiber's core. Used to reduce Fresnel reflection at fiber ends. *See also* index-matching fluid.

Index of Refraction Also refractive index. The ratio of the velocity of light in free space to the velocity of light in a fiber material. Symbolized by n. Always greater than or equal to one.

Infrared (IR) The region of the electromagnetic spectrum bounded by the long-wavelength extreme of the visible spectrum (about 0.7 μm) and the shortest microwaves about 0.1 mm). *See also* frequency, light.

Infrared Fiber Colloquially, optical fibers with best transmission at wavelengths of 2 μm or longer, made of materials other than silica glass. *See also* fluoride glasses.

InGaAs Indium gallium arsenide. Generally used to make high-performance long-wavelength detectors.

InGaAsP Indium gallium arsenide phosphide. Generally used for long-wavelength light emitters.

Injection Laser Diode (ILD) A laser employing a forward-biased semiconductor junction as the active medium. Stimulated emission of coherent light occurs at a pn junction where electrons and holes are driven into the junction.

Insertion Loss The loss of power that results from inserting a component, such as a connector or splice, into a previously continuous path.

Integrated Detector/preamplifier (IDP) A detector package containing a PIN photo-diode and transimpedance amplifier.

Integrated Services Digital Network (ISDN) An integrated digital network in which the same time-division switches and digital transmission paths are used to establish connections for services such as telephone, data, electronic mail and facsimile. How a connection is accomplished is often specified as a switched connection, nonswitched connection, exchange connection, ISDN connection, etc.

Intensity The square of the electric field strength of an electromagnetic wave. Intensity is proportional to irradiance and may get used in place of the term "irradiance" when only relative values are important.

Interchannel Isolation The ability to prevent undesired optical energy from appearing in one signal path as a result of coupling from another signal path. Also called crosstalk.

Interferometric Sensors Fiber optic sensors that rely on interferometric detection.

Intrinsic Losses Splice losses arising from differences in the fibers being spliced.

IPCEA Insulated Power Cable Engineers Association.

IPI Intelligent peripheral interface as defined by ANSI X3T9.3 document.

IR See infrared.

Irradiance Power per unit area.

ISA Instrument Society of America.

ISDN See integrated systems digital network.

ISO International Standards Organization.

Isolation See near-end crosstalk.

Jacket The outer, protective covering of the cable.

Jitter Small and rapid variations in the timing of a waveform due to noise, changes in component characteristics, supply voltages, imperfect synchronizing circuits, etc.

Jitter, Data Dependent (DDJ) Also called data dependent distortion. Jitter related to the transmitted symbol sequence. DDJ is caused by the limited bandwidth characteristics, non-ideal individual pulse responses, and imperfections in the optical channel components.

Jitter, Duty Cycle Distortion (DCD) Distortion usually caused by propagation delay differences between low-to-high and high-to-low transitions. DCD is manifested as a pulse width distortion of the nominal baud time.

Jitter, Random (RJ) Random jitter is due to thermal noise and may be modeled as a Gaussian process. The peak-to-peak value of RJ is of a probabilistic nature, and thus any specific value requires an associated probability.

JPEG Joint photographers expert group. International standard for compressing still photography.

Jumper A short fiber optic cable with connectors on both ends.

k Kilo. One thousand or 10^3.

K Kelvin. Measure of temperature where water freezes at 273° and boils at 373°.

kBaud One thousand bits of data per second.

kb/s See kBaud.

Kevlar® A very strong, very light, synthetic compound developed by Du Pont which is used to strengthen optical cables.

kg Kilogram. Approximately 2.2 pounds.

kHz One thousand cycles per second.

km Kilometer. 1 km = 3,280 feet or 0.62 miles.

Lambertian Emitter An emitter that radiates according to Lambert's cosine law. This law states that the radiance of certain idealized surfaces is dependent upon the angle from which the surface is viewed. The radiant intensity of such a surface is maximum normal to the surface and decreases in proportion to the cosine of the angle from the normal. Given by:

$$N = N_0 cosA$$

Where:

    N = The radiant intensity

    N_0 = The radiance normal (perpendicular) to an emitting surface.

A= The angle between the viewing direction and the normal to the surface.

LAN See local area network.

Large Core Fiber Usually, a fiber with a core of 200 μm or more.

Laser Acronym for l̲ight a̲mplification by s̲timulated e̲mission of r̲adiation A light source that produces, through stimulated emission, coherent, near monochromatic light. Lasers in fiber optics are usually solid-state semiconductor types.

Laser Diode A semiconductor that emits coherent light when forward biased.

Lateral Displacement Loss The loss of power that results from lateral displacement of optimum alignment between two fibers or between a fiber and an active device.

Launch Fiber An optical fiber used to couple and condition light from an optical source into an optical fiber. Often the launch fiber is used to create an equilibrium mode distribution in multimode fiber. Also called launching fiber.

LD See laser diode.

LED (Light-Emitting Diode) A semiconductor that emits incoherent light when forward biased.

LH Long-haul. A classification of video performance under RS-250B/C. Lower performance than medium-haul or short-haul.

L-I Curve The plot of optical output (L) as a function of current (I) which characterizes an electrical to optical converter.

Light In a strict sense, the region of the electromagnetic spectrum that can be perceived by human vision, designated the visible spectrum and nominally covering the wavelength range of 0.4 μm to 0.7 μm. In the laser and optical communication fields, custom and practice have extended usage of the term to include the much broader portion of the electromagnetic spectrum that can be handled by the basic optical techniques used for the visible spectrum. This region has not been clearly defined, but, as employed by most workers in the field, may be considered to extend from the near-ultraviolet region of approximately 0.3 μm, through the visible region, and into the mid-infrared region to 30 μm.

Light Piping Use of optical fibers to illuminate.

Lightguide *Synonym* optical fiber.

Lightwave The path of a point on a wavefront. The direction of the lightwave is generally normal to the wavefront.

Local Area Network (LAN) A communication link between two or more points within a small geographic area, such as between buildings.

Local Loop Synonym for Loop.

Long-Haul Telecommunications Long-distance telecommunications links such as cross-country or transoceanic.

Longitudinal Mode An optical waveguide mode with boundary condition determined along the length of the optical cavity.

Loop *1.* A communication channel from a switching center or an individual message distribution point to the user terminal. *2.* In telephone systems, a pair of wires from a central office to a subscribers's telephone. *3.* Go and return conductors of an electric circuit; a closed circuit. *4.* A closed path under measurement in a resistance test. *5.* A type of antenna used extensively in direction-finding equipment and in UHF reception. *6.* A sequence of instructions that may be executed iteratively while a certain condition prevails.

Loose-Tube A type of fiber optic cable construction where the fiber is contained within a loose tube in the cable jacket.

Loss The amount of a signal's power, expressed in dB, that is lost in connectors, splices, or fiber defects.

Loss Budget An accounting of overall attenuation in a system.

m Meter. 39.37″.

M Mega. One million or 10^6.

mA Milliamp. One thousandth of an Amp or 10^{-3} Amps.

MAC Multiplexed analog components. A video standard developed by the european community. An enhanced version, HD-MAC delivers 1250 lines at 50 frames per second, HDTV quality.

Macrobending In a fiber, all macroscopic deviations of the fiber's axis from a straight line.

MAN See metropolitan area network.

MAP Manufacturing automation protocol.

Margin Allowance for attenuation in addition to that explicitly accounted for in system design.

Mass Splicing Simultaneous splicing of many fibers in a cable.

Material Dispersion Dispersion resulting from the different velocities of each wavelength in a material.

MBaud One million bits of information per second. Also referred to as Mbps or Mb/s.

Mb/s See MBaud.

Mean Launched Power The average power for a continuous valid symbol sequence coupled into a fiber.

Mechanical Splice An optical fiber splice accomplished by fixtures or materials, rather than by thermal fusion.

Metropolitan Area Network A network covering an area larger than a local area network. A wide area network that covers a metropolitan area. Usually, an interconnection of two or more local area networks.

MFD See mode field diameter.

MH Medium-haul. A classification of video performance under RS-250B/C. Higher performance than long-haul and lower performance than short-haul.

MHz MegaHertz. One million Hertz (cycles per second).

Microbending Mechanical stress on a fiber may introduce local discontinuities called microbending. This results in light leaking from the core to the cladding by a process called mode coupling.

Micrometer One millionth of a meter or 10^{-6} meters. Abbreviated μm.

Microsecond One millionth of a second or 10^{-6} seconds. Abbreviated μs.

Microwatt One millionth of a Watt or 10^{-6} Watts. Abbreviated μW.

MIL-SPEC Military specification.

MIL-STD Military standard.

Misalignment Loss The loss of power resulting from angular misalignment, lateral displacement, and end separation.

mm Millimeter. One thousandth of a meter or 10^{-3} meters.

MM Abbreviation for multimode.

Modal Dispersion See multimode dispersion.

Modal Noise Modal noise occurs whenever the optical power propagates through mode-selective devices. It is usually only a factor with laser light sources.

Mode A single electromagnetic wave traveling in a fiber.

Mode Coupling The transfer of energy between modes. In a fiber, mode coupling occurs until equilibrium mode distribution (EMD) is reached.

Mode Evolution The dynamic process a multilongitudinal laser undergoes whereby the changing distribution of power among the modes creates a continuously changing envelope of the laser's spectrum.

Mode Field Diameter (MFD) A measure of distribution of optical power intensity across the end face of a single-mode fiber.

Mode Filter A device that removes higher-order modes to simulate equilibrium mode distribution.

Mode Scrambler A device that mixes modes to uniform power distribution.

Mode Stripper A device that removes cladding modes.

Modulation The process by which the characteristic of one wave (the carrier) is modified by another wave (the signal). Examples include amplitude modulation (AM), frequency modulation (FM), and pulse-coded modulation (PCM).

Modulation Index In an intensity-based system, the modulation index is a measure of how much the modulation signal affects the light output. It is defined as follows:

$$m = \frac{highlevel-lowlevel}{highlevel+lowlevel}$$

Monitor A television that receives its signal directly from a VCR, camera, or separate TV tuner for high-quality picture reproduction. Also a special type of television receiver designed for use with closed circuit TV equipment.

Monochrome Black and white TV signal.

MPEG Motion picture experts group. An international standard for compressing video that provides for high compression ratios. The standard has two recommendations: MPEG-1 compresses lower-resolution images for videoconferencing and lower-quality desktop video applications and transmits at around 1.5 Mb/s. MPEG-2 was devised primarily for delivering compressed television for home entertainment and is used at CCIR resolution when bit rates exceed 5.0 Mbits per second as in hard disk-based applications.

ms Milliseconds. One thousandth of a second or 10^{-3} seconds.

MTBF Mean time between failure. Time after which 50% of the units of interest will have failed.

MTTF Mean time to failure. See MTBF.

Multilongitudinal Mode Laser (MLM) An injection laser diode which has a number of longitudinal modes.

Multimode Dispersion Dispersion resulting from the different transit lengths of different propagating modes in a multimode optical fiber. Also called modal dispersion.

Multimode Fiber An optical fiber that has a core large enough to propagate more than one mode of light The typical diameter is 62.5 micrometers.

Multimode Laser Diode (MMLD) Synonym for Multilongitudinal mode laser.

Multiple Reflection Noise (MRN) The fiber optic receiver noise resulting from the interference of delayed signals from two or more reflection points in a fiber optic span. Also known as Multipath Interference.

Multiplexer A device that combines two or more signals into a single output.

Multiplexing The process by which two or more signals are transmitted over a single communications channel. Examples include time-division multiplexing and wavelength-division multiplexing.

MUSE Multiple sub-nyquist encoder. A high-definition standard developed in Europe that delivers 1125 lines at 60 frames per second.

mV Millivolt. One thousandth of a Volt or 10^{-3} Volts.

mW Milliwatt. One thousandth of a Watt or 10^{-3} Watts.

n Nano. One billionth or 10^{-9}.

N Newtons. Measure of force generally used to specify fiber optic cable strength.

nA Nanoamp. One billionth of an Amp or 10^{-9} Amps.

NA See numerical aperture.

NAB National Association of Broadcasters.

NA Mismatch Loss The loss of power at a joint that occurs when the transmitting half has a numerical aperture greater than the NA of the receiving half. The loss occurs when coupling light from a source to fiber, from fiber to fiber, or from fiber to detector.

National Electric Code® (NEC) A standard governing the use of electrical wire, cable and fixtures installed in buildings; developed by the NEC Committee of the American National Standards Institute (ANSI), sponsored by the National Fire Protection Association (NFPA), identified by the description ANSI/NFPA 70-1990.

Near-End Crosstalk (NEXT, RN) The optical power reflected from one or more input ports, back to another input port. Also known as isolation directivity.

Near Infrared The part of the infrared near the visible spectrum, typically 700 nm to 1500 nm or 2000 nm; it is not rigidly defined.

NEMA National Electrical Manufacturers Association.

NEP See noise equivalent power.

Network 1. An interconnection of three or more communicating entities and (usually) one or more nodes. 2. A combination of passive or active electronic components that serves a given purpose.

NFPA National Fire Protection Association.

nm Nanometer. One billionth of a meter or 10^{-9} meters.

Noise Equivalent Power (NEP) The noise of optical receivers, or of an entire transmission system, is often expressed in terms of noise equivalent optical power.

NRZ Nonreturn to zero. A common means of encoding data that has two states termed "zero" and "one" and no neutral or rest position.

ns Nanosecond. One billionth of a second or 10^{-9} seconds.

NTSC 1. National Television Systems Committee. The organization which formulated the NTSC system. 2. Standard used in the U.S. that delivers 525 lines at 60 frames per second.

Numerical Aperture (NA) The light-gathering ability of a fiber; the maximum angle to the fiber axis at which light will be accepted and propagated through the fiber. The measure of the light-acceptance angle of an optical fiber. NA = sin α, where α is the acceptance angle. NA is also used to describe the angular spread of light from a central axis, as in exiting a fiber, emitting from a source, or entering a detector.

$$NA = \sin\alpha = \sqrt{n_1^2 - n_2^2}$$

where:

α = Full acceptance angle.
n_1 = Core refractive index.
n_2 = Cladding refractive index.

nW Nanowatt. One billionth of a Watt or 10^{-9} Watts.

OC-x Optical carrier. A carrier rate specified in the SONET standard.

O/E Optical-to-electrical converter.

OEIC Opto-electronic integrated circuit.

OEM Original equipment manufacturer.

OLTS Optical loss test set.

1U One "U". "U" = 1.75 inches.

Open Standard Interconnect A seven-layer model defined by ISO for defining a communication network.

Optical Amplifier A device that amplifies an input optical signal without converting it into electrical form. The best developed are optical fibers doped with the rare earth element, erbium.

Optical Bandpass The range of optical wavelengths which can be transmitted through a component.

Optical Channel An optical wavelength band for WDM optical communications.

Optical Channel Spacing The wavelength separation between adjacent WDM channels.

Optical Channel Width The optical wavelength range of a channel.

Optical Continuous Wave Reflectometer (OCWR) An instrument used to characterize a fiber optic link wherein an unmodulated signal is transmitted through the link, and the resulting light scattered and reflected back to the input is measured. Useful in estimating component reflectance and link optical return loss.

Optical Directional Coupler (ODC) A component used to combine and separate optical power.

Optical Fall Time The time interval for the falling edge of an optical pulse to transition from 90% to 10% of the pulse amplitude. Alternatively, values of 80% and 20% may be used.

Optical Fiber A glass or plastic fiber that has the ability to guide light along its axis.

Optical Isolator A component used to block out reflected and unwanted light. Used in laser modules, for example. Also called an isolator.

Optical Link Loss Budget The range of optical loss over which a fiber optic link will operate and meet all specifications. The loss is relative to the transmitter output power.

Optical Loss Test Set (OLTS) A source and power meter combined to measure attenuation.

Optical Path Power Penalty The additional loss budget required to account for degradations due to reflections, and the combined effects of dispersion resulting from intersymbol interference, mode-partition noise, and laser chirp.

Optical Power Meter An instrument that measures the amount of optical power present at the end of a fiber or cable.

Optical Return Loss (ORL) The ratio (expressed in units of dB) of optical power reflected by a component or an assembly to the optical power incident on a component port when that component or assembly is introduced into a link or system.

Optical Rise Time The time interval for the rising edge of an optical pulse to transition from 10% to 90% of the pulse amplitude. Alternatively, values of 20% and 80% may be used.

Optical Time Domain Reflectometer (OTDR) An instrument that locates faults in optical fibers or infers attenuation by backscattered light measurements.

Optical Waveguide Another name for optical fiber.

OSI Open standards interconnect.

OTDR Optical time domain reflectometer.

p Pico. One trillionth or 10^{-12}.

pA Picoamp. One trillionth of an Amp or 10^{-12} Amps.

PABX Private automatic branch exchange.

PAL Phase alternation line. A composite color standard used in many parts of the world for TV broadcast. The phase alternation makes the signal relatively immune to certain distortions (compared to NTSC). Delivers 625 lines at 50 frames per second. PAL-plus is an enhanced-definition version.

Passive Branching Device A device which divides an optical input into two or more optical outputs.

PC Physical contact. Refers to an optical connector that allows the fiber ends to physically touch. Used to minimize backreflection and insertion loss.

PCB Printed circuit board.

PCM See pulse-code modulation.

PCS Fiber See plastic clad silica.

Peak Power Output The output power averaged over that cycle of an electromagnetic wave having the maximum peak value that can occur under any combination of signals transmitted.

PFM Pulse-frequency modulation. Also referred to as square wave FM.

Phase Constant The imaginary part of the axial propagation constant for a particular mode, usually expressed in radians per unit length. *See also* attenuation.

Phase Noise Rapid, short-term, random fluctuations in the phase of a wave caused by time-domain instabilities in an oscillator.

Photoconductive Losing an electrical charge on exposure to light.

Photodetector An optoelectronic transducer such as a PIN photodiode or avalanche photo-diode.

Photodiode A semiconductor device that converts light to electrical current.

Photon A quantum of electromagnetic energy. A particle of light.

Photonic A term coined for devices that work using photons, analogous to "electronic" for devices working with electrons.

Photovoltaic Providing an electric current under the influence of light or similar radiation.

Pigtail A short optical fiber permanently attached to a source, detector or other fiber optic device.

PINFET PIN detector plus a FET amplifier. Offers superior performance over a PIN alone.

PIN Photodiode See photodiode.

Planer Waveguide A waveguide fabricated in a flat material such as thin film.

Plastic Clad Silica (PCS) Also called hard clad silica (HCS). A step-index fiber with a glass core and plastic or polymer cladding instead of glass.

Plastic Fiber An optical fiber having a plastic core and plastic cladding.

Plenum The air handling space between walls, under structural floors, and above drop ceilings, which can be used to route intrabuilding cabling.

Plenum Cable A cable whose flammability and smoke characteristics allow it to be routed in a plenum area without being enclosed in a conduit.

Point-to-Point Transmission Transmission between two designated stations.

Polarization The direction of the electric field in the lightwave.

Polarization Maintaining Fiber Fiber that maintains the polarization of light that enters it.

Polarization Mode Dispersion (PMD) Polarization mode dispersion is an inherent property of all optical media. It is caused by the difference in the propagation velocities of light in the orthogonal principal polarization states of the transmission medium. The net effect is that if an optical pulse contains both polarization components, then the different polarization components will travel at different speeds and arrive at different times, smearing the received optical signal.

Port Hardware entity at each end of the link.

POS Point of sale.

POTS Plain old telephone system.

p-p Peak-to-peak. A peak-to-peak value is the algebraic difference between extreme values of a varying quantity.

PPM Pulse-position modulation. A method of encoding data.

Preform The glass rod from which optical fiber is drawn.

Profile Dispersion Dispersion attributed to the variation of refractive index contrast with wavelength.

ps Picosecond. One trillionth of a second or 10^{-12} seconds.

Pulse A current or voltage which changes abruptly from one value to another and back to the original value in a finite length of time. Used to describe one particular variation in a series of wave motions.

Pulse-Code Modulation (PCM) A technique in which an analog signal, such as a voice, is converted into a digital signal by sampling the signal's amplitude and expressing the different amplitudes as a binary number. The sampling rate must be at least twice the highest frequency in the signal.

Pulse Dispersion The spreading out of pulses as they travel along an optical fiber.

Pulse Spreading The dispersion of an optical signal as it propagates through an optical fiber.

pW Picowatt. One trillionth of a Watt or 10^{-12} Watts.

Quantum Efficiency In a photodiode, the ratio of primary carriers (electron-hole pairs) created to incident photons. A quantum efficiency of 70% means seven out of ten incident photons create a carrier.

Quaternary Signal A digital signal having four significant conditions. *See also* signal.

Radiation-Hardened Fiber An optical fiber made with core and cladding materials that are designed to recover their intrinsic value of attenuation coefficient, within an acceptable time period, after exposure to a radiation pulse.

Radiometer An instrument, distinct from a photometer, to measure power (Watts) of electromagnetic radiation.

Radiometry The science of radiation measurement.

Rayleigh Scattering The scattering of light that results from small inhomogeneities of material density or composition.

Rays Lines that represent the path taken by light.

Receiver A terminal device that includes a detector and signal processing electronics. It functions as an optical-to-electrical converter.

Receiver Overload The maximum acceptable value of average received power for an acceptable BER or performance.

Receiver Sensitivity The minimum acceptable value of received power needed to achieve an acceptable BER or performance. It takes into account power penalties caused by use of a transmitter with worst-case values of extinction ratio, jitter, pulse rise and fall times, optical return loss, receiver connector degradations, and measurement tolerances. The receiver sensitivity does not include power penalties associated with dispersion, jitter, or reflections from the optical path; these effects are specified separately in the allocation of maximum optical path penalty. Sensitivity usually takes into account worst-case operating and end-of-life (EOL) conditions.

Recombination Combination of an electron and a hole in a semiconductor that releases energy, sometimes leading to light emission.

Refraction The changing of direction of a wavefront in passing through a boundary between two dissimilar media, or in a graded-index medium where refractive index is a continuous function of position.

Refractive Index A property of optical materials that relates to the speed of light in the material.

Refractive Index Gradient The change in refractive index with distance from the axis of an optical fiber.

Refractive Index Profile The description of the value of the refractive index as a function of distance from the optical axis along an optical fiber diameter.

Regenerative Repeater A repeater, designed for digital transmission, in which digital signals are amplified, reshaped, retimed, and retransmitted.

Regenerator Synonym for regenerative repeater.

Repeater A receiver and transmitter set designed to regenerate attenuated signals. Used to extend operating range.

Residual Loss The loss of the attenuator at the minimum setting of the attenuator.

Responsivity The ratio of a photodetector's electrical output to its optical input in Amperes/Watt (A/W).

Return Loss See optical return loss.

RFI Radio frequency interference. Synonym of electromagnetic interference.

RGB Red, green, and blue. The basic parallel component set in which a signal is used for each primary color; or the related equipment or interconnect formats or standards.

Ribbon Cables Cables in which many fibers are embedded in a plastic material in parallel, forming a flat ribbon-like structure.

RIN Relative intensity noise. Often used to quantify the noise characteristics of a laser.

Ring A set of stations wherein information is passed sequentially between stations, each station in turn examining or copying the information, and finally returning it to the originating station.

Ring Network A network topology in which terminals are connected in a point-to-point serial fashion in an unbroken circular configuration.

Rise Time The time taken to make a transition from one state to another, usually measured between the 10% and 90% completion points of the transition. Alternatively the rise time may be specified at the 20% and 80% amplitudes. Shorter or faster rise times require more bandwidth in a transmission channel.

RMS Root mean square. Technique used to measure AC voltages.

RTS Request to send.

RZ Return to zero. A common means of encoding data that has two information states called "zero" and "one" in which the signal returns to a rest state during a portion of the bit period.

s Abbreviation for second.

SAE Society of Automotive Engineers.

SC Subscription channel connector. A push-pull type of optical connector that originated in Japan. Features high packing density, low loss, low backreflection, and low cost.

Scattering The change of direction of light rays or photons after striking small particles. It may also be regarded as the diffusion of a light beam caused by the inhomogeneity of the transmitting material.

SDH Synchronous digital hierarchy.

SECAM Système Èlectronique Couleur avec Mèmoire. A TV standard used in various parts of the world. Delivers 625 lines at 50 frames per second.

Selfoc Lens A trade name used by the Nippon Sheet Glass Company for a graded-index fiber lens; a segment of graded-index fiber made to serve as a lens.

Sensitivity See receiver sensitivity.

SH Short- haul. A classification of video performance under RS-250B/C. Higher performance than long-haul or medium-haul.

Sheath An outer protective layer of a fiber optic cable.

Shot Noise Noise caused by current fluctuations arising from the discrete nature of electrons.

Si Silicon. Generally used in detectors. Good for short wavelengths only (e.g., < 1000 nm).

Sideband Frequencies distributed above and below the carrier that contain energy resulting from amplitude modulation. The frequencies above the carrier are called upper sidebands, and the frequencies below the carrier are called lower sidebands.

Silica Glass Glass made mostly of silicon dioxide, SiO_2, used in conventional optical fibers.

Signal-to-Noise Ratio (SNR) The ratio of the total signal to the total noise which shows how much higher the signal level is than the level of the noise. A measure of signal quality.

Simplex Single element (e.g., a simplex connector is a single-fiber connector).

Simplex Cable A term sometimes used for a single-fiber cable.

Simplex Transmission Transmission in one direction only.

Single Attachment Concentrator A concentrator that offers one attachment to the FDDI network.

Single-Line Laser Synonym for single-longitudinal mode laser.

Single-Longitudinal Mode Laser (SLM) An injection laser diode which has a single dominant longitudinal mode. A single-mode laser with a side mode suppression ratio (SMSR)< 25 dB.

Single-mode (SM) Fiber A small-core optical fiber through which only one mode will propagate. The typical diameter is 8-9 microns.

Single-mode Laser Diode (SMLD) Synonym for single-longitudinal mode laser.

Single-mode Optical Loss Test Set (SMOLTS) An optical loss test set for use with single-mode fiber.

SI Units Abbreviation for International System of Units, commonly known as the metric system.

SLED See surface-emitting diode.

SM Abbreviation for single-mode.

SMA A threaded type of optical connector. One of the earliest optical connectors to be widely used. Offers poor repeatability and performance.

Smart Structures Also smart skins. Materials containing sensors (fiber optic or other types) to measure their properties during fabrication and use.

SMD Surface-mount device.

SMPTE Society of Motion Picture and Television Engineers.

SMT Surface-mount technology.

S/N See signal-to-noise ratio.

SNR See signal-to-noise ratio.

Soliton Pulse An optical pulse having a shape, spectral content, and power level designed to take advantage of nonlinear effects in an optical fiber waveguide, for the purpose of essentially negating dispersion over long distances.

SONET Synchronous optical network transport system. An interface standard for synchronous 2.488 Gb/s optical fiber transmission, developed by the Exchange Carriers Standards Association.

Source In fiber optics, a transmitting LED or laser diode, or an instrument that injects test signals into fibers.

Spectral Width A measure of the extent of a spectrum. For a source, the width of wavelengths contained in the output at one half of the wavelength of peak power. Typical spectral widths are 50 to 160 nm for an LED and 0.1-5 nm for a laser diode.

Spectral Width, Full Width, Half Maximum (FWHM) The absolute difference between the wavelengths at which the spectral radiant intensity is 50 percent of the maximum power.

Splice A permanent connection of two optical fibers through fusion or mechanical means.

Splitting Ratio The ratio of power emerging from two output ports of a coupler.

ST Straight tip connector. Popular fiber optic connector originally developed by AT&T.

Stabilized Light Source An LED or laser diode that emits light with a controlled and constant spectral width, central wavelength, and peak power with respect to time and temperature.

Star Coupler A coupler in which power at any input port is distributed to all output ports.

Star Network A network in which all terminals are connected through a single point, such as a star coupler or concentrator.

Step-Index Fiber Fiber that has a uniform index of refraction throughout the core.

Strength Member The part of a fiber optic cable composed of aramid yarn, steel strands, or fiberglass filaments that increase the tensile strength of the cable.

Submarine Cable A cable designed to be laid underwater.

Subscriber Loop Also called local loop. The link from the telephone company central office (CO) to the home or business (customer premises).

Supertrunk A cable that carries several video channels between facilities of a cable television company.

Surface-Emitting Diode An LED that emits light from its flat surface rather than its side. Simple and inexpensive, with emission spread over a wide angle.

Sync This signal is derived from the composite or combination of the horizontal and vertical drives. *See also* composite sync.

Synchronous A data signal that is sent along with a clock signal.

T Tera. One trillion or 10^{12}.

Tap Loss In a fiber optic coupler, the ratio of power at the tap port to the power at the input port.

Tap Port In a coupler where the splitting ratio between output ports is not equal, the output port containing the lesser power.

TAXI Transparent asynchronous transmitter-receiver interface. A chip used to transmit parallel data over a serial interface.

TBC Timebase corrector.

T-Carrier Generic designator for any of several digitally multiplexed telecommunications carrier systems.

TDM See time-division multiplexing.

TEC Abbreviation for thermoelectric cooler.

Tee Coupler A three-port optical coupler.

10BASE-F A fiber optic version of an IEEE 802.3 network.

10BASE-FB That portion of 10BASE-F that defines the requirements for a fiber backbone.

10BASE-FL That portion of 10BASE-F that defines a fiber optic link between a concentrator and a station.

10BASE-FP That portion of 10BASE-F that defines a passive star coupler.

10BASE-T A twisted-pair cable version of an IEEE 802.3 network.

10BASE-2 A thin-coaxial-cable version of an IEEE 802.3 network.

10BASE-5 A thick-coaxial-cable version of an IEEE 802.3 network; very similar to the original Ethernet specification.

Ternary A semiconductor compound made of three elements (e.g., GaAlAs).

TFOCA Tactical fiber optic cable assembly.

Thermal Noise Noise resulting from thermally induced random fluctuation in current in the receiver's load resistance.

Throughput Loss In a fiber optic coupler, the ratio of power at the throughput port to the power at the input port.

Throughput Port In a coupler where the splitting ratio between output ports is not equal, the output port containing the greater power.

Tight-Buffer A material tightly surrounding a fiber in a cable, holding it rigidly in place.

Time-Division Multiplexing (TDM) A transmission technique whereby several low-speed channels are multiplexed into a high-speed channel for transmission. Each low-speed channel is allocated a specific position based on time.

Token Ring A ring-based network scheme in which a token is used to control access to a network. Used by IEEE 802.5 and FDDI.

Total Internal Reflection The reflection that occurs when light strikes an interface at an angle of incidence (with respect to the normal) greater than the critical angle.

Transceiver A device that performs, within one chassis, both telecommunication transmitting and receiving functions.

Transducer A device for converting energy from one form to another, such as optical energy to electrical energy.

Transmitter A device that includes a source and driving electronics. It functions as an electrical-to-optical converter.

Tree A physical topology consisting of a hierarchy of master-slave connections between a concentrator and other FDDI nodes (including subordinate concentrators).

Trunk A physical loop topology, either open or closed, employing two optical fiber signal paths, one in each direction (i.e. counter-rotating) forming a sequence of peer connections between FDDI nodes. When the trunk forms a closed loop, it is sometimes called a trunk ring.

TTL Transistor-transistor logic.

UL Underwriter's Laboratory.

Unidirectional Operating in one direction only.

UV Ultraviolet.

V Volt.

VAC Volts, AC.

VDC Volts, DC.

Vestigial-Sideband (VSB) Transmission A modified double-sideband transmission in which one sideband, the carrier, and only a portion of the other sideband are transmitted. *See also* sideband.

VGA Video graphics array. A high-resolution color standard for computer monitors.

Videoconferencing Conducting conferences via a video telecommunications system.

Videophone A telephone-like service with a picture as well as sound.

Visible Light Electromagnetic radiation visible to the human eye; wavelengths of 400-700 nm.

Voice Circuit A circuit capable of carrying one telephone conversation or its equivalent; the standard subunit in which telecommunication capacity is counted. The U.S. analog equivalent is 4 kHz. The digital equivalent is 64 kbit/s in North America and in Europe.

VSB See vestigial-sideband transmission.

W See Watt.

WAN See wide area network.

Watt Linear measurement of optical power, usually expressed in milliwatts, microwatts, and nanowatts.

Waveguide A material medium that confines and guides a propagating electromagnetic wave. In the microwave regime, a waveguide normally consists of a hollow metallic conductor, generally rectangular, elliptical, or circular in cross-section. This type of waveguide may, under certain conditions, contain a solid or gaseous dielectric material. In the optical regime, a waveguide used as a long transmission line consists of a solid dielectric filament (optical fiber), usually circular in cross-section. In integrated optical circuits an optical waveguide may consist of a thin dielectric film. In the RF regime, ionized layers of the stratosphere and the refractive surfaces of the troposphere may also serve as a waveguide.

Waveguide Couplers A coupler in which light is transferred between planar waveguides.

Waveguide Dispersion The part of chromatic dispersion arising from the different speeds light travels in the core and cladding of a single-mode fiber (i.e., from the fiber's waveguide structure).

Wavelength The distance between points of corresponding phase of two consecutive cycles of a wave. The wavelength, is related to the propagation velocity, and the frequency, by:

$$\text{Wavelength} = \frac{\text{Propagation Velocity}}{\text{Frequency}}$$

Wavelength-Division Multiplexing (WDM) Sending several signals through one fiber with different wavelengths of light.

Wavelength Isolation A WDM's isolation of a light signal in the desired optical channel from the unwanted optical channels. Also called far-end crosstalk.

WDM See wavelength-division multiplexing.

Wide Area Network A physical or logical network that provides capabilities for a number of independent devices to communicate with each other over a common transmission-interconnected topology in geographic areas larger than those served by local area networks.

Wideband Possessing large bandwidth.

XT Abbreviation for crosstalk.

Y Coupler A variation on the tee coupler in which input light is split between two channels (typically planar waveguide) that branch out like a Y from the input.

Zero-Dispersion Wavelength (λ_0) In a single-mode optical fiber, the wavelength at which material dispersion and waveguide dispersion cancel one another. The wavelength of maximum bandwidth in the fiber.

FIBER OPTIC SYMBOLS

NATIONAL CABLE TELEVISION ASSOCIATION SYMBOLS

The following symbols were voted by the National Cable Television Association's Engineering Committee to become part of the NCTA Recommended Practice for Measurements on Cable Television Systems.

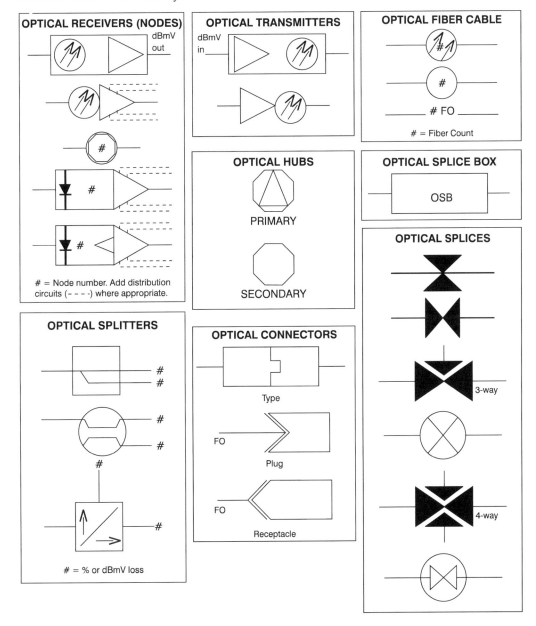

MILITARY STANDARD SYMBOLS

The following symbols are included in Military Standard MIL-STD-1864A. The purpose is to list the symbols for fiber optic parts for use on military drawings and wherever symbols for fiber optic parts are required.

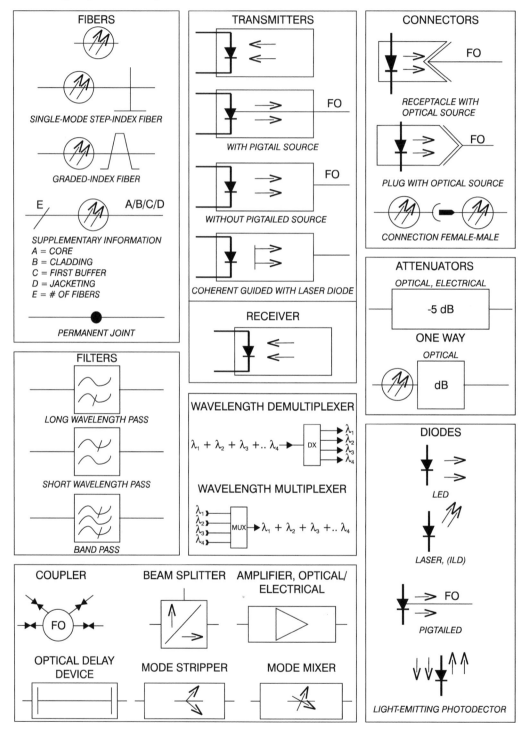

FIBER OPTIC SYSTEM CHECKLIST

Date		Contact	
Customer		Phone	
Address		Fax	

System Operational Requirements

APPLICATION

Video	CCTV		NTSC		HDTV	
	EDUCATION		CATV		# CHANNELS	
	BROADCAST		OTHER			

Audio	TELEPHONE	HIGH-QUALITY AUDIO		INTERCOM	
	STUDIO	OTHER			

Computer	HIPPI	FIBRE CHANNEL		FDDI	
	OTHER				

Other	

OPTICAL CHARACTERISTICS

Fiber Type	SM		50/125 μm		62.5/125 μm	
	100/140 μm		OTHER			

Optical Wavelength	PRIMARY	nm	SECONDARY	nm	OTHER	nm

Attenuation	850 nm	dB/km	1300 nm	dB/km	1550 nm	dB/km

Connector Type	ST		SC		FC	
	BICONIC		SMA 906		OTHER	

Minimum End-to-End Optical Loss		dB
Maximum End-to-End Optical Loss		dB

FIBER OPTIC SYSTEM CHECKLIST

Type of Signals

ANALOG SIGNALS

System Bandwidth		MHz
Low-Frequency Response		MHz
System Signal-to-Noise Ratio		dB
Linearity Requirements		

DIGITAL SIGNALS

Coding Scheme	NRZ		RZ		OTHER
Data Rate		bps			
Bit Error Rate	10^{-8}		10^{-9}		OTHER
Logic Format	TTL		ECL		PECL
	OTHER				

OTHER SIGNALS

Min. Required Receiver Optical Power	dB		AVG		PEAK
Receiver Dynamic Range		dB			
Max Optical Power Allowed at Receiver	dBm		AVG		PEAK

Terminal Equipment

SPACE AVAILABLE FOR:

	LENGTH			WIDTH			HEIGHT
Transmitter		" x			" x		"
Receiver		" x			" x		"
Repeater		" x			" x		"
Terminal Equipment Connections	RS-232			BNC		OTHER	
Terminal Equipment Mounting	PC Board			Rack		OTHER	

POWER SUPPLY REQUIREMENTS

Voltages	AC		DC		VOLTS
Current		mA			
Frequency		Hz			
Power		Watts			

System Layout

SYSTEM LOCATION

Locations of Equipment	BUILDING _____	OTHER _____
Distance Between Stations	_____ m	

SYSTEM ENVIRONMENT

For Terminals and Repeaters	BUILDING _____	OUTDOOR _____	
For Cables (routing)	DUCTS _____	BURRIED _____	AERIAL _____
	OTHER _____		
Operating Temp. Range	_____ °C	to _____ °C	
Storage Temp. Range	_____ °C	to _____ °C	
Vibration Requirements	_____		
Other Environmental Conditions	_____		

Notes _____
_____
_____
_____
_____
_____
_____
_____
_____
_____
_____
_____
_____
_____

INDUSTRIAL, MILITARY, & BELLCORE STANDARDS

The following is a listing of the industrial, military, and Bellcore standards that apply to the field of fiber optics. The industrial and military standards are available from the EIA Standards Sales Office by writing or calling:

> EIA Standards Sales Office (Global Engineering)
> 2500 Wilson Boulevard
> Arlington, VA 22201-3834
> Phone: (800) 854-7179

Bellcore standards are available by writing or calling:

> Information Exchange Management
> Bellcore
> 445 South Street, Room 2J-125 (PO Box 1910)
> Morristown, NJ 07962-1910
> Phone: (800) 521-CORE (2673)
> FAX: (908) 336-2259
> For Foreign Calls: (908) 336-2559

INDUSTRIAL STANDARDS

TSB-62	Informative Test Methods (ITMs) for Fiber Optic Fibers, Cables, Opto-Electronic Sources and Detectors, Sensors, Connecting and Terminating Devices, and Other Fiber Optic Components
TSB-62-1	Characterization of Large Flaws in Optical Fibers by Dynamic Tensile Testing with Sensoring Category 3
TSB-62-2	Method for Measurement of Hydrogen Evolved from Coated Optical Fiber
TSB-63	Reference Guide for Fiber Optic Test Procedures
EIA-440-A	Fiber Optic Terminology (ANSI/EIA-440-A-88)
EIA-455-A	Standard Test Procedure for Fiber Optic Fibers, Cables, Transducers, Sensors, Connecting and Terminating Devices, and Other Components
EIA-455-1A	FOTP-1 Cable Flexing for Fiber Optic Interconnecting Devices
EIA/TIA-455-2B	FOTP-2 Impact Test Measurements for Fiber Optic Devices
EIA/TIA-455-3A	FOTP-3 Procedure to Measure Temperature Cycling Effects on Optical Test Fibers, Optical Cable, and Other Passive Fiber Optic Components
TIA/EIA-455-4B	FOTP-4 Fiber Optic Component Temperature Life
TIA/EIA-455-5B	FOTP-5 Humidity Test Procedure for Fiber Optic Components
EIA/TIA-455-6B	FOTP-6 Cable Retention Test Procedure for Fiber Optic Cable Interconnecting Devices
EIA/TIA-455-7	FOTP-7 Numerical Aperture of Step-Index Multimode Optical Fibers by Output Far-Field Radiation Pattern Measurement
EIA/TIA-455-10	FOTP-10 Procedure for Measuring the Amount of Extractable Material in Coatings Applied to Optical Fiber
TIA/EIA-455-11B	FOTP-11 Vibration Test Procedure for Fiber Optic Components and Cables
EIA/TIA-455-12A	FOTP-12 Fluid Immersion Test for Fiber Optic Components
EIA/TIA-455-13	FOTP-13 Visual and Mechanical Inspection of Fibers, Cables, Connectors and/or Other Fiber Optic Devices
EIA/TIA-455-14A	FOTP-14 Fiber Optic Shock Test (Specified Pulse)
EIA/TIA-455-15A	FOTP-15 Altitude Immersion
EIA/TIA-455-16A	FOTP-16 Salt Spray (Corrosion) Test for Fiber Optic Components
EIA-455-17A	FOTP-17 Maintenance Aging of Fiber Optic Connectors and Terminated Cable Assemblies
EIA-455-18A	FOTP-18 Acceleration Testing of Fiber Optic Components and Assemblies
TSB-19	Optical Fiber Digital Transmission Systems: Considerations for Users and Suppliers
EIA-455-20	FOTP-20 Measurement of Change in Optical Transmittance
EIA-455-21A	FOTP-21 Mating Durability for Fiber Optic Interconnecting Devices
EIA/TIA-455-22B	FOTP-22 Ambient Light Susceptibility of Fiber Optic Components
EIA-455-23A	FOTP-23 Air Leakage Testing of Fiber Optic Components Seals
EIA/TIA-455-24	FOTP-24 Water Peak Attenuation Measurement of Single-mode Fibers
EIA/TIA-455-25A	FOTP-25 Repeated Impact Testing of Fiber Optic Cables and Cable Assemblies
EIA-455-26A	FOTP-26 Crush Resistance of Fiber Optic Interconnecting Devices

EIA/TIA-455-28B	FOTP-28 Method for Measuring Dynamic Tensile Strength of Optical Fibers
EIA/TIA-455-29A	FOTP-29 Refractive Index Profile (Transverse Interference Method)
EIA/TIA-455-30B	FOTP-30 Frequency Domain Measurement of Multimode Optical Fiber Information Transmission Capacity
EIA/TIA-455-31B	FOTP-31 Fiber Tensile Proof Test Method
EIA/TIA-455-32A	FOTP-32 Fiber Optic Circuit Discontinuities
EIA-455-33A	FOTP-33 Fiber Optic Cable Tensile Loading and Bending Test
EIA-455-34	FOTP-34 Interconnection Device Insertion Loss Test
EIA/TIA-455-35A	FOTP-35 Fiber Optic Component Dust (Fine Sand) Test
EIA-455-36A	FOTP-36 Twist Test for Fiber Optic Connecting Devices
EIA/TIA-455-37A	FOTP-37 Low or High Temperature Bend Test for Fiber Optic Cable
EIA/TIA-455-39A	FOTP-39 Fiber Optic Cable Water Wicking Test
TIA/EIA-455-41A	FOTP-41 Compressive Loading Resistance of Fiber Optic Cables
EIA/TIA-455-42A	FOTP-42 Optical Crosstalk in Components
EIA-455-43	FOTP-43 Output Near-Field Radiation Pattern Measurement of Optical Waveguide Fibers
TIA/EIA-455-44B	FOTP-44 Refractive Index Profile (Refracted Ray Method)
EIA-455-45B	FOTP-45 Microscopic Method for Measuring Fiber Geometry of Optical Waveguide Fibers
EIA/TIA-455-46A	FOTP-46 Spectral Attenuation Measurement for Long-Length Graded-Index Fibers
EIA/TIA-455-47B	FOTP-47 Output Far-Field Radiation Pattern Measurement
EIA/TIA-455-48B	FOTP-48 Measurement of Optical Fiber Cladding Diameter Using Laser-Based Instruments
EIA/TIA-455-49A	FOTP-49 Measurement for Gamma Irradiation Effects on Optical Fibers and Cables
EIA-455-50A	FOTP-50 Light Launch Conditions for Long-Length Graded-Index Optical Fibers Spectral Attenuation Measurements
EIA/TIA-455-51A	FOTP-51 Pulse Distortion Measurement of Multimode Glass Optical Fiber Information Capacity
EIA/TIA-455-53A	FOTP-53 Attenuation by Substitution Measurement for Multimode Graded-Index Optical Fibers or Fiber Assemblies Used in Long-Length Communications Systems
EIA/TIA-455-54A	FOTP-54 Mode Scrambler Requirements for Overfilled Launching Conditions to Multimode Fibers
EIA/TIA-455-55B	FOTP-55 End-View Methods for Measuring Coating and Buffer Geometry of Optical Fibers
EIA-455-56A	FOTP-56 Test Method for Evaluating Fungus Resistance of Optical Waveguide Fibers and Cables
EIA/TIA-455-57A	FOTP-57 Optical Fiber End Preparation and Examination
EIA/TIA-455-58A	FOTP-58 Core Diameter Measurements of Graded-Index Optical Fibers
EIA/TIA-455-59	FOTP-59 Measurement of Fiber Point Defects Using an OTDR
EIA/TIA-455-60	FOTP-60 Measurement of Fiber or Cable Length Using an OTDR
EIA/TIA-455-61	FOTP-61 Measurement of Fiber or Cable Attenuation Using an OTDR
EIA/TIA-455-62A	FOTP-62 Measurement of Optical Fiber Macrobend Attenuation
EIA-455-65	FOTP-65 Optical Fiber Flexure Test
EIA/TIA-455-69A	FOTP-69 Test Procedure for Evaluating Minimum and Maximum Exposure Temperature on the Optical Performance of Optical Fiber
EIA/TIA-455-71	FOTP-71 Procedure to Measure Temperature Shock Effects on Fiber Optic Components
EIA/TIA-455-75	FOTP-75 Fluid Immersion Test for Optical Waveguide Fibers
EIA/TIA-455-76	FOTP-76 Method for Measuring Dynamic Fatigue of Optical Fibers by Tension
EIA/TIA-455-77	FOTP-77 Procedures to Qualify a Higher-Order Mode Filter for Measurements of Single-mode Fibers
EIA/TIA-455-78A	FOTP-78 Spectral Attenuation Cutback Measurement for Single-mode Optical Fibers
EIA-455-80	FOTP-80 Cutoff Wavelength of Uncabled Single-mode Fiber by Transmitted Power
EIA/TIA-455-81A	FOTP-81 Compound Flow (Drip) Test for Filled Fiber Optic Cable
EIA-455-82B	FOTP-82 Fluid Penetration Test for Fluid-Blocked Fiber Optic Cable
EIA-455-83A	FOTP-83 Cable to Interconnecting Device Axial Compressive Loading
EIA-455-84B	FOTP-84 Jacket Self-Adhesion (Blocking) Test for Fiber Optic Cable
TIA/EIA-455-85A	FOTP-85 Fiber Optic Cable Twist Test
EIA-455-86	FOTP-86 Fiber Optic Cable Jacket Shrinkage
EIA/TIA-455-87B	FOTP-87 Fiber Optic Cable Knot Test
EIA-455-88	FOTP-88 Fiber Optic Cable Bend Test
EIA-455-89A	FOTP-89 Fiber Optic Cable Jacket Elongation and Tensile Strength Test
EIA-455-91	FOTP-91 Fiber Optic Cable Twist-Bend Test
TIA/EIA-455-92	FOTP-92 Optical Fiber Cladding Diameter and Noncircularity Measurement by Fizeau Interferometry

TIA/EIA-455-93	FOTP-93 Test Method for Optical Fiber Cladding Diameter and Noncircularity by Noncontacting Michelson Interferometry
EIA-455-94	FOTP-94 Fiber Optic Cable Stuffing Tube Compression
EIA-455-95	FOTP-95 Absolute Optical Power Test for Optical Fibers and Cables
EIA/TIA-455-96	FOTP-96 Fiber Optic Cable Long-Term Storage Temperature Test for Extreme Environments
EIA/TIA-455-98A	FOTP-98 Fiber Optic Cable External Freezing Test
TIA/EIA-455-99	FOTP-99 Gas Flame Test for Special Purpose Fiber Optic Cable
EIA/TIA-455-100A	FOTP-100 Gas Leakage Test for Gas-Blocked Fiber Optic Cable
EIA-455-102	FOTP-102 Water Pressure Cycling
TIA/EIA-455-104A	FOTP-104 Fiber Optic Cable Cyclic Flexing Test
TIA/EIA-455-106	FOTP-106 Procedure for Measuring the Near Infrared Absorbance of Fiber Optic Coating Materials
EIA/TIA-455-107	FOTP-107 Return Loss for Fiber Optic Components
EIA/TIA-455-127	FOTP-127 Spectral Characterization of Multimode Laser Diodes, Performance of Optical Fibers
EIA-455-162	FOTP-162 Fiber Optic Cable Temperature-Humidity Cycling
EIA/TIA-455-164A	FOTP-164 Single-mode Fiber, Measurement of Mode Field Diameter by Far-Field Scanning
EIA/TIA-455-165A	FOTP-165 Mode Field Diameter Measurement by Near-Field Scanning (Single-mode)
EIA/TIA-455-167A	FOTP-167 Mode Field Diameter Measurement, Variable Aperture Method in Far-Field
EIA/TIA-455-168A	FOTP-168 Chromatic Dispersion Measurement of Multimode Graded-Index and Single-mode Optical Fibers by Spectral Group Delay Measurement in the Time Domain
EIA/TIA-455-169A	FOTP-169 Chromatic Dispersion Measurement of Optical Fibers by Phase-Shift Method
EIA/TIA-455-170	FOTP-170 Cable Cutoff Wavelength of Single-mode Fiber by Transmitted Power
EIA-455-171	FOTP-171 Attenuation by Substitution Measurement Short-Length Multimode Graded-Index and Single-mode Fiber Optic Cable Assemblies
EIA-455-172	FOTP-172 Flame Resistance of Firewall Connector
EIA/TIA-455-173	FOTP-173 Coating Geometry Measurement of Optical Fiber, Side-View Method
EIA-455-174	FOTP-174 Mode Field Diameter of Single-mode Fiber by Knife-Edge Scanning in Far-Field
TIA/EIA-455-175A	FOTP-175 Chromatic Dispersion Measurement of Single-mode Optical Fibers by the Differential Phase-Shift Method
TIA/EIA-455-176	Method for Measuring Optical Fiber Cross-Sectional Geometry by Automated Gray-Scale Analysis
TIA/EIA-455-177A	FOTP-177 Numerical Aperture Measurement of Graded-Index Optical Fibers
EIA/TIA-455-178	FOTP-178 Measurements of Strip Force Required for Mechanically Removing Coatings from Optical Fibers
EIA-455-179	FOTP-179 Inspection of Cleaved Fiber End Faces by Interferometry
EIA/TIA-455-180	FOTP-180 Measurement of Optical Transfer Coefficients of a Passive Branching Device
TIA/EIA-455-181	FOTP-181 Lightning Damage Susceptibility Test for Fiber Optic Cables with Metallic Components
EIA/TIA-455-184	FOTP-184 Coupling Proof Overload Test for Fiber Optic Interconnecting Devices
EIA/TIA-455-185	FOTP-185 Strength of Coupling Mechanism for Fiber Optic Interconnecting Devices
EIA/TIA-455-186	FOTP-186 Gauge Retention Force Measurement for Components
EIA/TIA-455-187	FOTP-187 Engagement and Separation Force of Fiber Optic Connector Sets
EIA/TIA-455-188	FOTP-188 Low-Temperature Testing of Fiber Optic Components
EIA/TIA-455-189	FOTP-189 Ozone Exposure Test of Fiber Optic Components
EIA/TIA-455-190	FOTP-190 Low Air Pressure (High Altitude) Testing of Fiber Optic Components
EIA/TIA-458-B	Standard Optical Fiber Material Classes and Preferred Sizes
TIA/EIA-4720000-A	General Specification for Fiber Optic Cables
EIA-472A000	Sectional Specification for Fiber Optic Communication Cables for Outside Aerial Use
EIA-472B000	Sectional Specification for Fiber Optic Communication Cables for Underground and Buried Use
TIA/EIA-472C000-A	Sectional Specification for Fiber Optic Communication Cables for Indoor Use
EIA-472D000-A	Sectional Specification for Fiber Optic Communication Cables for Outside Plant Use
EIA/TIA-4750000-B	Generic Specification for Fiber Optic Connectors
EIA/TIA-475C000	Sectional Specification for Type FSMA Connectors
EIA/TIA-475CA00	Blank Detail Specification for Optical Fibers and Cable Type FSMA, Environmental Category I
EIA/TIA-475CB00	Blank Detail Specification for Optical Fibers and Cables Type FSMA, Environmental Category II
EIA/TIA-475CC00	Blank Detail Specification Connector Set for Optical Fibers and Cables Type FSMA, Environmental Category III.
TIA/EIA-475E000	Sectional Specification for Fiber Optic Connectors Type BFOC/2.5

TIA/EIA-475EA00	Blank Detail Specification for Connector Set for Optical Fibers and Cables, Type BFOC/2.5, Environmental Category I
TIA/EIA-475EB00	Blank Detail Specification for Connector Set for Optical Fibers and Cables, Type BFOC/2.5, Environmental Category II
TIA/EIA-475EC00	Blank Detail Specification for Connector Set for Optical Fibers and Cables, Type BFOC/2.5, Environmental Category III
EIA-4920000	Generic Specification for Optical Waveguide Fibers
EIA-492A000	Sectional Specification for Class Ia Multimode, Graded-Index Optical Waveguide Fibers
EIA/TIA-492AAAA	Detail Specification for 62.5-μm Core Diameter/125-μm Cladding Diameter Class Ia Multimode, Graded-Index Optical Waveguide Fibers
EIA-492B000	Sectional Specification for Class IV Single-mode Optical Waveguide Fibers
EIA-492BA00	Blank Detail Specification for Class IVa Dispersion, Unshifted Single-mode Optical Waveguide Fibers
EIA/TIA-492BB00	Blank Detail Specification for Class IVb Dispersion, Shifted Single-mode Optical Waveguide Fibers
EIA-5090000	Generic Specification for Fiber Optic Terminal Devices
EIA-5150000	Generic Specification for Optical Fiber and Cable Splices
EIA-515B0000	Sectional Specification for Splice Closures for Pressurized Aerial, Buried, and Underground Fiber Optic Cables
EIA/TIA-526	Standard Test Procedures for Fiber Optic Systems
EIA/TIA-526-2	OFSTP-2, Effective Transmitter Output Power Coupled into Single-mode Fiber Optic Cable
EIA/TIA-526-3	OFSTP-3, Fiber Optic Terminal Equipment Receiver Sensitivity and Maximum Receiver Input
TIA/EIA-526-10	OFSTP-10, Measurement of Dispersion Power Penalty in Digital Single-mode Systems
EIA/TIA-526-11	OFSTP-11, Measurement of Single-reflection Power Penalty for Fiber Optic Terminal Equipment
EIA/TIA-526-14	OFSTP-14, Optical Power Loss Measurements of Installed Multimode Fiber Cable Plant
EIA/TIA-526-15	OFSTP-15, Jitter Tolerance Measurement
TIA/EIA-526-16	OFSTP-16, Jitter Transfer Function Measurement
TIA/EIA-526-17	OFSTP-17, Output Jitter Measurement
TIA/EIA-526-18	OFSTP-18, Systematic Jitter Generation Measurement
EIA/TIA-559	Single-mode Fiber Optic System Transmission Design

EIA/TIA-559-1	Single-mode Fiber Optic System Transmission Design
EIA/TIA-587	Fiber Optic Graphic Symbols
EIA/TIA-590	Standard for Physical Location and Protection of Below-ground Fiber Optic Cable Plant
EIA/TIA-598	Color Coding of Fiber Optic Cables
EIA/TIA-604	Fiber Optic Connector Intermateablity Standards

MILITARY STANDARDS

MIL-STD-188-111A	Interoperability and Performance Standards for Fiber Optic Communication Systems
MIL-HDBK-415	Design Handbook for Fiber Optic Communication Systems
FED-STD-1037B	Glossary of Telecommunication Terms (Includes fiber optic terms.)
DOD-STD-1678	Fiber Optic Test Methods and Instrumentation
MIL-STD-1773	Fiber Optic Mechanization of an Aircraft Internal Time Division
MIL-STD-1863A	Interface Designs and Dimensions for Fiber Optic Interconnection Devices
MIL-STD-2163C	Insert Arrangements for MIL-C-28876 Connectors, Circular, Plug and Receptacle Style, Multiple Removable Termini
MIL-C-22520/10D	Crimping Tools, Terminal, Hand-Held, Wire, for Coaxial, Shielded, Contacts, Ferrules, Terminal Lugs, Splices, and End Caps
MIL-D-24620A	Detectors, PIN, and APD, Fiber Optic, General Specification
MIL-D-24620/2	Detector, PIN Type A and B, Class 1, Style 3
MIL-D-24620/3	Detector, APD, Type A and B, Class 1, Style 3
MIL-D-24620/4A	Detector, PINFET, 1100-1600 nm, Glass Pigtailed, Hermetically Sealed, Dual In-Line Package (DIP) Couplers, Passive
MIL-C-24621B	Couplers, Passive, General Specification
MIL-C-24621/1	Couplers, Cable Splitter, Passive Glass Connectorized Output, Type I, 820-910 nm 2 x 2, Transmission Star, 50/125 μm Fiber, Graded-Index, MM, Bidirectional
DOD-C-24621/2	Couplers, Cable Splitter, Passive Glass Pigtailed Output, Type II, 820-910 nm, 4 x 4, Transmission Star, 50/125 μm Fiber, Graded-Index, MM, Bidirectional
DOD-C-24621/3	Couplers, Cable Splitter Passive Glass Pigtailed Output, Type II, 820-910 nm, 20 x 20 Transmission Star, 100/140 μm, 50/125 μm Fiber
DOD-C-24621/4	Couplers, Cable Splitter Passive Glass Pigtailed Output, Type II, 820-910 nm 5 x 5, Transmission Star, 100/140 μm, 50/125 μm Fiber
MIL-C-24621/5A	Couplers, Cable Splitter Passive Glass Pigtailed Output, Type II, 1280-1360 nm, 2 x 2 Transmission Star, 62.5/125 μm Fiber, Bidirectional

MIL-S-24622A	Sources, LED Type General Specification
MIL-S-24622/2	LED, Type A, Class 1, Style 3
MIL-S-24622/3A	LED, 1290 nm, Pigtailed, Hermetically Sealed, Dual In-line Package
MIL-S-24623B	Splice, Fiber Optic Cable, General Specifications
DOD-S-24623/1	Splice, Fiber Optic Cable, Fiber Splice
DOD-S-24623/2	Splice, Fiber Optic Cable, Fiber Splice Enclosure
DOD-S-24623/3	Splice, Fiber Optic Cable, Splice, Cable/Fiber
MIL-S-24623/4A	Splice, Fiber Optic, Housing, Fiber
MIL-S-24623/5A	Splice, Fiber Optic, Housing, Cable
MIL-H-24626	(Navy) Cable Harness Assemblies, Pressure Proof, Fiber Optic, General Specification
MIL-H-24626/1	(Navy) Cable Harness, Pressure Proof
MIL-P-24628	(Navy) Penetrators, Hull Connectorized, Connectors, Pressure-Proof
MIL-P-24628/1	Penetrators, Hull, Connectorized, Pressure Proof, Submarine Plug Connector, 32 Channel, Type 1
MIL-R-24720	Receivers, Digital, Fiber Optic Shipboard, General Specification for VSMF
MIL-R-24720/1	Receiver, Digital, Fiber Optic, Shipboard, 0.5 to 16 Mbps, Manchester Encoded
MIL-T-24721	Transmitters, Digital, Fiber Optic, Shipboard, General Specification for VSMF
MIL-T-24721/1	Transmitters, Digital, Fiber Optic, Shipboard, DC to 16 Mbps, Manchester Encoded
MIL-S-24725A	Switches, Fiber Optic, Shipboard, General Specification for VSMF
MIL-S-24725/1A	Switches, Fiber Optic, Shipboard, Electrical, Nonlatching, bypass, Multimode Cable, Stand-alone
MIL-S-24725/2	Switches, Fiber Optic, Nonhard Mounted, Electrical, Nonlatching, Bypass, Multimode Cable, Stand-alone (Metric)
MIL-A-24726	Attenuators, Fiber Optic, Shipboard, General Specifications for VSMF
MIL-A-24726/1	Attenuator, Fiber Optic, Shipboard, Fixed, Connectorized, Single-mode, Stand-alone
MIL-R-24727	Rotary Joints, Fiber Optic, Shipboard, General Specification for VSMF
MIL-R-24727/1	Rotary Joints, Fiber Optic, Shipboard, Single Fiber, On-axis, Multimode Cable Pigtail
MIL-R-24727/2	Rotary Joints, Fiber Optic, Shipboard, Multiple Fibers, On-Axis
MIL-I-24728A	Interconnection Box, Fiber Optic, General Specification for VSMF
MIL-I-24728/2A	Interconnection Box, Fiber Optic, Submersible
MIL-M-24731	Multiplexers, Demultiplexers, Frequency-Division, Fiber Optic Interfaceable, Shipboard, Metric, General Specification for VSMF
MIL-M-24731/1	Multiplexers, Demultiplexers, Frequency-Division, Fiber Optic, Shipboard, 4:1 Channels, 30 MHz (Double Sideband)
MIL-M-24731/2	Multiplexers, Demultiplexers, Frequency-Division, Fiber Optic, Shipboard, 4:1 Channels, 30 MHz (Double Sideband)
MIL-L-24732	Light Source, Rigid and Flexible Fiberscope, Fiber Optic, General Specification for VSMF
MIL-L-24732/1	Light Source, Fiberscope, Fiber Optic, 150 Watt
MIL-C-24733	Controller Interface Unit, Fiber Optic, General Specification for VSMF
MIL-C-24733/1	Controller Interface Unit, Fiber Optic, 2 Fiber Channels, Multimode Fiber
MIL-F-24734	Fiberscope, Fiber Optic, General Specification for VSMF
MIL-F-24734/1	Fiberscope, Fiber Optic, Type F, Style 1, Class 1, Options F
MIL-F-24734/2	Fiberscope, Fiber Optic, Type R, Style 4, Class 1, Options F
MIL-T-24735	Transmitters, Light Signal, Analog, Fiber Optic, Shipboard, General Specification for VSMF
MIL-T-24735/1	Transmitter, Analog, Fiber Optic, Shipboard, 0.5 to 60 MHz, (0.5 dB Passband)
MIL-M-24736	Multiplexers, Demultiplexers, Time-Division, Fiber Optic Interfaceable, Shipboard, General Specification for VSMF
MIL-M-24736/1	Multiplexers, Demultiplexers, Time-Division, Fiber Optic Interfaceable, Shipboard, 8:1 Channels, 311 Mbps per Input Channel
MIL-M-24736/2	Multiplexers, Demultiplexers, Time-Division, Fiber Optic Interfaceable, Shipboard, 1:8 Channels, 311 Mbps per Output Channel
MIL-R-24737	Receivers, Light Signal, Analog, Fiber Optic, Shipboard, General Specification for VSMF
MIL-R-24737/1	Receivers, Light Signal, Analog, Fiber Optic, Shipboard, 0.5 to 60 MHz (0.5 dB Passband)
MIL-C-28688	Cable, Fiber Optic, Packaging of
MIL-C-28876D	Connectors, Circular Plug and Receptacle
MIL-C-28876/1D	Connectors, Circular, Plug and Receptacle Style, Screw Threads, Multiple Removable Termini, General Specifications
MIL-T-29504A	Termini, Connector, Removable, General Specification
MIL-T-29504/1A	Termini, Connector, Removable, Environmental, Class 2, Type II, Style A, Pin Terminus (For MIL-C-28876 and MIL-C-83526 Connectors)
MIL-F-49291B	Optical Fiber, General Specification
MIL-F-49291/1A	Optical Fiber, 50/125 μm, Radiation Hardened
MIL-F-49291/2A	Optical Fiber, 100/140 μm, Radiation Hardened

MIL-F-49291/3A	Optical Fiber, 50/125 μm
MIL-F-49291/4A	Optical Fiber, 100/140 μm
MIL-F-49291/6B	Optical Fiber, 62.5/125 μm, Radiation Hardened
MIL-F-49291/7B	Optical Fiber, Single-mode, Dispersion Unshifted, Radiation Hardened
MIL-F-49291/08	Optical Fiber, 400/430 μm, Avionic Rated
MIL-F-49291/09	Optical Fiber, 400/430 μm, Radiation Hardened, Avionic Rated
MIL-C-49292A	Cable Assemblies, Non-Pressure Proof, Fiber Optic, Metric, General Specification
MIL-C-49292/1	Cable Assemblies, Non-Pressure Proof, Fiber Optic, Metric, Branched, MIL-C-83522. MIL-C-83526, Connectors, and DOD-C-85045 Cables
MIL-C-49292/2	Cable Assemblies, Non-Pressure Proof, Fiber Optic, Metric, Single Bundle, 6-Position, MIL-C-83526 Connectors, and DOD-C-85045 Cables
MIL-C-49292/3	Cable Assemblies, Non-Pressure Proof, Fiber Optic, Metric, Connector, 2-Position, Hermaphroditic, Jam-Nut, MIL-C-83526 Connectors, and DOD-C-85045 Cables
MIL-C-49292/4A	Cable Assemblies, Non-Pressure Proof, Fiber Optic, Metric
MIL-C-49292/6	Cable Assemblies, Non-Pressure Proof, Fiber Optic, Metric, Single Fiber, Glass, MIL-C-83522 Connectors, and DOD-C-85045 Cables
MIL-C-49292/7A	Cable Assemblies, Non-Pressure Proof, Fiber Optic, Test, Metric
MIL-F-50533	Fiber, Acrylic, Fibrillated
MIL-F-50809	Fiber, Acrylic, Fibrillatable
MIL-I-81969/46A	Installing and Removal Tools, Type I, Class 2, Composition C, Size 16 Termini
MIL-I-81969/47A	Installing and Removal Tools, Type II, Class 2, Composition C, Size 16 Termini
MIL-I-81969/48A	Installing and Removal Tools, Type III, Alignment, Sleeve, Class 1, Composition C, Size 16 Termini
MIL-I-81969/49B	Installing and Removal Tools, Hand, Right Angle, Type 1, Class 2, Composition A, Size 16 Termini
MIL-C-83522D	Connector, Single Terminus, General Specifications
MIL-C-83522/1E	Connector, Plug, Single Terminus, Threaded (Step-Nose Down Interface). Lensless, Epoxy, for 50/125 μm, 62.5/125 μm, and 100/140 μm
MIL-C-83522/2F	Connector, Plug, Single Terminus, Threaded (Straight Nose Interface) Lensless, Epoxy, For 50/125 μm, 62.5/125 μm, and 100/140 μm
MIL-C-83522/3F	Connector, Plug-Receptacle--Adapter Style, Fixed Single Terminus, Threaded, Adapter Bulkhead Mount
MIL-C-83522/4E	Connector, Plug-Receptacle--Adapter Style, Fixed Single Terminus, Threaded, Receptacle, Low Profile Parallel PC Mount
MIL-C-83522/5D	Connector, Plug-Receptacle--Adapter Style, Fixed Single Terminus, Threaded, Receptacle, Bulkhead Mount
MIL-C-83522/6D	Connector, Plug-Receptacle--Adapter Style, Fixed Single Terminus, Threaded, Plug, Expanded Beam Lens, Epoxy, for 50/125 μm, and 100/140 μm fiber
MIL-C-83522/7B	Connector, Plug-Receptacle-Adapter Style, Fixed Single Terminus, Threaded, Receptacle, Parallel PC Mount
MIL-C-83522/8B	Connector, Plug-Receptacle--Adapter Style, Fixed Single Terminus, Threaded, Receptacle, Bulkhead, Hex Nut Mount
MIL-C-83522/9	Connector, Receptacle, PC Mount, Active
MIL-C-83522/10	Connector, Receptacle, Flange Mount Active (Used with MIL-T-29504/4 Pin Terminus, Size 16)
MIL-C-83522/11	Connector, Adapter, In-Line-Cable Panel Mount (Used with MIL-T-29504/4 Pin Terminus, Size 16)
MIL-C-83255/15	Connectors, Plug-Receptacle-Adapter Style, Fixed Single Terminus, Environmental, Crimp and Cleave
MIL-C-83522/16A	Connector, Single Terminus Plug, Adapter Style, 2.5 mm, Bayonet Coupling, Epoxy
MIL-C-83522/17A	Connector, Single Terminus, Adapter, 2.5 mm, Bayonet Coupling, Bulkhead Panel Mount
MIL-C-83522/18A	Connector, Single Terminus, Adapter 2.5 mm, Bayonet Coupling PC Mount
MIL-T-83523	Tools, Fiber Optic, General Specifications
MIL-T-83523/1	Tools, Fiber Optic, Hand Terminating, Type II
MIL-T-83523/2A	Tools, Fiber Optic, Polishing Bushing Assembly, Type IV
MIL-T-83523/3A	Tools, Fiber Optic, Scribe, Carbide, Type I
MIL-T-83523/4A	Tools, Fiber Optic, Scribe, Diamond or Sapphire, Type I
MIL-T-83523/5	Tools, Fiber Optic, Retaining Band Strain Relief, Type VIII
MIL-T-83523/6	Tools, Fiber Optic, Polishing Fixture Assembly, Type IV
MIL-T-83523/7	Tools, Fiber Optic, Wrench, Spanner, Type V
MIL-T-83523/8	Tools, Fiber Optic, Wrench, Spanner, Type V
MIL-T-83523/9	Tools, Fiber Optic, Scribe, Conical and Wedge, Type I
MIL-T-83523/10	Tools, Fiber Optic, Hand, Type VII, Carbide/Diamond Composition C
MIL-T-83523/16A	Tools, Fiber Optic, Stripping, (50/125 μm), (100/140 μm), (400/430 μm), (600/630 μm), (1000/1040 μm), Type X, Composition C
MIL-M-83524	Microscope, Optical, Monocular, Hand-Held, Portable, Militarized, 200X Magnification
MIL-K-83525	Kit, Portable, Optical Microscope, Militarized, 200X Magnification

MIL-C-83526A	Connectors, Hermaphroditic, Circular, Environment Resistant
MIL-C-85045E	Fiber Optic Cables (Metric), General Specification
DOD-C-85045/2B	Cable, Fiber Optic,1, 2, 4, and 6 Fiber, Heavy Duty, Metric
DOD-C-85045/3A	Cables, Fiber Optic, Heavy Duty, With Gel Filling and Flooding Compound
DOD-C85045/4A	Cables, Fiber Optic, Heavy Duty, Ruggedized with Steel Sheathing Rodent Protection, Gel Filling and Flooding Compound
DOD-C-85045/5A	Cable, Fiber Optic, Heavy Duty, Ruggedized with Non-Metallic Sheathing Rodent Protection, Gel Filling and Flooding Compound
DOD-C-85045/6C	Cable, Fiber Optic, Environmental, Type II
DOD-C-85045/8	Cable, Fiber Optic, Ruggedized, Radiation Hardened
DOD-C-85045/9	Cable, Fiber Optic, Break Out Individually Jacketed Fibers
DOD-C-85045/10	Cables, Fiber Optic, with Gel Filling and Flooding Compound
DOD-C-85045/11	Cable, Fiber Optic, Steel Sheathing Rodent Protection, Gel Filling and Flooding Compound
MIL-C-85045/12	(NASA) High Reliability, Single Fiber, Multimode, Graded Index, Out-Gassing Resistant
MIL-C-85045/13A	Cable, Cross-Linked, Eight Fibers, Cable Configuration Type 2, Loose Tube, Cable Class SM and MM
MIL-C-85045/14A	Cable, Cross-Linked, One Fiber, Cable Configuration Type 2, Application B, Cable Class SM and MM
MIL-C-85045/15	Cable, Cross-Linked, Four Fibers, Cable Configuration Type 2, Application B, Cable Class SM and MM
MIL-C-85045/16	Cable, Cross-Linked, One Fiber, Cable Configuration Type 2, Tight Buffered, Cable Class SM and MM
MIL-C-85045/17	Cable, Cross-Linked, Eight Fibers, Cable Configuration Type 2, Application B, Cable Class SM and MM
MIL-C-85045/18	Cable, Cross-Linked, Four Fibers, Cable Configuration Type 2, Application B, Cable Class SM and MM

BELLCORE STANDARDS

GR-20-CORE	Generic Requirements of Optical Fiber and Fiber Optic Cable, Issue1, Sept. 1994
TR-TSY-000020	Generic Requirements for Optical Fiber and Optical Fiber Cable, Issue 4, Mar. 1989
GR-26-CORE	Generic Requirements for Optical Fiber and Optical Fiber Cable, Issue 1, Sept. 1994
TA-TSY-000038	Digital Fiber Optic Systems, Requirements & Objectives, Issue 3, April 1986
ST-TEC-000051	Telecommunications Transmission Engineering Textbook - Volume 1: Principles, Third Edition, Issue 1, Aug. 1987
ST-TEC-000052	Telecommunications Transmission Engineering Textbook - Volume 2: Facilities, Third Edition, Issue 1, May 1989
ST-TEC-000053	Telecommunications Transmission Engineering Textbook - Volume 3: Networks and Services, Third Edition, Issue 1, May 1989
ST-CSP-000054	Trademarks, Acronyms and Abbreviations Commonly Used in the Telecommunications Industry, Issue 1, Oct. 1987
TR-TSY-000187	Optical Cable Placing Winces, Issue 1, Mar. 1985
TR-TSY-000196	Generic Criteria for Optical Time Domain Reflectometers, Issue 2, Sept. 1989
TR-NWT-000198	Generic Requirements for Optical Loss Test Sets (OLTSs), Issue 4, Sept. 1993
TA-NWT-000199	Specification of Memory Administration Messages at the OS/NE Interface, Issue 7, Jan. 1993
TR-NWT-000199	Operations Application Messages-Memory Administration messages (OTGR), Issue 2, Dec. 1992
TA-NWT-000233	Wideband and Broadband Digital Cross-Connect Systems Generic Criteria, Issue 4, Nov. 1992
TR-NWT-000233	Wideband and Broadband Digital Cross-Connect Systems Generic Requirements and Objectives (A Module of TSGR, FR-440), Issue 3, Nov. 1993
GR-253-CORE	Synchronous Optical network (SONET) Transport Systems: Common Generic Criteria, Issue 1, Dec. 1994
TR-NWT-000264	Generic requirements for Optical Fiber Cleaves, Issue2, Dec. 1993
TR-TSY-000264	Optical Fiber Cleaving Tools, Issue 1, Dec. 1986
TR-TSY-000266	Optical Patch Panels, Issue 1, Oct. 1985
GR-326-CORE	Generic Requirements for Single-mode Optical Fiber Connectors, Issue 1, Dec. 1994
TR-NWT-000332	Reliability Prediction Procedure for Electronic Equipment (A module of RQGR, FR-NWT-000796), Issue 3, Sept. 1990
TR-NWT-000356	Generic Requirements for Optical Cable Interduct, Issue 2, Oct. 1992
TA-TSY-000384	Preliminary Generic Requirements for PFM Video and Data Transmission on Single-mode Fiber, Issue 1, April 1986
GR-409-CORE	Generic Requirements for Premises Fiber Optic Cable, Issue 1, May 1994
TA-NWT-000418	Generic Reliability Assurance Requirements for Fiber Optic Transport Systems (A module of RQGR, FR-NWT-000796), Issue 1, May 1988

TR-TSY-000441	Submarine Splice Closures for Fiber Optic Cable, Issue 1, June 1989
TR-OPT-000449	Generic Requirements and Design Considerations for Fiber Distributing Frames, Issue 1, Dec. 1991
TA-NPL-000464	Generic Requirements and Design Considerations for Optical Digital Signal Cross-Connect Systems, Issue 1, September, 1987
TR-NWT-000468	Reliability Assurance Practices for Optoelectronic Devices in Central Office Applications, Issue 1, Dec. 1991
SR-TSY-000686	Cost Comparison of 45 Mb/s Video Over Fiber with Digital Alternate Technologies, Issue 1, Aug. 1987
TR-TSY-000761	Generic Criteria for Chromatic Dispersion Test Sets, Issue 1, Feb. 1990
TR-NWT-000764	Generic Criteria for Optical Fiber Identifiers, Issue 1, Aug. 1990
TR-TSY-000765	Splicing Systems for Single-mode Optical Fibers, Issue 2, June 1992
GR-771-CORE	Generic Requirements for Fiber Optic Splice Closures, Issue 1, July 1994
TR-TSV-000774	SMDS Operations Technology Network Element Generic Requirements, Issue 1, March 1992
TA-NWT-000782	SONET Digital Switch Trunk Interface Criteria, Issue 2, Oct. 1992
TR-TSY-000782	SONET Digital Switch Trunk Interface Criteria (A Module of TSGR FR-440 and of LSSGR, FR-64), Issue 2, Sept. 1989
TR-TSY-000786	Optical Source Module Generic Requirements for Subscriber Loop Distribution, Issue 2, Dec. 1991
GR-820-CORE	OTGR Section%.1: Generic Transmission Surveillance, Issue 1, Nov. 1994
SR-NWT-000821	Field Reliability Performance Study Handbook (A module of RQGR, FR-NWT-000796), Issue 3, Dec. 1990
GR-826-CORE	User Interface Generic Requirements for Supporting Network Element Operations, Issue 1, June 1994
TR-TSY-000842	Generic Requirements for SONET Compatible Digital Radio, Issue 1, July 1988
TR-TSY-000843	Generic Requirements for Optical and Optical/Metallic Buried Service Cable, Issue 1, January 1989
SR-TSY-000857	Preliminary Special Report on Broadband ISDN Access, Issue 1, Dec. 1987
TR-TSY-000886	Generic Criteria for Optical Power Meters, Issue 1, March 1991
TR-TSY-000887	Generic Criteria for Fiber Optic Stabilized Light Sources, Issue 1, March 1990
TR-TSY-000901	Generic Requirements for WDM (Wavelength Division Multiplexing) Components, Issue 1, Aug. 1989

TR-TSY-000902	Generic Requirements for Non-Concrete Splice Enclosures, Issue 1, Dec. 1988
TA-NWT-000909	Generic Requirements and Objectives for Fiber-In-The-Loop Systems, Issue 2, Dec. 1993
TR-NWT-000909	Generic Requirements and Objectives for Fiber-In-The-Loop Systems, Issue 1, Dec. 1991
TA-NWT-000910	Generic Requirements for Fiber Optic Attenuators, Issue 2, Dec. 1992
TR-NWT-000917	SONET Regenerator (SONET RGTR) Equipment Generic Criteria (A Module of TSGR, FR-440), Issue 1, Dec. 1990
TR-TSY-00944	Generic Requirements for Optical Distribution Cable, Issue 1, July 1990
TR-TSY-000949	Generic Requirements for Service Terminal Closures with Optical Cable, Issue 1, June 1990
GR-950-CORE	Generic Requirements for Optical Network Unit (ONU) Closures, Issue 1, Dec. 1994
TR-NWT-000955	Generic Requirements for Single and Multi-Fiber Strippers, Issue 2, Jan. 1994
TR-NWT-000983	Reliability Assurance Practices for Optoelectronic Devices in Loop Applications, Issue 2, Dec. 1993
GR-1009-CORE	Generic Requirements for Fiber Optic Clip-on Test Sets, Issue 1, Mar. 1995
TR-TSY-001028	Generic Criteria for Optical Continuous Wave Reflectometers, Issue 1, May 1990
TA-TSY-001040	SONET Test Sets for Acceptance and Maintenance Testing: Generic Criteria, Issue 1, July 1990
GR-1042-CORE	Generic Requirements for Operations Interfaces Using OSI Tools - Information Model Overview: Synchronous Optical Network (SONET) Transport Information Model, Issue 1, Oct. 1994
GR-1042-IMD	Generic Requirements for Operations Interfaces Using OSI Tools - Information Model Details: Synchronous Optical Network (SONET) Transport Information Model, Issue 1, Oct. 1991
TR-NWT-001073	Generic Requirements for Fiber Optic Switches, Issue 1, Dec. 1993
GR-1081-CORE	Generic Requirements for Field-Mountable Optical Fiber Connectors, Issue 1, Jan. 1995
GR-1095-CORE	Generic Requirements of Multi-Fiber Splicing Systems for Single-Mode Optical Fibers, Issue 1, May 1994
TR-NWT-001121	Generic Requirements for Self-Supporting Optical Fiber Cable, Issue 1, Oct. 1991
SR-TSY-001128	Communicating by Light Fiber Optics Slide Package, Issue 2, June 1990
SR-TSY-001129	Communicating by Light Fiber Optics Slide Package, Issue 2, June 1990

TR-NWT-001137	Generic Requirements for Hand-held Optical Power Meters, Issue 1, Dec. 1991
TR-NWT-001138	Generic Requirements for Mini-OTDRs and Fiber Break Locators, Issue 1, Oct. 1992
SR-TSY-001171	Methods and Procedures for System Reliability Analysis (A module of RQGR, FR-NWT-000796), Issue 1, January 1989
TR-NWT-001190	Generic Requirements for Fiber Optic Cable Locators, Issue 1, July 1993
TR-NWT-001196	Generic Requirements for Splice Verification Sets, Issue 1, Nov. 1991
TR-NWT-001197	Generic Requirements For Locally Powering ONUs for Fiber-In-The-Loop Systems, Issue 1, July 1992
GR-1209-CORE	Generic Requirements for Fiber Optic Branching Components, Issue 1, Nov. 1994
TR-NWT-001222	Generic Requirements for Optical Terminators, Issue 1, June 1992
GR-1230-CORE	SONET Bidirectional Line-Switched Ring Equipment Generic Criteria, Issue 1, Dec. 1993
TA-NWT-001250	Generic Requirements for SONET File Transfer, Issue 2, June 1992
TR-NWT-001295	Generic Requirements for Remote Fiber Testing Systems (RFTSs), Issue 2, Sept. 1993
TR-NWT-001309	TSC/RTU Generic Requirements for Remote Optical Fiber Testing and Related Access and Test Messages, Issue 2, Sept. 1993
TA-NWT-001312	Generic Requirements for Optical Fiber Amplifier Performance, Issue 2, Dec. 1993
TR-NWT-001319	Generic Requirements for Fiber Optic Visual Fault Finders, Issue 1, May 1993
TR-NWT-001322	Generic Requirements for Steam Resistant Cables, Issue 1, Jan. 1994
SR-NPL-001324	Timing Characteristics of DTMF to DP Converters, Issue 1, Feb. 1989
GR-1332-CORE	Generic Requirements for Data Communication Network Security, Issue 1, Jan. 1994
FA-NWT-001345	Framework Generic Requirements for Element Manager (EM) Applications for SONET Subnetworks, Issue 1, Sept. 1992
GR-1365-CORE	SONET Private Line Service Interface Generic Criteria for End Users, Issue 1, Dec. 1994
SR-TSY-001369	Introduction to Reliability of Laser Diodes and Modules, Issue 1, June 1989
GR-1374-CORE	SONET Inter-Carrier Interface Physical Layer Generic Criteria for Carriers, Issue 1, Dec. 1994
GR-1375-CORE	Self-Healing Ring-Functionality in Digital Cross-Connect Systems, Issue 1, Aug. 1994
GR-1377-CORE	SONET OC-192 Transport System Generic Criteria, Issue 1, Nov. 1994
GR-1380-CORE	Generic Requirements for Fusion Splice Protectors, Issue 1, May 1994
TA-NWT-001385	Generic Requirements for Opto-electronic Devices in Fiber Optic Systems, Issue 1, April 1993
GR-1400-CORE	SONET Dual-Fed Unidirectional Path Switch Ring (UPSR) Equipment Generic Criteria, Issue 1, March 1994
TA-NWT-001402	DS# HCDS TSC/RTU and DTAU Functional Requirements and TL1 Access and Testing Messages, Issue 1, May 1993
FA-NWT-001413	Framework Generic Requirements for Element Management Layer Applications for Management of Fiber-In-The-Loop Systems and Remote Digital Terminals, Issue 1, May 1993
SR-TSY-001425	Temperature Cycling Test of Laser Modules Used for Interoffice Applications, Issue 1, June 1989
GR-1435-CORE	Generic Requirements for Multi-Fiber Optical Connectors, Issue 1, October 1994
SR-TSY-001468	A Compilation of Results from Recent Bellcore Studies on Fiber-to-the-Home (FTTH), Issue 1, Dec. 1989
TA-NWT-001500	Generic Requirements for Powering Optical Network Units in Fiber-in-the-Loop Systems, Issue 1, Dec. 1993
SR-TSY-001681	Bellcore Fiber-in-the-Loop (FITL) Architecture Summary Report, Issue 1, June 1990
SR-NWT-001756	Automatic Protection Switching For SONET, Issue 1, Oct. 1990
SR-NWT-002014	Suggested Optical Cable Code (SOCC), Issue 1, April 1992
SR-NWT-002041	Transport Surveillance for Time Division Multiple Access (TDMA) Based Point-to-Multipoint Fiber-in-the-Loop Systems, Issue 1, Jan. 1992
SR-OPT-002104	TIRKS® Time Slot Numbering Schemes for SONET Add/Drop Multiplex Equipment (ADM), Issue 1, Nov. 1991
SR-NWT-002224	SONET Synchronization Planning Guidelines, Issue 1, Feb. 1992
SR-NWT-002287	INA Cycle 1 Managed Object Specification, Issue 2, April 1993
SR-TSV-002387	SONET Network and Operations Plan: Feature, Functions, and Support, Issue 1, Aug. 1992
SR-NWT-002439	Interface Functions and Information Model for Initial Support of SONET Operations Using OSI Tools, Issue 1, Dec. 1992
SR-TSV-002672	EML Applications for Fault Management: Intelligent Alarm Filtering for SONET, Issue 1, Mar. 1994

SR-NWT-002723	Applicable TL1 Messages for SONET Network Elements, Issue 1, June 1993
SR-ARH-002744	Single-Mode Fiber Connectors Technology, Issue 1, Aug. 1993
SR-STS-002751	OCS OS-NE Interface Support for the CMISE/OSI Protocol Stack, Issue 1, June 1994
GR-2833-CORE	Generic Operations Interfaces Using OSI Tools: Information Model for Integrated Digital Loop Carrier and Fiber-In-The-Loop Systems, Issue 1, Feb. 1994
GR-2837-CORE	ATM Virtual Path Functionality in SONET Rings - Generic Criteria, Issue 1, Dec. 1994
SR-3317	OPS/INE to Network Element Generic OSI/CMISE Interface Support, Issue 1, Dec. 1994
TR-73536	Technical Requirements for Optical Connectors, Oct. 1989
TR-73539	BellSouth Generic Requirements for Optical Fiber and Optical Fiber Cable, Issue A, Nov. 1989
TR-73540	BellSouth Technical Requirements for Fiber Optic Splice Closures, Issue A, Nov. 1989
TR-73541	BellSouth Technical Requirements for Fiber Optic Mechanical Splice Systems, Issue A, Nov. 1990
TR-73542	Technical Requirements for Fiber Terminating Equipment for Remote Cabinets and Customer Premise Locations, Issue A, Nov. 1989

SOCIETIES, CONFERENCE SPONSORS & TRADE MAGAZINES

SOCIETIES & CONFERENCE SPONSORS

Armed Forces Communication and Electrotonic Association (AFCEA)
4400 Fair Lakes Court
Fairfax, VA 22033
TEL:(703) 631-6125

American Electronics Association
5201 Great American Parkway
Santa Clara, CA 95054
TEL:(408) 987-4200

American Federation of Information Processing Societies (AFIPS)
1899 Preston White Drive
Reston, VA 22091
TEL:(703) 620-8900

American Institute of Physics (AIP)
11 W. 42nd Street
New York, NY 10036
TEL:(212) 642-4900

American National Standards Institute (ANSI)
1430 Broadway
New York, NY 10018
TEL:(212) 354-3300

American Physical Society (APS)
1 Physics Ellipse
College Park, MD 20740-3844
TEL:(301) 209-3200

American Society for Industrial Security (ASIS)
1655 North Fort Myer Drive - Suite 1200
Arlington, VA 22209

Architecture Technology Corp.
PO Box 24344
Minneapolis, MN 55424
TEL:(612) 935-2035

Building Industry Consulting Service International Inc. (BICSI)
10500 University Center Dr. - Suite 100
Tampa, FL 33621-6415
TEL:(813) 979-1991

Cahners Exposition Group
999 Summer Street
Stanford, CT 06905-9990
TEL:(203) 964-8287

Competitive Telecommunication Association
1140 Connecticut Ave. — Suite 220
Washington, D.C. 20036
TEL:(202) 296-6650

CW Communications Conference Management Group
375 Cochituate Road
PO Box 9171
Framingham, MA 01701
TEL:(508) 879-0700

ECOC/IOOC, SDSA
65 avenue Edouard Vaillant
92100 Boulogne Billancourt
FRANCE
TEL:(33-1) 46 08 5661

Electronic Industries Association (EIA)
2500 Wilson Boulevard
Arlington, VA 22201
TEL:(703) 907-7500

ElectroniCast Corp.
800 S. Claremont St.
San Mateo, CA 94402
TEL:(415) 343-1398

E.J. Krause & Associates, Inc.
3 Bethesda Metro Center - Suite 510
Bethesda, MD 20814
TEL:(301) 986-7800

European Physical Society (EPS)
PO Box 69
CH-1213, Petit-Lancy 2
SWITZERLAND
TEL:(22) 93 11 30

Events Management International, Inc.
737 Webster St.
Marshfield, MA 02050
TEL:(508) 879-6700

Hannover Fairs USA Inc.
103 Carnegie Center
PO Box 7066
Princeton, NJ 08540
TEL:(609) 987-1202

Information Gatekeepers Inc. (IGI)
214 Harvard Avenue
Boston, MA 02134
TEL:(617) 232-3111

Institute of Electrical and Electronic Engineers Inc. (IEEE)
345 East 47th Street
New York, NY 10017
TEL:(212) 705-7900

Institute of Electronics and Communication Engineers of Japan (IECEJ)
Business Center for Academic Societies Japan
Hamazaki Building, 4F
40-14 Hongo 2-chome, Bunkyo-ku
Tokyo, 113
JAPAN
TEL:(03) 817-5831

Institute of Electrical Engineers (IEE)
PO Box 8
Southgate House
Suitevenage, Herts, SG1 1GH
UK
TEL:(0438) 313311

Instrument Society of America (ISA)
67 Alexander Drive
PO Box 12277
Research Triangle Park, NC 27709
TEL:(919) 549-8411

The Interface Group
300 First Ave.
Needham, MA 02194
TEL:(617) 449-6600

International Commission for Optics (ICO)
PO Box 4413
61044 Tel Aviv
ISRAEL

International Communications Association (ICA)

12750 Merit Drive - Suite 710, LB89
Dallas, TX 75251
TEL:(214) 233-3889

International Communications Industries Association

3150 Spring Street
Fairfax, VA 22031
TEL:(703) 273-7200

International Telecommunication Union (ITU)

Places Des Nations, CH-1211
Geneva, 20
SWITZERLAND
TEL:(022) 995190

International Television Association

6311 North O'Connor Road, LB-51
Irving, TX 75039
TEL:(214) 869-1112

International Wire & Cable Symposium (IWCS)

174 Main Street
Eatontown, NJ 07724
TEL:(908) 389-0990

Interop, Inc.

480 San Antonio Rd. Suite 100
Mountain View, CA 94040-1219
TEL:(415) 578-6900

Kallman Associates

20 Harrison Ave.
Waldwick, NJ 07463
TEL: (201) 652-7070

Kessler Marketing Intelligence (KMI)

31 Bridge Street
Newport, RI 02840
TEL:(401) 849-6771

E.J. Krause & Associates, Inc.

7315 Wisconsin Ave. Suite 450N
Bethesda, MD 20814
TEL: (301) 986-7800

Laser Association of America (LAA)

72 Mar Street
San Francisco, CA
TEL:(415) 621-5776

Lasers and Electro-Optics Society (LEOS)

445 Hoes Lane
PO Box 1331
Piscataway, NJ 08855
TEL:(201) 562-3892

Laser Institute of America (LIA)

12424 Research Parkway - Suite 130
Orlando, FL 32826
TEL:(407) 380-1553

National Association of Broadcasters

1771 N St., NW
Washington, D.C. 20036
TEL:(202) 429-5350

National Association of State Telecommunications Directors (NASTD)

Iron Works Pike
PO Box 11910
Lexington, KY 40578
TEL:(606) 231-1939

National Cable Television Association (NCTA)

1724 Massachusetts Avenue N.W.
Washington, D.C. 20036
TEL:(202) 775-3606

National Electrical Contractors Association (NECA)

3 Bethesda Metro Center, Suite 1100
Bethesda, MD 20814
TEL:(301) 657-3110

National Engineering Consortium (NEC)

303 Wacker Drive - Suite 740
Chicago, IL 60601
TEL:(312) 938-3500

National Institute of Standards Technology (NIST)

724.02 325 Broadway
Boulder, CO 80303
TEL:(303) 497-3300

National Research Council of Canada, Electrical Engineering Division

Building M50, Montreal Road Ottawa
Ontario, K1A 0R6
CANADA
TEL:(613) 993-9009

National Sound and Communication Association (NSCA)

10400 South Roberts Road
Palos Hills, IL 60465

North American Telecommunications Association (NATA)

2000 M Street N.W, - Suite 550
Washington, D.C. 20036
TEL:(202) 296-9800

Optoelectronic Industry and Technology Development Association (OITDA)

20th Mori Building
7-4 Nishi Shimbashi 2-chome
Minato-ku Tokyo, 105
JAPAN
TEL:(03) 508-2091

Optical Sensors Collaborative Association (OSCA)

c/o Sira Limited
South Hill Chislehurt, Kent BR7 5EH UK
TEL:(01) 467 2636

Optical Society of America (OSA)

2010 Massachusetts Avenue N.W.
Washington, D.C. 20036
TEL:(202) 416-1950

Pacific Telecommunications Council (PTC)

1110 University Avenue - Suite 308
Honolulu, HI 96826
TEL:(808) 941- 3789

Power and Communications Contractors Association (PCCA)

6301 Stevenson Avenue - Suite One
Alexandria, VA 22304
TEL:(703) 823-1555

Reed Exhibition Companies

International Security Conference
Jacob Javits Convention Center
New York, NY

Society of Cable Television Engineers (SCTE)

669 Exton Commons
Exton, PA
TEL:(610) 363-6888

Society of Optical and Quantum Electronics (SOQE)

PO Box 245
McLean, VA 22101
TEL:(703) 642-5835

Society of Photo-Optical Instrumentation Engineers (SPIE)

PO Box 10
Bellingham, WA 98227-0010
TEL:(206) 676-3290

Society of Telecommunications Consultants

1841 Broadway - Suite 1203
New York, NY 10023
TEL:(212) 582-3909

Telecommunications Association (TCA)

2001 Pennsylvania Ave. NW Suite 800
Washington, D.C. 20006
TEL:(202) 457-4912

Telecommunications Industry Association (TIA)

150 N. Michigan Avenue - Suite 600
Chicago, IL 60601-7524
TEL:(312) 782-8596

Trade Associates

6001 Montrose Rd. Suite 900
Rockville, MD 20852
TEL:(301) 468-3210

U.S. Army Communications - Electronics Command (CECOM)

AMSELCOM D-4
Fort Monmouth, NJ 07703
TEL:(908) 544-3163

United States Telephone Association (USTA)

1401 H St. NW
Washington, D.C. 20005
TEL:(202) 326-7300

United States Telecommunications Suppliers Association (USTSA)

150 N. Michigan Avenue - Suite 600
Chicago, IL 60601
TEL:(312) 782-8597

Wire Association International

1570 Boston Post Road - Box H
Guildford, CT 06437
TEL:(203) 453-2777

Wire Industry Suppliers Association

7297 Lee Highway - Suite N
Falls Church, VA 22042
TEL:(703) 533-9530

World Expo Corporation

111 Speen St.
PO Box 9107
Farmingham, MA 01701-9107
TEL:(508) 879-6700

FIBER OPTIC & RELATED TRADE MAGAZINES

Access Control

5 Penn Plaza
New York, NY 10001

Applied Optics

Optical Society of America
2010 Massachusetts Ave. NW
Washington, D.C. 20036
TEL:(202) 416-1950

Broadcast Engineering

Intertec Publishing Corporation
PO Box 12901
Overland Park, KS 66282-2901
TEL:(913) 888-4664
FAX:(913) 541-6697

Canadian Security

Security Publishing, Ltd.
PO Box 430, Station O
Toronto, Ontario M4A 2P1
CANADA

CCTV Applications & Technology

Burke Publishing Co.
15825 Shady Grove Rd. - Suite 130
Rockville, MD 20850

Communications Technology

50 South Steele - Suite 500
Denver, CO 80209
TEL:(303) 355-2101
FAX:(303) 355-2144

Communications Week

CMP Publications, Inc.
600 Community Drive
Manhasset, NY 11030
TEL:(516) 562-5530
FAX:(703) 562-5055

ComNet

111 Speen Street
PO Box 9107
Framingham, MA 01701-9515

EDN

Reed Publishing USA
275 Washington St.
Newton, MA 02158-1630

EE Product News

Intertec Publishing Corp.
707 Westchester Ave.
White Plains, NY 10604
TEL:(914) 949-8500
FAX:(914) 682-0922

Electro Optics

Milton Publishing Company Ltd.
5 Tranquil Passage, Blackheath, London SE3 OBY
ENGLAND

Electronic Design

Penton Publishing Inc.
611 Route # 46 West
Hasbrouck Heights, NJ 07604
TEL:(201) 393-6060
FAX:(201) 393-0204

Electronic Engineering Times

600 Community Drive
Manhasset, NY 11030

Electronic Product News
Rue Verte 216
1210 Brussels 21
BELGIUM

Electronic Product News
Elsevier Publishing
One Penn Plaza, 26th Floor
New York, NY 10119

Fiber Optics and Communication
Information Gatekeepers, Inc.
214 Harvard Avenue
Boston, MA 02134

Fiberoptic Product News
301 Gibraltar Drive
Box 650
Morris Plains, NJ 07950-0650

Fiber Optic News
Phillips Publishing, Inc.
7811 Montrose Road
Potomac, MD 20854

Fiber Optics International
4th Floor, Britannia House
960 High Road
London N12 9RY
ENGLAND
TEL:(44)81-446-5141

LAN
Editorial Office
12 West 21 Street
New York, NY 10010

LAN TIMES
McGraw-Hill
7050 Union Park Center, Suite 240
Midvale, Utah 84047

Laser Focus World
Pennwell Publishing Co.
1421 S. Sheridan
Tulsa, OK 74112
TEL:(800) 331-4463

Lasers & Optics Bulletin
Opto-Laser Info
7 Placid Harbor
Dana Point, CA 92629

Lasers & Optronics
301 Gibralter Drive
PO Box 650
Morris Plains, NJ 07950-0650
TEL:(201) 292-5100
FAX:(201) 605-1220

Lightwave
Pennwell Publishing
Advanced Technology Group
5th Floor
Ten Tara Blvd.
Nashua, NH 03062-2801

Microwaves & RF
Penton Publishing, Inc.
1100 Superior Ave.
Cleveland, OH 44114
TEL:(216)-696-7000

Military & Aerospace Electronics
Sentry Publishing
1900 West Park Drive
Westboro, MA 01581

Optics & Photonics News
Optical Society of America
2010 Massachusetts Ave., NW
Washington, D.C. 20036

Outside Plant
PO Box 183
Cary, IL 60013

Photonics Spectra
Laurin Publishing Co.
Editorial Offices
2 South Street
Berkshire Common
Pittsfield, MA 01202-1146

RF Design
Cardiff Publishing Company
6300 S. Syracuse Way
Suite 650
Englewood, CO 80111
TEL:(303) 220-0600

Security
Cahners Publishing Co.
1350 E. Touhy Avenue
Des Plaines, IL 60018-3358
TEL:(708) 635-8800
FAX:(708) 635-9950

Security Management
American Society for Industrial Security (ASIS)
1655 North Fort Myer Drive - Suite 1200
Arlington, VA 22209

Sound & Video Contractor
Intertec Publishing Corporation
9221 Quivira
Overland Park, KS 66215

Telemanagement
1400 Bayley Street
Office Mall Two, Suite 3
Pickering, Ontario L1W 3R2
CANADA

Television Broadcast
PSN Publications, Inc.
2 Park Avenue
New York, NY 10016
TEL:(212)-779-1919
FAX:(212)-213-3484

Video Systems
Intertec Publishing Corporation
9221 Quivira Road
Overland Park, KS 66215
TEL:(913) 888-4664
FAX: (913) 541-6697

INDEX

My Thali

My Thali

A Simple Indian Kitchen

Joe Thottungal

with Anne DesBrisay

Figure.1

Vancouver / Toronto / Berkeley

For Suma

Recipes are chef tested.

23 24 25 26 27 5 4 3 2 1

Cataloguing data is available from Library and Archives Canada
ISBN 978-1-77327-195-8 (hbk.)

Design by Naomi MacDougall
Photography by Christian Lalonde, Photoluxstudio.com,
except images on pages 6, 9 photographed by Christo Raju.
Prop styling by Irene Garavelli
Food styling by Sylvie Benoit
Endsheets: Vecteezy.com

Editing by Michelle Meade
Copy editing by Pam Robertson
Proofreading by Renate Preuss
Indexing by Iva Cheung

Figure 1 Publishing Inc.
Vancouver BC Canada
www.figure1publishing.com
Printed and bound in China by C&C Offset Printing Co.
Distributed internationally by Publishers Group West

Figure 1 Publishing works in the traditional, unceded territory of the xʷməθkʷəy̓əm (Musqueam), Sḵwx̱wú7mesh (Squamish), and səlilwətaɬ (Tsleil-Waututh) peoples.

RECIPE NOTES

Unless stated otherwise:

Vegetables are medium-sized.

Milk is whole.

All herbs are fresh.

Curry leaves are always fresh.

Mustard seeds are always black.

Black pepper is always freshly ground.

Contents

Recipes

The Genesis of Thali, the Restaurant

Our first Ottawa restaurant, Coconut Lagoon, was well established, busy, and always full. Over its first decade, the restaurant had received lovely press, hosted several festivals and food events, contributed to dozens of community causes, and been honored with awards. I'd been incredibly lucky with my staff. Most of them, including my chef de cuisine Rajesh Gopi and my brothers Majoe and Thomas, had been with me since we first opened our doors in 2004. But Coconut Lagoon pretty much ran itself. I was looking for the challenge, for me and for my staff, of doing something quite different in a quite different setting.

Just as Coconut Lagoon introduced the city to the southern coastal cuisine of Kerala, I wanted to showcase the experience of thali. Essentially, to open a second restaurant that served a daily tasting menu only, thereby giving our team the freedom to be creative every day, and for our diners to experience a complete meal of many parts, a mini buffet on the plate, served at communal tables. A guest had only to decide on the type of thali—vegan, vegetarian, beef, lamb, chicken, fish, or seafood—and within a few minutes of ordering, the food would arrive.

When the downtown Ottawa space that is now Thali became available, it felt perfect. On a corner lot at street level, wrapped in tall windows with tall walls, the room was big and streamed with light. Above, and all around the restaurant, were thousands of office workers. And the space had never been a restaurant, which gave us a blank slate to design something splendid. Thali did so well in that first year we thrived on lunch service alone, bolstered with catering. And then, of course, the pandemic hit like a sledgehammer. During its turbulence, the towers emptied, the streets became desolate, and those office workers who streamed through our doors every day were all working from home. To add to the heartache, in May 2020 our beloved Coconut Lagoon restaurant was destroyed by fire.

And so, the pandemic years have been ones of rethinking and rebuilding and redefining our priorities. As they were for so many in our industry. But they also granted me more time to cook at home with my family, and more time to write this second cookbook, to share these recipes—and that's been a privilege and a pleasure.

Introduction

As a Kerala schoolboy my lunch box was a banana leaf, plucked from our garden, filled with food, folded over, and wrapped in a page of yesterday's newspaper. It made a tidy, perfect parcel to tuck into a schoolbag.

Unwrapped at lunch break, there would be the pleasure of a still-warm meal, its flavor ramped up by the long contact with the fragrant leaf. Inside, perhaps an egg omelette, some cabbage thoran, a curry of black chickpeas in a thick tomato gravy, a piece of last night's fried fish. One of my mum's pickles would always be there, along with a spoonful of raita. And rice, of course, always rice. I would wolf it all down, then toss the leaf on the side of the road to decompose or, more likely, to be eaten by the temple cows that wandered free in our town. No plastic left behind, no food waste, and no bulky tiffin box to lug to after-school cricket practice.

As a kid, I would never have called this tasty muddle of food a *thali*—it was just my school lunch. But that's what it was. Because a thali is a meal of many parts, which makes the contents of that leaf, modest as they were, count as one.

The format of a meal of many parts is simply the way we eat in India, the way we have eaten for centuries. Not to be confused with the more European style of a meal of many courses, but one that involves many elements, served together. The dishes are always hyper-regional and -seasonal, usually culturally or religiously significant, and typically presented on a round, rimmed plate (also called a thali) holding small metal bowls. Or, in my part of India, more often presented on a banana leaf.

Cooking thali might sound like a lot of work. And it sometimes is if the occasion calls for an eye-popping all-out feast in full-on splendor. But everyday thalis can be quite simple. Many dishes can be whipped up in fifteen minutes, recipes can be doubled, allowing leftovers for another day, and almost all can be made ahead, leaving the cook with little to do but make rice while the curries warm up for supper. With one or two anchor dishes, plus rice, the balance of a thali is made up of the small stuff—the condiments and accompaniments and sides that give the meal flavor, character, zing, and balm, usually found in jars or bowls in the fridge. No sweat, really.

As an Indian chef, the tradition of thali dominates my life. It is the way I eat at home, it is the way I celebrate life's big events, and it is the style of cooking we offer at my second Ottawa restaurant, Thali.

My first cookbook, *Coconut Lagoon*, was a tribute to my home state of Kerala, an introduction to its unique cuisine, essential ingredients, and traditional dishes. It was written for all those unfamiliar with the particular pleasures of its particular cuisine.

I wanted this cookbook to showcase south Indian home cooking—the dishes my wife Suma prepares for our family, defined by seasonality, affordability, nutrition, and tradition, recipes passed down from grandmother to mother to daughter—and the ways we create complete meals, those we serve on thalis at the restaurant.

Meals Ready

Like most kids, what my mother packed for school lunch was always of interest to me, but nothing compared with the thrill of a travel thali. Growing up in the 1980s, in the southern city of Thrissur, journeys were great fun. Mostly because they were rare, and because they were defined by food. Eating out was an expensive treat for our family of six, so a restaurant meal only happened when we'd visit grandparents.

During the layovers between the bus rides required for these trips, there would be lunch, either in the busy station restaurant or at a small nearby hotel. Sometimes the rickshaw driver would be queried for the best place for an affordable family meal. My parents, brothers, and I would then join the queue outside that place, waiting for the doors to open. "Meals Ready" (written in both English and Malayalam) would be the sign to show up—sometimes an ornate wooden sign, sometimes just a torn piece of cardboard in the eatery's window. The doors would then open, and the magic would begin. We'd wash hands at communal sinks, sit at communal tables, and, within seconds, the feast was on. There might be sambar to start, served with rice, and then into small containers would be spooned the day's offerings: a meat or fish curry, a dal, a vegetable stir-fry. There would be complementary side dishes—chutneys, pickles, raitas, and pappadum. Something sweet—a rice pudding, perhaps—and always unlimited sambaram, the refreshing buttermilk drink. If we wanted more of something—more aviyal, more sambar, more pickles—we'd signal with a hand or a little whistle and a watchful server carrying replenishments in metal buckets would fill us up. There was no lingering in these eateries: food would arrive quickly, be consumed quickly, and be paid for quickly with cash. Others would take our seats and we'd get on with the next leg of the journey. Granted, there was nothing glamorous about these restaurant thalis. But for a kid, there was the fun of food that was different from what we'd get at home, served

in unlimited quantities, and eaten in a bustling place with strangers around us.

And then there were the spreads of all spreads, those eye-popping "big occasion" thalis. At baptisms, weddings, funerals, housewarmings, and birthdays, and at all manner of religious holidays, there would be feasts, none more elaborate in Kerala than the banana leaf thali, the Onam sadya (page 81), held at the close of the ten-day festival of the harvest. My brothers and I would be given new shirts to wear. There would be flowers everywhere, dancing, parades, and fireworks. But it was the lunch feast I loved best: dozens and dozens of delicious dishes and condiments, all to honor the bounty, each placed in its proper spot on the long green leaf, with rice down the center. (Rice is always at the heart of any thali, modest or grand.)

Those childhood pleasures fueled my desire to introduce my city to the joys of thalis, both simple and homey, and grand and glorious.

How to Use This Book

While it may have been easier to follow a conventional order for the chapters, I wanted the book to reflect the importance of the ingredients in a thali, of the ways in which a thali is constructed. As a result, you will find the chapter on "Rice" near the beginning of the book and the one for "Condiments" near the end.

Each recipe in this book can be joined with other recipes to create a thali, a complete and balanced meal. I've provided inspiration for what those complete meals might be on pages 83–91. And though it's not typically Indian, a recipe can also be enjoyed on its own, as the hero of supper, or with a small supporting cast of friends: a chutney, a crunchy topping, an herby ending, or a soothing raita. Ideas for those supports are also supplied in these pages. Just keep in mind that most recipe quantities are based on multiple elements coming together as a thali.

Some of these recipes can be prepared in thirty minutes in a single pot (Kale Ularthiyathu, page 60, or Vegetable Uppumavu, page 36). Others are more ambitious dishes you turn to when you want to spend the day in the kitchen working on a celebration feast (Duck Biryani, page 38, or Mango-Coconut Mousse, page 170).

Whatever suits your time and temperament, the finest cooking begins with the finest ingredients. In Indian cuisine, that begins and ends with the spice box. If your spices are ancient, I'm sorry, but toss them and start fresh! I recommend you buy spices whole, in small quantities as required, and store them in airtight containers away from the light.

Most of these recipes require spices and other aromatics to be cooked in rapid succession, so always measure out the ingredients and have them on hand by the stove before you start cooking. I cannot emphasize this enough!

If you're lucky to live near an Indian or South Asian grocery shop, source spices there—they tend to be the freshest and of the finest quality. Otherwise, purchase your spices from a reputable source online (page 181).

I have provided you with recipes for our pickles, chutneys, spice blends, breads, and fried snacks that lend flavor, fire, and crunch to meals, but many of these extras can be found at food shops, restaurants, and bakeries specializing in Indian and South Asian food. Experiment with brands to find ones you like.

A note on cooking time: Ovens and stoves are variable, as are pots and pans, so be aware that cooking times are approximate.

A note on serving size: "Serves 4" means the recipe will serve four if considered a side dish, as part of a thali.

A note on dietary symbols: The recipes throughout this book feature dietary symbols for common dietary requirements: gluten-free (GF), dairy-free (DF), and vegan (VG). Please note that the symbols pertain to the recipes only and not accompanying "To serve" dishes.

Equipment

There are only two pieces of kitchen equipment essential to preparing these recipes. One is a good, sharp chef's knife (there's a lot of chopping in Indian cooking!) and the other a heavy-bottomed frying pan or a sturdy wok with a handle (or an Indian vessel like a *kadai* or *uruli*) for roasting spices and building curries. Otherwise, this list of tools ranges from convenient-to-have to a wish list for the future. Many you will have in your kitchen; others can be sourced online (page 181) or at an Indian or South Asian grocery.

BLENDER OR HAND BLENDER For making drinks.

BOX GRATER To grate fresh coconut into shavings.

CHAKLA AND BELAN If you're going to make Chapati (page 124) on a regular basis, invest in this tidy set: a small, round chapati board (*chakla*) and thin, light rolling pin (*belan*). A standard cutting board and rolling pin make acceptable substitutes.

COCONUT SCRAPER The traditional coconut scraper is called a *chirava* (page 17). Its spike-edged blade makes quick work of any coconut shaving job. Alternatively, use the large holes of a box grater or pulse the coconut meat in a food processor.

FOOD PROCESSOR For use when making ginger-garlic paste, puréeing soups, and grinding coconut.

KADAI OR URULI Traditional Indian cooking vessels, round-bottomed and high-sided, for building stews and curries and for deep-frying. A Chinese wok with a handle would be the next best thing.

MORTAR AND PESTLE For cracking or grinding spices, and making pastes.

PRESSURE COOKER A core piece of equipment in any Indian home, it cooks food faster than stovetop methods, while conserving precious fuel and preserving the aromas

and nutrients in the food. There are many styles of pressure cookers, so it is difficult to give precise instructions without knowing the style you are using. I suggest you refer to the cooker's manual for guidance. If you don't have a pressure cooker, a heavy-bottomed saucepan with a tight-fitting lid is the best alternative.

PUTTU MAKER Every Kerala kitchen would have a puttu maker, a steamer with a tall cylindrical tube (called a *kutty*) used for making *puttu*, a south Indian steamed rice flour and coconut log that is a popular breakfast dish (see Oats Puttu, page 35). If you don't have a puttu maker, a steamer or a double boiler with a steamer insert will do.

SPICE MILL Manual or electric. Alternatively, many use a coffee grinder dedicated to spices.

SPLASH GUARD This round, perforated screen is recommended as a portable "lid" for tempering spices safely in hot oil. Can be purchased at any dollar store.

TADKA *Tadka* is the Hindi word for tempering spices in oil or ghee as well as the name given to the small pan with high sides and a flat bottom used for this purpose. A small frying pan is the next best thing.

THALIS AND KATORIS Of course, you can use any large plate or platter, of any size and shape, and any small bowls you have around the house to contain the various components of a thali—you can even use a banana leaf as we do in south India. But if you want to go traditional, there are dedicated thalis and bowls (commonly called katoris) widely available and varying wildly in price from the simplest stainless steel to copper, brass, bronze, and even pure silver.

Food for Thought

There's a recipe in this book called Food for Thought Curry (page 49). It was the dish we made when the world shut down in March 2020, and we ramped up the work of a program called Food for Thought, or FFT, when facing a locked-down walk-in filled with fresh food about to rot, a bored staff eager to help, and the wish to do some good for our community. We had the physical space at Thali to spread out, mask up, and cook. And so, we did that. We made a curry with our veggies and some bags of red lentils donated by our friends at the Ottawa Mission. We packaged it up with rice and bread and gave it away to anyone hungry and able to come to our downtown doors.

Word spread of what was going on at Thali and, within days, more food was donated, more hands were offered to cook, more volunteers arrived to package up meals, and drivers began to deliver.

Two years later and FFT now has its own commercial kitchen, a dozen paid staff, dozens more volunteers, government grants, and a much-appreciated charitable status. We've been able to escalate production and we now put out thousands of meals a week, focusing, still, on feeding the food insecure in our community—homebound seniors, new Canadians housed in emergency shelters, refugee families in motels with no way of cooking a hot, nourishing meal. We are grateful this wretched pandemic has given us, our team, and our marvelous army of volunteers the opportunity to shine a light on the challenge of food insecurity in our city, and to feel useful.

For more information about the program, please visit www.foodforthought.cafe.

JOE THOTTUNGAL, Culinary Director, Food for Thought

SYLVAIN DE MARGERIE, Founder and President, Food for Thought

The Spice Box

Most of the spices you'll encounter in these pages will be familiar to you and easily available in well-stocked supermarkets. There are a few, however, that will either require some specialized shopping at an Indian or Asian grocery store or a commitment to ordering spices online. See page 181 for a list of reputable suppliers.

I recommend you buy spices whole, as required, and in small quantities, and store them in airtight containers away from the light.

ASAFETIDA POWDER It literally means fetid, or putrid, so use this odorous extract from the rhizome of the ferula plant liberally at your peril: a little goes a long way. In small amounts, though, it makes a lovely umami difference, in fish dishes and dals. *Bitter, Pungent*

BLACK MUSTARD SEEDS A foundational spice in south Indian cooking, the flavor of these seeds is released when tempered in hot coconut oil. If you can't source black mustard seeds, use brown mustard seeds, and use a little more as the brown seeds aren't as pungent. *Pungent, Spicy, Tangy, Warm*

BLACK PEPPER This king of spices has been highly prized since ancient times for its pungent, fiery taste. *Earthy, Peppery, Piney, Pungent, Spicy, Warm, Wood*

CARDAMOM The so-called queen of spices, and my favorite, cardamom was the warm scent of my childhood. Our native variety in Kerala is slim but packs an aromatic punch. It's pale green in color, with a thin, papery outer shell and small black seeds, and both the husks and the seeds are used. *Citrusy, Peppery, Pungent, Warm*

CHAAT MASALA (page 21) Composed of black salt, chili powder, cumin seeds, and dry mango powder, chaat masala is a ground spice blend used for marinades and in salad dressings. *Salty, Spicy, Tangy*

CHILES, DRIED, RED When I use red chiles, they are typically dried, and snapped into two or three pieces before being added to a dish. Source chiles that are deep red and unbroken if you can. *Earthy, Grassy, Spicy*

CINNAMON Used to bring a sweet perfume to curries, puddings, and sweets, cinnamon is also essential to our house Garam Masala (page 20). *Aromatic, Fragrant, Spicy, Sweet, Warm, Woody*

CLOVES Sweet, pungent, and powerfully aromatic, cloves are used whole as flavoring or blended into spice mixes. *Aromatic, Bitter, Pungent, Sweet, Warm*

CORIANDER Aromatic with citrus tones, whole coriander seeds are dry-roasted to bring out their perfume. Ground coriander, in addition to flavoring dishes, is also a natural thickener. *Aromatic, Citrusy, Earthy*

CUMIN Nutty, distinctively perfumed, and a common ingredient in Indian cuisine. Toasting the seeds brings out their flavor. For a health boost, drink water infused with cumin seeds. *Aromatic, Earthy, Pungent, Sharp, Sweet, Warm*

FENNEL SEEDS The dried seeds of the fennel plant, they look like green cumin seeds and have a sweet anise aroma. Fennel seeds are used as an aid in digestion. *Sweet, Warm*

FENUGREEK Fenugreek seeds (*methi*) can be added to brines, curries, and soups. You'll often need to temper them to remove the bitter edge. Earthy and strongly perfumed, dried fenugreek leaves (*kasuri methi*) are used most commonly in butter chicken. *Bitter, Nutty, Rich, Sweet*

GARAM MASALA A masala is the name given to a mixture of freshly ground seasonings. We make our house Garam Masala (page 20) from whole spices, toasted to release their perfume, then ground in a spice mill. *Aromatic, Floral, Savory, Spicy, Sweet, Warm*

The Art of Tempering Spices

Tempering, blooming, or crackling (called *tadka*) is a cooking method in which spices are added to hot fat, to extract their fragrance, perfume the oil, and add flavor to a dish. Spices are either tempered at the beginning of a recipe for a curry or a stir-fry, or at the end to finish a dish—a dal or a raita—with a flourish of sputtering spices. The process is not difficult, but tempering is a skill requiring a good nose and an eagle eye, best learned through practice and from making mistakes. The biggest mistake being that you may burn the spices. In which case, there is no remedy other than to start again with fresh spices and to be more vigilant.

As tempering can take only a few seconds, ingredients are added in rapid succession, and timing and temperature are everything, it's critically important that you have a mise en place of measured spices and aromatics at the ready, near the stove. Also important is the right pan. If you are tempering spices at the beginning of a recipe, you'll need a larger pan, preferably with a flat bottom and high sides, to continue building the dish.

Tempering spices at the end of a dish will require a small frying pan or a tadka pan designed specifically for the purpose.

Start with hot oil or ghee and reduce heat to medium before adding the spices to help ensure they don't burn. If spices do burn, toss them and start fresh.

You can briefly cover the pan with a splash guard or a lid to keep the spices from popping out of the pan but be prepared to remove it fast to continue with the recipe.

INDIAN BAY LEAVES (*tej patta*) Not to be confused with the Mediterranean bay laurel leaves you might have in your pantry, our bay leaves are longer and wider, with a sweeter, cinnamon-clove aroma. Source them at a South Asian grocer. *Aromatic, Fragrant, Warm*

KASHMIRI CHILI POWDER Bright red and with a tempered heat, this is the powdered form of the dried Kashmiri red chile. Kashmiri powder also adds a beautiful color to a curry. *Smoky, Tangy*

MACE The source of mace is the precious, scarlet-colored lacy covering of the nutmeg kernel, produced by an evergreen tree. I use it dried to add flavor and aroma to gravies and biryanis. *Citrusy, Pungent, Warm*

MUSTARD SEEDS *See black mustard seeds.*

RED CHILI POWDER Ground dried red chiles are added to just about all of our dishes. It is not to be confused with Kashmiri chili powder (see above), which imparts color more than heat. You may substitute cayenne powder, though it's hotter than our red chili powder, so use a little less. *Spicy, Warm*

STAR ANISE One of the star spices in our Garam Masala (page 20), star anise has sweet, musky, licorice notes and can be used sparingly in biryanis and meat curries. *Bitter, Pungent, Sweet, Warm*

TURMERIC POWDER Used universally in our cuisine, turmeric has myriad health benefits and lends a golden color and an earthy, pungent flavor to many of our curries. Buy it already ground or source the dried root and grate it yourself. *Bitter, Citrusy, Earthy, Pungent*

Pantry

Many of these ingredients are available in major supermarkets or can be sourced from South Asian or Indian grocers or online.

ATTA A hard, refined wheat flour with a high gluten content used for making Chapati (page 124).

BOONDI A crispy fried gram flour Indian snack used in making Boondi Mango Raita (page 155).

COCONUT OIL Buy a good-quality, pure virgin oil and store it in a dark, dry place.

GRAM FLOUR Also known as besan, this pulse flour is made from ground chickpeas.

JAGGERY A toffee-flavored sugar, widely used in India and made from the unrefined sap of date palms or coconut palms or from the juice of sugarcane. Sold in blocks, jaggery is typically grated over food.

MALABAR TAMARIND (KUDAMPULI) *Puli* means sour in Malayalam, and kudampuli is a key ingredient for adding a sour note to dishes, particularly to fish curries. The fruit, which looks like a mini pumpkin, is seeded, sun-dried, and then slowly smoked for weeks until black and shriveled. It needs rinsing and soaking before using, and both the soaked fruit and its soaking liquid are used in recipes. Stored in an airtight container away from light, Malabar tamarind will keep for years.

MOR MILAGAI (BUTTERMILK CHILES) These are chiles repeatedly marinated in salted yogurt and sun-dried over a few days. They are popular snacks to add fire and flavor to any south Indian thali. As the warmth of the sun is essential to the process, mor milagai are rarely made from scratch—especially in a northern climate! You buy them packaged and shallow-fry them up to order. They can be added to any thali menu.

PULSES *See page 42.*

RAVA (OR RAWA) South Indian word for soft wheat semolina, a coarse flour made from ground durum wheat, also known as *sooji* in north India.

RICE *Basmati* is a long, slender-grained, aromatic rice widely available in supermarkets, though the better varieties are often found in South Asian groceries.

Kaima (or *jeerakasala*), a small-grained, highly flavored basmati varietal, is commonly used in the Malabar area for making biryani (page 38). *Matta* (or Kerala red rice) is grown in the Palakkad district of Kerala and is the state's most iconic varietal, protected with its own geographical indication. *Patna*, a mild-flavored, long-grain white rice, can be used mainly for rice puddings (page 166), dosa (page 127), and chapati.

TAMARIND PULP AND PASTE A principal souring agent in many of our dishes. Recipes that call for tamarind paste (page 22) begin with tamarind pulp, usually sold in rectangular blocks wrapped in plastic. To make tamarind paste, you need to soak small chunks of the pulp in a little hot water to soften it, breaking it down further with your fingers as it cools, and then push the soaked pulp through a strainer into a bowl. Discard the contents of the strainer and use the brown sticky paste in the recipe. You can find bags or jars of premade tamarind paste, but I don't recommend their use: they tend to be too thin and watery, and can have an unpleasant metallic taste.

Produce

Curry leaves, tamarind, plantain, mango, jackfruit, snake gourd, fresh green chiles, moringa drumsticks, ginger root, and certainly coconut are the native fruits and vegetables I grew up with, and they were near impossible to source when I first landed in Canada. Today, I can find all I need in supermarkets, produce shops, and Indian groceries.

BANANA LEAVES Used for cooking, wrapping, and as plates, particularly during special occasion feasts. The leaf imparts subtle and sweet aromas.

COCONUT In Kerala, where I was raised, the coconut is the backdrop to everything, the king of our cultural life and foundational in our cuisine. We use its oil, its milk, its meat, and its sap in our kitchens and homes. The leaves are made into brooms and baskets, the shells into cooking utensils, and the dried bark into fuel for the fire.

When selecting a coconut, it should feel heavy in your hand. Its three "eyes" should be dry and clean, without moisture marks or black spots. And when you shake it, it should make a sloshy sound, indicating there is water inside.

CURRY LEAVES Along with coconuts, curry leaves are at the heart of south Indian cooking, giving our dishes a unique citrus, herby flavor. Buy them fresh, bright green, and still on the stem. Stored in the fridge, they will keep for about a week. You can also freeze them, still on the stem, in an airtight container.

GINGER ROOT Whenever possible, choose young, fresh ginger, pale in color and with a papery thin skin, usually sold as "organic." If you can source it from an Asian grocer, rather than the supermarket, where the ginger tends to be older, darker, and woodier, you are golden.

INDIAN GREEN CHILES Fresh and slender, these add a bright heat to much of our cooking. Easily found in Indian groceries, though less common in North American supermarkets. The more widely available Thai green (bird's eye) chile is a good substitute.

MORINGA DRUMSTICKS Also known as *Moringa oleifera* pods, these tender fruit pods of the moringa tree are a staple vegetable in south Indian cooking, high in vitamins, minerals, and fiber.

SHALLOTS (OR MADRAS ONIONS) Indian shallots are used extensively in our cuisine and tend to be smaller and sweeter than European shallots (which may be substituted).

TAPIOCA Also known as kappa, yuca, cassava, and manioc, tapioca is a potassium-rich, vital food for millions in the tropics. The edible bits are the tuberous root and the leaves. Cooked and mashed, the tuber makes my favourite side for a juicy steak.

TURMERIC ROOT You will use ground turmeric in most of these recipes, but for turmeric (or "golden") milk, we recommend buying the root (or the rhizome, the underground stem) and grating it fresh. It looks a bit like ginger root (indeed, it's in the same family) and, as you would for sourcing ginger, look for organic turmeric that's fresh looking and thin skinned.

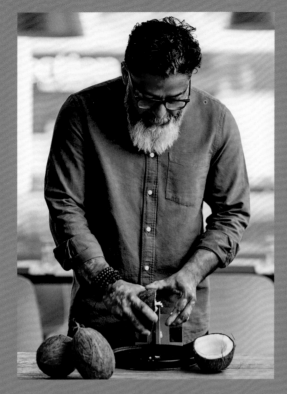

Coconut: Cracking and Grating

My technique for opening a coconut is to use the blunt side of a sturdy butcher's knife, giving firm whacks around the equator of the coconut, rotating as I go, until the shell eventually cracks. With a bowl beneath to catch the water, I insert the knife tip into the crack, then wiggle it around until the coconut splits in half. The coconut flesh can then be pried out in chunks, peeled away from the fine brown skin.

For grated coconut, use the coarse grinder of a food processor, a box grater, or invest in an authentic coconut scraper, called a *chirava*.

Grated coconut freezes well. You can buy unsweetened frozen grated coconut in most supermarkets (look for frozen coconut from India, and better yet from Kerala, if you can source it), or you can process a fresh coconut yourself. In our home, we buy a few coconuts at a time, grate them all, and freeze them in airtight containers for future use.

Essentials

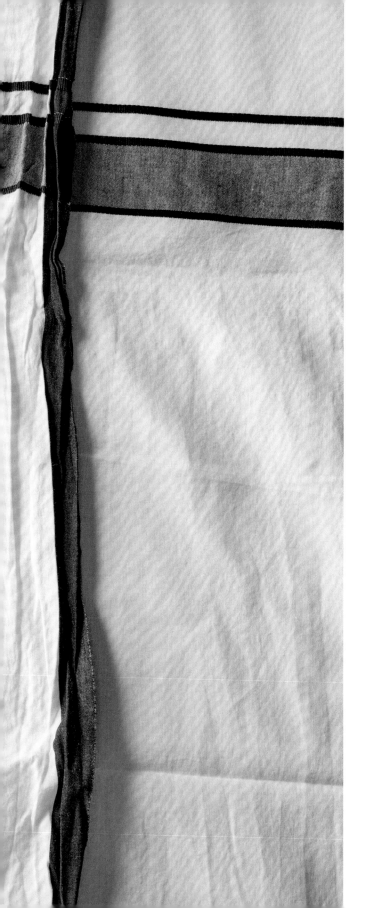

These are the building blocks: the spice blends, gravies, pastes, and fried flavor bombs that will make the recipes in this book come together with ease! You will want to begin with a shopping trip to an Indian or South Asian grocery store or to the "ethnic" aisle of a well-stocked supermarket for the spices. If these are not available to you, I encourage you to shop online and buy from a reputable spice merchant (page 181).

Garam Masala

MAKES ½ CUP

Once you make your own garam masala you will never go back to buying the supermarket stuff. This is our house blend, and you can whizz it up in three minutes. Start with fresh spices!

2 star anise

1 (3-inch) cinnamon stick, broken into smaller pieces

3 Tbsp fennel seeds

1 Tbsp cloves

1 Tbsp green cardamom pods

In a frying pan over medium heat, dry-roast all spices for 2 minutes, until toasty, crispy, and fragrant. Set aside to cool.

Using a spice mill, small food processor, or mortar and pestle, grind into a fine powder.

Store unused garam masala in an airtight container in a dark place for up to a couple of weeks. After that, it starts to lose its perfume.

House Spice Blend

MAKES 1 CUP

Every cook has their own top secret spice blend and this one belongs to our chef de cuisine, Rajesh. We have twisted his arm, gently, and he's agreed to share it.

15 cloves

15 green cardamom pods

10 dried red chiles, snapped in half

2 star anise

½ cinnamon stick

½ cup coriander seeds

2 Tbsp fennel seeds

1 Tbsp black peppercorns

1½ tsp ground turmeric

In a frying pan over medium heat, dry-roast all spices except turmeric for 2 minutes, until toasty, crispy, and fragrant. Set aside to cool.

Using a spice mill, small food processor, or mortar and pestle, grind into a fine powder. Stir in turmeric.

Store unused spice blend in an airtight container in a dark place for up to a week. After that, it starts to lose its perfume.

Chaat Masala

MAKES ½ CUP

An essential spice blend in north Indian cooking, particularly in street food, we use chaat masala to give a sour-salty kick to all manner of snacks, curries, and drinks.

1½ Tbsp cumin seeds

1 Tbsp coriander seeds

1 Tbsp dried pomegranate seeds

3 Tbsp mango powder

1 tsp black salt

½ tsp asafetida powder

¼ tsp red chili powder

In a frying pan over medium heat, dry-roast cumin and coriander seeds for 2 minutes, until toasty, crispy, and fragrant. Set aside to cool slightly.

Using a spice mill, small food processor, or mortar and pestle, grind cumin, coriander, and pomegranate seeds into a fine powder. Add the remaining ingredients and mix well.

Store unused chaat masala in an airtight container in a dark place for a couple of weeks. After that, it starts to lose its perfume.

Sambar Powder

MAKES ½ CUP

This blend of roasted spices, seeds, and pulses is always added to a south Indian sambar.

10 dried red chiles, snapped in half

2 Tbsp white lentils (*urad dal*)

2 Tbsp split and skinned chickpeas (*chana dal*)

3 Tbsp coriander seeds

1 Tbsp cumin seeds

1½ tsp fenugreek seeds

1 Tbsp ground turmeric

1½ tsp asafetida powder

In a frying pan over medium-low heat, dry-roast chiles, lentils, chickpeas, and the coriander, cumin, and fenugreek seeds. Stir spices and pulses for about 2 minutes, until they darken slightly, turning crispy and fragrant. Set aside to cool a little.

Using a spice mill, small food processor, or mortar and pestle, grind into a fine powder. Stir in turmeric and asafetida.

Store unused sambar powder in an airtight container in a dark place for a couple of weeks. After that, it starts to lose its perfume.

Tamarind Paste

MAKES ABOUT 2 CUPS

Recipes that call for tamarind paste begin with tamarind pulp, usually sold in dark, rectangular blocks wrapped in plastic. As a general rule, to make one tablespoon of tamarind paste, tear off about two table-spoons of the pulp.

1 (14-oz) pkg pure
seedless tamarind pulp

Divide tamarind pulp into small chunks and soak the chunks in enough hot water to cover for about 15 minutes, breaking down the chunks further with your fingers once the water has cooled a little to help separate the pulp from its tough membranes, and then push the soaked pulp through a strainer into a bowl. Discard the contents of the strainer and use the brown sticky paste in the recipe. Tamarind paste will keep in a sealed jar for about a week in the fridge and can also be frozen. I recommend freezing tamarind paste in ice cube trays, then storing the cubes in resealable bags.

Ghee

MAKES ABOUT 1½ CUPS

Clarified and lightly caramelized butter, ghee is sweet and nutty flavored. It is shelf stable, has a high smoke point, and is the foundational cooking oil in northern India.

2 cups (4 sticks)
unsalted butter

Melt butter in a small heavy-bottomed saucepan over medium heat, until butter separates, milk solids sink, and a layer of creamy white foam rises to the top. Reduce heat and simmer butter for 15 minutes, periodically skimming off the foam, until it turns to a golden liquid and no more solids are rising to the surface. (If necessary, adjust the heat so the solids don't burn.) Remove from heat. Set aside for 15 minutes to cool and allow the flavors to further infuse.

Line a fine-mesh strainer with a double layer of cheese-cloth or place a coffee filter over a glass jar. Strain ghee, discarding any browned milk solids (or better yet, keep them to add to biryanis, vegetables, or pulse dishes). What's left is ghee. It can be stored in a dark cupboard for 2 months or up to a year in the fridge.

Ginger-Garlic Paste

MAKES ½ CUP

This essential aromatic is featured in many of our dishes, and it's dead easy to make if you have a food processor. Otherwise, a mortar and pestle will do the trick beautifully.

1 (4-inch) knob ginger, peeled, coarsely chopped, and patted dry

10 cloves garlic

Transfer ginger, garlic, and 2 Tbsp water to a mortar and pestle or mini food processor. Process until smooth, scraping down the sides as required.

The paste can be stored in an airtight container in the fridge for up to 3 days.

Basic Gravy

MAKES ABOUT 3 CUPS

Make a batch of this fragrant gravy and store it in the fridge. It will keep, gathering deeper flavors, for about a week.

5 Tbsp vegetable oil

9 cloves

7 green cardamom pods

1 cinnamon stick, broken in half

1 star anise

8 cloves garlic, finely chopped

1 (1-inch) piece ginger, peeled and finely chopped

3 onions, finely chopped

1 Tbsp salt

2 Tbsp ground coriander

1 Tbsp ground turmeric

1½ tsp red chili powder

Pinch of mace

2 large ripe tomatoes, chopped

Heat oil in a heavy-bottomed frying pan or wok over medium heat, until oil is shimmering. Add cloves, cardamom, cinnamon, and star anise and roast for 1–2 minutes until spices release their fragrance. Add garlic and ginger and sauté for another 2 minutes.

Stir in onions and salt and sauté for 10 minutes, until onions are soft and translucent. Stir in coriander, turmeric, chili powder, and mace and cook, stirring well, for another minute.

Add tomatoes and sauté for 2 minutes. Remove from heat and set aside.

The gravy can be stored in an airtight container in the fridge for up to a week.

Fried Onions

MAKES ABOUT 1 CUP

Nothing like a final flourish of crispy onions as a topping to a dish!

1 cup vegetable oil

2 cooking onions, very thinly sliced

Heat oil in a heavy-bottomed frying pan or wok over medium heat, until it reaches a temperature of 350°F. To check, add a slice of onion to the oil. If it begins to sizzle right away, the oil is ready.

Sauté onions for 15–20 minutes, until golden brown. Using a slotted spoon, transfer onions to a baking sheet lined with paper towels and drain excess oil. Set aside to cool.

Fried onions will keep in an airtight container at room temperature for a week.

Rice

Kerala may be the land of coconuts, but it is also the land of rice. Vast swaths of our countryside are covered in paddy fields. Rice is what we eat every day, a vital part of our diet, and significant in many of our customs and celebrations. There is no thali without rice, and usually, the rice in a complete thali would be plain—basmati or matta—to be a support to its companion dishes, rather than a dominant flavor. We've provided recipes for perfectly cooked plain rice, and for ways of adding aroma, texture, color, and nutrition to a rice dish, either with freshly steamed rice or rice leftovers.

JEERA RICE

Jeera Rice

SERVES 3–4

1 Tbsp Ghee (page 22)

10 curry leaves (optional)

2 tsp cumin seeds

3 cups cooked
basmati rice

2 Tbsp Fried Onions
(page 24)

Salt, to taste

Chopped cilantro,
for garnish (optional)

Heat ghee in a large frying pan or wok over medium-high heat. Reduce heat to medium and toss in curry leaves (if using) and cumin seeds. Sauté for 20 seconds, until seeds are fragrant and leaves have crisped and curled. Add rice and fried onions and stir-fry, until warmed through. Season with salt and garnish with cilantro (if using).

Lemon Rice

SERVES 3–4

1 Tbsp Ghee (page 22)

1 tsp black mustard seeds

10 curry leaves

2 dried red chiles,
snapped in half

1 tsp white lentils (*urad dal*)

½ tsp ground turmeric

3 cups freshly cooked
basmati rice

1 Tbsp fresh lemon juice

Salt, to taste

Cilantro leaves, for garnish

Fried Onions (page 24),
for garnish

With measured spices nearby, heat ghee in a small frying pan (or tadka pan) over medium-high heat, until shimmering. (To test the heat, add a couple of mustard seeds. If they start to sizzle, the oil is ready.) Reduce heat to medium, add mustard seeds, and crackle them for just a few seconds until popping subsides, taking care they don't burn. Immediately add curry leaves, chiles, and lentils and roast for 1 minute, until dal has turned golden brown and leaves have crisped and curled. Add turmeric and cook for 1 minute, until the smell of raw spice disappears.

Put hot rice in a serving bowl, and mix in lemon juice and contents of pan. Season with salt to taste, garnish with cilantro and fried onions, and serve.

Rice Pulao

SERVES 4–6

Clear your fridge of vegetables for this dish. And feel free to improvise: substitute zucchini for mushrooms, snap peas for green beans, parsnips for carrots.

¼ cup Ghee (page 22), plus extra if needed

4 cloves

4 green cardamom pods

2 star anise

½ cinnamon stick

1 onion, finely chopped

3 cloves garlic, finely chopped

2 tsp salt (divided)

4 green beans, cut into 1-inch lengths

2 mushrooms, quartered

1 small carrot, cut into 1-inch batons

¼ cauliflower, cut into small florets

¼ cup fresh or frozen peas

2 cups basmati rice, washed, rinsed, and soaked (page 31)

Fried Onions (page 24), for garnish (optional)

Heat ghee in a heavy-bottomed saucepan with a tight-fitting lid over medium-high heat, until it begins to sizzle. Reduce heat to medium, then add cloves, cardamom, star anise, and cinnamon. Stir for 1–2 minutes, until fragrant and lightly roasted.

Add onions and sauté for 4–5 minutes, until golden brown. Add garlic, 1 tsp salt, and all vegetables except the peas. Cook until vegetables are al dente. Stir in peas.

Using a slotted spoon, transfer vegetables and spices to a plate, keeping as much ghee in the pan as possible. Add the rice to the pan and stir to coat in ghee. (Add more ghee if you wish.) Add the remaining 1 tsp salt and 3 cups water, stir, and bring to a boil. Cover and cook over low heat for 12–15 minutes. Remove pan from heat and set aside, covered, to rest for 10 minutes. Fluff rice with a fork.

Combine rice and vegetables in a serving bowl. Garnish with fried onions (if using).

Basmati

Basmati rice is like wine: each harvest is unique, giving the rice its own flavor profile and temperament. Something often overlooked in any discussion about cooking perfect rice is the advice to find your favorite brand. Experiment with a few to discover the one you like best (I prefer the basmati rice grown in Pakistan) and once you've found it, commit to it. You'll know how it behaves—how much water it likes, how much time it takes to steam to perfection, how long it likes to rest.

Critically important in preparing basmati is to first remove the starch or you'll end up with a big gloopy mess. Measure out the quantity of rice you need, add it to a bowl or saucepan, and wash it well. This means to cover the rice with cold water, swish it around with your hands, and drain out the first bathwater. Repeat three to four times until the water remains clear. Then drain, add a fresh bath of cold water, and soak the rice for 20–30 minutes. Drain the rice again and add it to a saucepan with a tight-fitting lid and the correct amount of water. The ratio of rice to water is generally 1 to 1½, depending on the brand. In addition, the smaller the grain, the less water is required.

Bring rice and water to a boil. Stir well, reduce heat to low, cover, and cook for 12–15 minutes. Remove from heat, keep the lid on, and set aside to rest for 10 minutes before fluffing with a fork.

Curd Rice

SERVES 4

What the West calls "yogurt," we in India call "curd." This cold dish has equal parts rice mixed with yogurt (curd), fired up with ginger, chiles, and a pinch of asafetida powder, then doused with a pan of tempered spices and dal. Cool and fiery, it's a protein-packed vegetarian go-to dish for a hot summer day, traditionally served in a clay pot called a *chatti*. For curd rice, we want the rice to be cooked to a more porridge-like consistency.

2 cups basmati or patna rice, washed, rinsed, and soaked (page 31)

2 cups plain yogurt

3 Indian or Thai green chiles, finely chopped

2 Tbsp finely chopped ginger

2 tsp salt

Pinch of asafetida powder

2 Tbsp coconut oil

1 Tbsp black mustard seeds

1 Tbsp white lentils (*urad dal*)

20 curry leaves

Chopped cilantro, for garnish

Shredded carrot, for garnish

Combine rice and 4 cups water in a saucepan and bring to a boil. Cover, reduce heat to low, and cook for 15–18 minutes. Remove pan from heat and set aside, covered, to rest for 10 minutes.

Keep whatever water remains and slightly mash the rice. Stir in yogurt, chiles, ginger, salt, and asafetida. Chill until ready to serve.

With measured spices nearby, heat oil in a small frying pan (or tadka pan) over medium-high heat, until oil is shimmering. (To test the heat, add a couple of mustard seeds. If they start to sizzle, the oil is ready.) Reduce heat to medium, add mustard seeds, and crackle them for just a few seconds until popping subsides, taking care they don't burn. Immediately add lentils and curry leaves and roast for 1 minute, until dal is golden brown and leaves have crisped.

Pour contents of frying pan over the curd rice. Garnish with cilantro and carrots, then serve.

Matta Rice

SERVES 4–6

Kerala's indigenous rice, matta rice is chubbier and coarser than basmati, and with a pinkish shell that turns beige as it cooks. Grown in the black soils of the Palakkad region, matta is highly nutritious and rich in fiber. Note that it requires the same washing, rinsing, and soaking as basmati, and it takes longer to cook.

2 cups matta rice, washed, rinsed, and soaked

1½ tsp salt

Put rice, salt, and 10 cups water in a large heavy-bottomed saucepan. Bring to a boil. Cover, reduce heat to low, and simmer for 45 minutes, adding more water to ensure the rice is constantly covered. Cook until rice is soft but still holds its shape. Drain out water, fluff with a fork, and serve warm.

Oats Puttu

SERVES 4

My family eats this variation on the classic south Indian rice puttu pretty much every morning, topped with banana slices and a bit of jaggery or brown sugar, or sometimes as a savory late-night snack with a leftover black chickpea curry (page 50). Our twist is to use Canadian oats in lieu of rice flour. Fresh grated coconut will make a big difference to the flavor and texture. Every Kerala kitchen has a puttu maker—a steamer attached to a cylindrical metal tube, which makes puttu's traditional log shape—but the recipe can easily be made without one. See below for instructions. If using a puttu maker, this recipe will fill the cylinder twice.

2½ cups large flake (not instant) oats

¼ tsp salt

1 cup fresh grated coconut (divided)

TO SERVE (OPTIONAL)
Curry of your choice or banana slices and a sweetener such as jaggery, brown sugar, or honey

---

NO PUTTU MAKER? NO PROBLEM! Simply combine all of the soaked oats with all the grated coconut and use a steamer with a lid or a double boiler with a perforated steamer insert to cook the mixture for about 10–12 minutes. When the puttu is light and fluffy, it is cooked.

Combine oats and salt in a bowl. Slowly sprinkle in about ¾ cup water, mixing it in with your fingertips, until the mixture begins to form a clump when you squeeze it in the palm of your hand, then un-clumps when you release the squeeze. (The amount of water required will vary depending on the quality and cut of the oats.) Set aside to soften for 10 minutes.

Transfer oats to a food processor and pulse two or three times, just to break down the oats to a coarse meal. Transfer to a bowl, then stir in half the coconut, reserving the other half in a separate bowl. For the first puttu, you will use about half of each component.

Using a puttu maker, add the thin metal perforated disk to the bottom of the cylindrical tube (called a *puttu kutty*). Then fill the tube with alternating layers of grated coconut and the oats-coconut mixture, beginning with a thin layer of coconut, then a thicker layer of the oats-coconut mixture, a thin layer of coconut, a thicker layer of oats-coconut mixture, and finishing with a final layer of coconut.

Fill the steamer a quarter full with water. Attach puttu kutty and its perforated lid and set the puttu maker over high heat. Once water is boiling, reduce heat to medium and cook for about 5 minutes more, or until steam rises from the holes in the top, indicating the puttu is cooked. Using a towel or oven mitts to protect your hands, push the steaming oats puttu out of its cylinder using the puttu "stick" (or the handle of a wooden spoon). Repeat with remaining coconut and oats-coconut mixture to make a second puttu. This will take less time as the water is already steaming hot.

Serve oats puttu savory with a curry or sweet, topped with banana slices and your preferred sweetener.

Vegetable Uppumavu

SERVES 4

A fine and filling morning porridge made with roasted semolina, grated coconut, and pretty much any vegetable on hand. Bonus: Uppumavu has staying power not only in the belly but out of the fridge—vegetarian and dairy-free, there is nothing to spoil. That means it was often in my travel bag on the overnight train to cooking school in Chennai. Eaten in the morning in my dorm room, it was the taste of home.

Heat oil in a heavy-bottomed saucepan over medium-high heat, until oil is shimmering. Reduce heat to medium, add mustard seeds, and let them splutter and pop for a few seconds. Add chiles, ginger, and onions and sauté for 1 minute.

Add carrots, beans, and peas and sauté for 2 minutes. Stir in curry leaves and turmeric and cook for another minute. Add coconut, salt, and 3 cups water. Increase to high heat and bring to a boil. Slowly add semolina, stirring constantly to prevent lumps forming. Cook for 3 minutes, or until semolina is soft. Remove from heat, cover, and set aside to rest for 5 minutes. Serve.

1 Tbsp coconut oil

1½ tsp black mustard seeds

2 Indian or Thai green chiles, finely chopped

1 (½-inch) piece ginger, peeled and finely chopped

1 onion, finely chopped

2 carrots, finely chopped

10 green beans, wax beans, or purple string beans, finely chopped

¼ cup fresh or frozen peas

20 curry leaves

¼ tsp ground turmeric

1 cup fresh or frozen (thawed) grated coconut

1½ tsp salt

2 cups semolina (*rava* or *sooji*)

Duck Biryani

SERVES 4

A biryani reminds me of my time as a young cook in Saudi Arabia. It's a celebration rice dish of layered flavors and textures, deeply aromatic, and yes, a bit of work ... but so completely rewarding. Full credit for this recipe goes to my sous chef Renju, who came to Thali as a culinary student from Algonquin College and has proven such a talent that this most festive of dishes is now his domain.

DUCK MASALA

2 Tbsp Ghee (page 22)

2 cloves

2 green cardamom pods

1 star anise

½ cinnamon stick

½ tsp fennel seeds

3 Indian or Thai green chiles, finely chopped

1 Tbsp Ginger-Garlic Paste (page 23)

2 large onions, thinly sliced

1 large tomato, chopped

1 lb duck breasts (about 3 small) or duck legs, cubed, or a whole duck, sectioned into 16 pieces

2 Tbsp House Spice Blend (page 20)

1½ tsp salt

½ cup plain yogurt

DUCK MASALA Heat ghee in a heavy-bottomed frying pan or wok over medium heat, until it begins to sizzle. Add cloves, cardamom, star anise, cinnamon, and fennel seeds and stir for 30 seconds to infuse flavors.

Add chiles and ginger-garlic paste and sauté for 1–2 minutes. Add onions and sauté for 2 minutes, then stir in tomatoes. Add duck meat, spice blend, salt, and yogurt and mix well.

Pour in 1 cup water and bring to a gentle boil. Reduce heat to medium-low, cover, and simmer for 20 minutes (40 minutes if using the leg meat or whole duck), until meat is tender. Remove from heat and set aside.

RICE In a heavy-bottomed saucepan, combine all ingredients except rice. Add 4 cups water and bring to a boil. Stir in rice. Cover, reduce heat to low, and cook for 8 minutes, or until rice is about 70% cooked. (It will finish cooking in the oven.) Remove from heat and set aside.

ASSEMBLY Preheat oven to 350°F.

Heat ghee in a frying pan over medium-low heat. Add nuts and sauté for 3–5 minutes, until golden brown. Add raisins and sauté for 30 seconds. Using a slotted spoon, transfer nuts and raisins to a plate, keeping as much ghee in the pan as possible. Set aside.

Add half the rice to a Dutch oven or casserole with a tight-fitting lid. Spread the duck masala evenly on top. Scatter half the fried onions, mint, and cilantro over the duck masala layer, then top with the remaining rice. Drizzle ghee on top (you can use the leftover ghee from roasting the nuts and raisins). Cover with a layer of parchment paper, then the lid. Bake for 20 minutes, until rice is fully cooked and entire dish is warmed through.

Sprinkle nuts and raisins on top. Top with the remaining fried onions, mint, and cilantro.

Serve with pineapple achar, cucumber-tomato raita, and pappadum, if you like.

RICE

2 cloves

2 green cardamom pods

1 star anise

½ cinnamon stick

½ tsp salt

1½ Tbsp Ghee (page 22)

Juice of ½ lime

2 cups high-quality kaima or basmati rice, rinsed well and soaked (page 31)

ASSEMBLY

2 Tbsp Ghee (page 22), plus extra for drizzling

Handful of raw cashew nuts or nuts of your choice

Handful of raisins or dried fruit of your choice

1 qty Rice (see here)

1 qty Duck Masala (see here)

½ cup Fried Onions (page 24, divided)

Small bunch of mint leaves, coarsely chopped (divided)

Small bunch of cilantro, coarsely chopped (divided)

TO SERVE (OPTIONAL)

Pineapple Achar (page 145)

Cucumber-Tomato Raita (page 155)

Pappadum

Dals

Since forever, dried beans, legumes, pulses, split peas (collectively known as dals) have been an integral part of Indian cuisine and an affordable and essential source of everyday protein for a vegetarian. Whether cooked plainly and jazzed up with a sizzling pan of tempered spices, mixed with other dals in a slow-cooked stew, ground into flour for deep fried fritters, or transformed into a breakfast pancake, dals are marvelously versatile. Indeed, India is both the largest producer and consumer of dals in the world. In my home (and in my childhood home) there are always pots of cowpeas, black chickpeas, or *toor dal* soaking, and no meal would be complete without at least one dal dish. As with everything in India, there are infinite and wonderful uses of dals, and regional variations are extensive and immense!

On the Pulse

Indian pulses, or dals, can be categorized into three types: the whole dal, the split dal with the skins on, and the split dal with the skins off. In fact, a great many of the dals we use are farmed on the Canadian Prairies. (Canada is the world's leading producer of lentils.)

A pressure cooker takes much of the work, time, and energy out of cooking most dals and every Indian kitchen has one. If yours doesn't, dals can be cooked on the stovetop and most will benefit from presoaking.

ADZUKI BEANS (RED COWPEAS) These small red beans have a nutty, neutral flavor.

BLACK-EYED PEAS (COWPEAS) Black-eyed peas are not peas but rather an earthy white bean with a creamy consistency.

BLACK GRAM A black bean with a creamy-colored interior prized for its superior nutritive qualities and for its versatility in our cuisine. Black gram comes whole (*urad*) or split (*urad dal chic*) or split and hulled, known as *urad dal* or white lentils. It's this latter, all-creamy-white version of black gram we use most often in the recipes in this book.

CHICKPEAS, WHITE (KABULI) OR BLACK (KALA CHANA) AND SPLIT AND SKINNED (CHANA DAL) We use both the white chickpea and the smaller—and, to my mind, much more delicious—black chickpea in our cuisine. The black chickpea is actually more brown in color, has a rough outer coating, retains its shape when cooked, and is a terrific source of protein.

KIDNEY BEANS These hearty beans have a dense structure and mild sweetness. One of India's most popular kidney bean recipes is *rajma*, a fragrant and nourishing stew of kidney beans, onions, tomatoes, and spices.

MUNG BEANS (GREEN GRAM) Small, ovoid in shape, and green in color (as the name would suggest), this bean is closely related to the adzuki bean.

PIGEON PEAS, SPLIT (TOOR DAL) This popular pulse is hulled and split, which translates to quick and easy cooking. They can also be used to thicken curries, soups, and stews.

RED LENTILS (MASOOR DAL) This quick-cooking pulse is a south Indian staple and used in many curries.

Black-Eyed Pea Curry

SERVES 4

There were always watery pots filled with beans or lentils in my childhood home, a soaking task my mother took on just before bed to speed up the morning tasks of breakfast and lunch curries. Here, the creamy soft peas soak up the lovely flavors of a rich tomato-coconut masala, and are finished with our house garam masala.

1 cup black-eyed peas

1½ Tbsp salt (divided)

1½ Tbsp ground coriander (divided)

¼ tsp ground turmeric

2 Tbsp coconut oil

15 curry leaves

6 Indian or small shallots, finely chopped

5 cloves garlic, finely chopped

1 ripe tomato, finely chopped

1½ tsp red chili powder

1½ tsp Garam Masala (page 20)

½ cup coconut milk

TO SERVE (OPTIONAL)

Basmati rice or Lemon Rice (page 29)

Ginger Pachadi (page 144)

Soak the black-eyed peas in cold water for at least 8 hours but preferably overnight. Drain and rinse well.

In a large saucepan, combine black-eyed peas, 1 Tbsp salt, 1½ tsp coriander, and turmeric. Add 4 cups water and bring to a boil. Reduce heat to medium-low and simmer for 1 hour, or until beans are soft. If needed, add more water to keep peas covered.

Meanwhile, heat oil in a heavy-bottomed saucepan over medium heat, until oil is shimmering. Add curry leaves, shallots, and garlic and sauté for 3–4 minutes, until shallots are deep golden brown. Add tomatoes and cook for 1 minute. Add the remaining 1 Tbsp coriander, the remaining 1½ tsp salt, chili powder, and garam masala and cook for 2–3 minutes, until the oils from the spices rise to the surface. Add black-eyed peas (do not drain) and coconut milk and bring to a boil. Remove from heat.

Serve with rice and, if you wish, a spoon of ginger pachadi for zing.

MAKE IT QUICK! The pressure cooker is an essential piece of kitchen equipment in every Indian household and can save you significant prep and cook time. If you'd like to cook the curry in a pressure cooker, follow the manufacturer's instructions for suggested cooking time.

Chickpea Sundal

SERVES 4

This savory street food snack beats movie popcorn, hands down. Made in a jiffy, chickpeas are warmed up in hot, fragrant oil and topped with curry leaves, crispy urad dal, mustard seeds, and roasted coconut. The pinch (just a pinch!) of asafetida powder lends a bit of umami joy, and the lime juice brings it all back to balance.

4 cups canned chickpeas, well washed, drained, and patted dry

½ cup fresh or frozen (thawed) grated coconut

1½ Tbsp coconut oil

2 tsp black mustard seeds

2 tsp white lentils (*urad dal*)

15 curry leaves

3 dried red chiles, snapped in half

Pinch of asafetida powder

½ tsp salt

Juice of 1 lime

TO SERVE (OPTIONAL)
Chapati (page 124)
Ginger Pickle (page 151)

Combine chickpeas and coconut in a large frying pan or wok over low heat.

With measured spices nearby, heat oil in a small frying pan over medium-high heat, until oil is shimmering. (To test the heat, add a couple of mustard seeds. If they start to sizzle, the oil is ready.) Reduce heat to medium, add mustard seeds, and crackle them for just a few seconds until popping subsides, taking care they don't burn. Immediately add lentils and sauté until golden brown. Stir in curry leaves and chiles, then sprinkle in asafetida.

Add the seasoning mixture to the pan of chickpeas. Increase heat to high and sauté for 3–4 minutes to dry the chickpeas and give them a bit of crunch. Remove from heat. season with salt, and sprinkle lime juice on top.

Serve on their own as a snack, or tucked into a chapati with a dollop of ginger pickle, if you like.

Dal Makhani

SERVES 4

Makhani means "buttery" in Hindi and, for vegetarians, this dal dish is the closest thing to the seduction of butter chicken … all those soothing, creamy, and rich flavors minus the meat. This is our family's Friday night supper, with a side of raita and a stack of parathas for sopping up every drop of sauce.

1½ cups black gram (*urad*), washed and soaked overnight

½ cup red kidney beans, washed and soaked overnight

1 Tbsp salt

¼ cup (½ stick) butter, plus extra for garnish

8 green cardamom pods

3 black cardamom pods

5 cloves

4 Indian bay leaves

3 Indian or Thai green chiles, finely chopped

1 cinnamon stick

3 tsp Ginger-Garlic Paste (page 23)

2 onions, coarsely chopped

2 Tbsp Kashmiri chili powder

1 tsp red chili powder

4 large tomatoes, puréed (3 cups)

½ tsp ground cumin

1 cup whipping cream (35%)

1 Tbsp dried fenugreek leaves (*kasuri methi*), for garnish

¼ cup chopped cilantro, for garnish

TO SERVE (OPTIONAL)

Malabar Parathas (page 126)

Raita of your choice (page 152)

Drain and rinse the soaked beans, then add them to a large saucepan. Add salt and enough cold water to cover by an inch. Bring to a boil, then reduce heat to medium-low and simmer for 1 hour, or until beans are very soft. If necessary, top up with more water.

Heat butter in a heavy-bottomed frying pan or wok over medium heat until sizzling. Add cardamom, cloves, bay leaves, chiles, cinnamon, and ginger-garlic paste. Roast for 2 minutes, or until fragrant and softened. Add onions and sauté for 3–4 minutes, until onions are brown. Add chili powders and cook for 1 minute, until the smell of raw spice disappears.

Pour in tomato purée and simmer for 2–3 minutes, until the oil from the spices rises to the surface. Add the cooked beans along with their cooking water and ½ cup hot water. Simmer, stirring occasionally, for 10 minutes. Add cumin and cream and simmer for another 5 minutes.

Garnish with fenugreek leaves and cilantro, and add another dollop of butter to melt slowly in the heat. This dish goes well with parathas and raita.

TRY TRADITIONAL! Considered one of the Punjab's most beloved dishes, the keys to a great dal makhani are to never skimp on the butter or cream and to cook it low and slow—in fact, the more it simmers, the more fabulous the flavor. If you have time, and to really make the most of this dish, after adding the cumin and cream, simmer for another 30 minutes to 1 hour on low heat, stirring occasionally and adding more glugs of cream and butter.

Food for Thought Curry

SERVES 4

When the pandemic first shut us down, we were left with fridges filled with fresh produce. We called our friends at the Ottawa Mission and their response was to donate many bags of red lentils. What else to do but feed people? This was the first curry we made for the Food for Thought charity (page 12), a program that serves the food insecure in our community who suffered so much from the lockdowns.

1 cup red lentils, rinsed

1 Tbsp salt, plus extra to taste

1 tsp ground turmeric (divided)

3 Tbsp vegetable oil

1½ tsp black mustard seeds

1½ tsp cumin seeds

3 dried red chiles, snapped in half

15 curry leaves

5 cloves garlic, chopped

2 Indian or Thai green chiles, finely chopped

1 onion, finely chopped

1 carrot, cut into ½-inch cubes

1 red or yellow bell pepper, seeded and cut into ½-inch cubes

½ Chinese or Japanese eggplant, cut into ½-inch cubes

½ zucchini, cut into ½-inch cubes

¼ cauliflower, cut into small florets

Juice of ½ lemon

Coarsely chopped cilantro, for garnish

TO SERVE (OPTIONAL)
Pappadum or Malabar Parathas (page 126)

Heirloom Tomato Chutney (page 143)

In a heavy-bottomed saucepan, combine lentils, 1 Tbsp salt, and ½ tsp turmeric. Pour in 4 cups water and bring to a boil over high heat. Reduce heat to medium and simmer, uncovered, for 8–10 minutes, until lentils are soft but still hold their shape. Remove from heat and set aside.

While lentils are cooking, prepare the seasoning. With measured spices and aromatics nearby, heat oil in a large frying pan or wok over medium-high heat, until oil is shimmering. (To test the heat, add a couple of mustard seeds. If they start to sizzle, the oil is ready.) Reduce heat to medium, add mustard seeds, and crackle them for just a few seconds until popping subsides, taking care they don't burn. Immediately add cumin seeds, red chiles, and curry leaves and sauté for another minute. Add garlic, green chiles, and onions and sauté for 7 minutes, or until onions are softened and translucent. Add the remaining ½ tsp turmeric and mix well.

Add vegetables and sauté for 8–10 minutes, or until softened. Add mixture to pot of lentils and heat through over medium heat. Stir in lemon juice and cilantro and season with salt to taste.

Serve with pappadum or parathas and a chutney, if you wish.

Nadan Kadala Masala

SERVES 4

Nadan means "homestyle," and this classic breakfast curry, thick and nourishing, would often reappear as lunch in my schoolbag, served with rice, some raita, and my mum's lemon pickle, packaged up in a banana leaf. They're called black chickpeas but really, they are brown—and truly more nutritious, and delicious, than their white cousins.

1 cup black chickpeas, rinsed and soaked overnight

1 tsp salt

2 tsp ground coriander (divided)

1 Tbsp coconut oil

1 small onion, chopped

1 (1-inch) piece ginger, peeled and crushed

5 cloves garlic, crushed

15 curry leaves

1 tsp Garam Masala (page 20)

½ tsp red chili powder

½ cup fresh or frozen (thawed) grated coconut

TO SERVE (OPTIONAL)

Oats Puttu (page 35)

Pickle of your choice (page 149 or page 151)

Raita of your choice (page 152)

Discard any chickpeas that have floated to the surface. Drain chickpeas. Combine chickpeas, salt, and 1 tsp coriander in a large saucepan. Add 5 cups water and bring to a boil. Reduce heat to medium-low and simmer for 1½ hours, until tender. If needed, add more water to keep chickpeas covered. Drain, then set pan aside.

Heat oil in a frying pan or wok over medium heat. Add onions and sauté for 1 minute. Add ginger, garlic, and curry leaves and sauté for another 2 minutes. Add garam masala, chili powder, and remaining 1 tsp coriander and cook for 1 minute, until the smell of raw spice disappears. Stir in grated coconut.

Transfer mixture to the pan of chickpeas and stir to combine. Heat through over medium heat, season to taste, and serve on its own or alongside oats puttu and sides of pickle and raita.

MAKE IT QUICK! The pressure cooker is an essential piece of kitchen equipment in every Indian household and can save you significant prep and cook time. Cook the chickpeas for 45 minutes in a pressure cooker instead of a saucepan, if you like.

Tadka Dal

SERVES 4

I could measure my life in dal recipes! This one is a family favorite and a recipe my children can easily manage. Simple, quick, and bright yet soothing… a supper (or breakfast!) of tadka dal with rice and chapati tops the list of comfort dishes. Red lentils, farmed on the Canadian Prairies, are a fast cook, and once you've mastered the art of tempering spices this dish will become a standard, budget-friendly go-to on a cold winter night.

2 cups dried red lentils, rinsed

½ tsp salt, plus extra to taste

1 Tbsp coconut oil

1 Tbsp black mustard seeds

1 Tbsp cumin seeds

2 dried red chiles, snapped in half

5 cloves garlic, finely chopped

6 Indian or small shallots, finely chopped

2 Indian or Thai green chiles, finely chopped

About 20 curry leaves

2 ripe tomatoes, diced

TO SERVE (OPTIONAL)
Basmati rice

Chapati (page 124)

Basic Raita (page 154)

In a heavy-bottomed saucepan, combine lentils and ½ tsp salt. Pour in 4 cups water and bring to a boil over high heat. Reduce heat to medium and simmer, uncovered, for 8–10 minutes, until lentils are soft but still hold their shape. Remove from heat and set aside.

With measured spices and aromatics nearby, heat oil in a medium frying pan over medium-high heat, until oil is shimmering. (To test the heat, add a couple of mustard seeds. If they start to sizzle, the oil is ready.) Reduce heat to medium, add mustard seeds, and crackle them for just a few seconds until popping subsides. Add cumin seeds and red chiles and stir for 30 seconds, taking care they do not burn. Immediately add garlic, shallots, and green chiles and sauté for 3 minutes. Stir in curry leaves and tomatoes and sauté for another 3 minutes.

Pour mixture over the dal and stir to combine. Season with salt to taste and serve hot with basmati rice, a basket of chapati, and basic raita (if using).

DRIED CHILES

Vegetables

Vegetables are critical in any Indian thali, and so is the matter of texture when composing one. You would never serve two or three vegetable dishes, all wet or saucy, without the balance of a dry dish—a stir-fry, a thoran, or a lovely thick masala. And so, we have divided this chapter into two sections: "dry" vegetable dishes and "saucy" ones, and encourage you to select from both when building your thali.

Egg Roast Masala

SERVES 4

Hard-cooked eggs are bathed in a tomato masala and finished with a touch of cream. This classic egg dish is typical breakfast fare, served with chapati, but it also makes a quick, easy, and affordable vegetarian dinner.

¼ cup coconut oil

1 tsp fennel seeds

1 (1-inch) piece ginger, peeled and cut into matchsticks

4 cloves garlic, chopped

3 Indian or Thai green chiles, chopped

3 small onions, sliced

1½ tsp salt

2 large tomatoes, thinly sliced

1 Tbsp ground coriander

2 tsp Kashmiri chili powder

1 tsp ground turmeric

½ tsp Garam Masala (page 20)

½ tsp black pepper

¼ cup whipping cream (35%)

4 warm hard-boiled eggs, peeled

Cilantro, for garnish (optional)

TO SERVE (OPTIONAL)

Chapati (page 124) or Malabar Parathas (page 126)

Baby Arugula and Sunflower Seed Raita (page 155)

Heat oil in a heavy-bottomed frying pan or wok over medium heat. Add fennel seeds and roast for about 30 seconds. Add ginger and garlic and sauté for 2–3 minutes, until garlic starts to color.

Add chiles, onions, and salt and sauté for another 2–3 minutes, until onions are light brown. Add tomatoes and cook for 1–2 minutes. Stir in coriander, chili powder, turmeric, garam masala, and pepper and cook for 1 minute, until the smell of raw spice disappears. Pour in cream and bring back to a simmer.

Quarter the eggs and place on top of the masala. (Alternatively, leave the eggs whole and simmer them in the masala for 1 minute to warm them up.) Garnish with cilantro (if using).

Serve with chapati or parathas and perhaps a dollop of raita.

TRY TRADITIONAL! If you want to be fancy about it, make shallow cuts in the eggs before adding them to allow the masala to enter and flavor the eggs.

Roasted Beet Salad

SERVES 4

Roasting is not an Indian thing. Ovens simply weren't available, so everything was boiled, steamed, or fried. But when Suma's mother was visiting our Canadian home, she discovered her favorite beet salad recipe took on an entirely new depth of flavor when the beets were roasted. We often serve this for a late summer lunch, when young beets are abundant, with a stack of chapatis.

8 assorted small beets (ruby red, golden, Chioggia, and/or rainbow), scrubbed and patted dry

2 tsp salt (divided)

2 Tbsp vegetable oil

1 Tbsp coconut oil

3 Tbsp white wine vinegar

15 curry leaves, thinly sliced

3 Indian or Thai green chiles, finely chopped

2 red onions, thinly sliced

TO SERVE (OPTIONAL)
Jeera Rice (page 29)
Chapati (page 124)

Preheat oven to 400°F.

In a bowl, combine beets, 1½ tsp salt, and vegetable oil and mix well. Transfer to a large piece of aluminum foil and wrap them tightly. Roast for 45 minutes, or until they can be easily pierced with a knife. Set aside until cool enough to handle.

Peel off beet skins, slice into wedges, and place warm beets in a serving bowl. Stir in coconut oil. Then add vinegar, most of the curry leaves (reserving some for garnish if you wish), chiles, onions, and the remaining ½ tsp salt and mix well. Chill in the fridge.

Garnish with reserved curry leaves and serve with jeera rice and chapati (if using).

Kale Ularthiyathu

SERVES 6

The rough translation for *ularthiyathu* (or, more commonly, *ularthu*) is "to fry" and though kale is not a green we would typically fry up in India (it likes a northern climate), we grow it in abundance in our Ottawa garden, so this dish is often on our table. The white lentil (*urad dal*) gives a nutty flavor and a bit of crunch, along with a boost of protein and iron, to an already nutritious side dish. Omit or reduce the chiles if you want a milder version.

2 small onions, finely chopped

2 Indian or Thai green chiles, finely chopped

1½ cups fresh or frozen (thawed) grated coconut

1½ tsp salt

1 tsp red chili flakes

1 tsp ground turmeric

2 Tbsp coconut oil

2 Tbsp white lentils (*urad dal*)

2–3 bunches kale, spines removed and leaves finely chopped

TO SERVE (OPTIONAL)
Lemon Rice (page 29)

Cranberry-Coconut Chutney (page 140)

In a bowl, combine onions, chiles, coconut, salt, chili flakes, and turmeric and set near the stove.

Heat oil in a large frying pan or wok over medium-high heat, until oil is shimmering. Add lentils and sauté for 1 minute, or until toasty and crunchy. Add the coconut mixture, toss to coat, and sauté for 3 minutes. Add kale and sauté for another 2–3 minutes, until leaves are wilted but still have some crunch.

Serve with lemon rice and a side of cranberry-coconut chutney, if you wish.

Kerala-Style Aviyal

SERVES 4

A rich and hearty vegetable stew, cooked with coconut, chiles, and curry leaves and finished with the creamy tang of yogurt. There are no hard rules about which vegetables to use for aviyal, so raid the fridge and improvise.

5 Indian or Thai green chiles, halved lengthwise

2 moringa drumsticks, peeled, halved lengthwise, and cut into 2-inch batons

1 onion, sliced

1 carrot, cut into 2-inch batons

1 potato, cut into 2-inch batons

1 Chinese or Japanese eggplant, cut into 2-inch batons

1 green plantain, cut into 2-inch batons

¼ yellow yam, cut into 2-inch batons

Handful of long beans, cut into 2-inch pieces

2 tsp salt (divided)

1 Tbsp ground turmeric (divided)

3 Indian or small shallots, coarsely chopped

2 cups fresh or frozen (thawed) grated coconut

2 tsp cumin seeds

3 Tbsp plain yogurt

20 curry leaves

3 Tbsp coconut oil

TO SERVE (OPTIONAL)
Matta Rice (page 33)

Pickle of your choice (page 149 or page 151)

In a large saucepan, combine all the cut vegetables except the shallots with 1 tsp salt, 1½ tsp turmeric, and ¼ cup water. Stir, then cover. Steam over medium-low heat for 10 minutes, or until vegetables are tender.

Using a mortar and pestle or food processor, combine shallots, coconut, cumin seeds, remaining 1½ tsp turmeric, and enough water to make a rough paste. Add paste to the pan of vegetables, along with yogurt, remaining 1 tsp salt, and curry leaves, and stir to combine. Cook over low heat for 3 minutes to warm through. Stir in coconut oil.

Serve hot with matta rice and any pickle you'd like.

MORINGA DRUMSTICKS

KAPPA

Kappa Mash

SERVES 8

Kappa—the poor man's starch and my favorite mash with a steak or a roast. (It's also known as yuca, manioc, cassava, or, in India, tapioca.) When I was a kid, sent to buy kappa at the market, I was taught to slyly break off the tip of the tuber to see if the flesh was indeed white and tender. If it was blue-veined or dry looking, there would be trouble if I brought it home. So, there's my trick for sourcing a beautiful piece of fresh kappa.

2 lbs kappa, peeled and cut into ½-inch cubes

1½ tsp salt (divided)

1½ tsp ground turmeric (divided)

2 cups fresh grated coconut

4 Indian or Thai green chiles, halved lengthwise

½ tsp black pepper

2 Tbsp coconut oil

1 Tbsp black mustard seeds

8–12 curry leaves

4 Indian or small shallots, finely chopped

In a large saucepan, combine kappa, 1 tsp salt, and ½ tsp turmeric. Add enough cold water to cover and bring to a boil. Reduce heat to medium-low and simmer, covered, for 10 minutes, or until kappa is tender.

Meanwhile, combine coconut, chiles, and remaining 1 tsp turmeric in a medium bowl.

Drain kappa, then return to the pan. Place the coconut mixture on top. Cover and set aside for 3 minutes to steam in residual heat. Season with remaining ½ tsp salt and pepper, stir with a wooden spoon to combine and make a rough mash of the kappa, then transfer to a serving bowl.

With measured spices and aromatics nearby, heat oil in a small frying pan (or tadka pan) over medium-high heat, until oil is shimmering. (To test the heat, add a couple of mustard seeds. If they start to sizzle, the oil is ready.) Reduce heat to medium, add mustard seeds, and crackle them for just a few seconds until popping subsides, taking care they don't burn. Immediately add curry leaves and shallots and cook for 2–3 minutes, until shallots are golden brown and curry leaves are crispy and curled.

Pour the sizzling spices and shallots over the kappa mash and serve immediately.

Eggplant Mezhukkupuratti

SERVES 4

A long word for a quick dish! *Mezhukkupu-ratti* is a traditional south Indian stir-fry, and here we've used what we call snake eggplant, those long purple pretties that fry up fast. Once you've measured these spices and sliced the eggplant, this dish will come together in five minutes.

4 Chinese or Japanese eggplants

3 Tbsp coconut oil (divided)

1 tsp black mustard seeds

12 curry leaves

10 cloves garlic, crushed

6 Indian or small shallots, coarsely crushed

4 dried red chiles, snapped in half

1 Tbsp salt

1 tsp ground turmeric

1 tsp black pepper

½ tsp red chili powder

Juice of ½ lemon

½ tsp Garam Masala (page 20)

TO SERVE (OPTIONAL)
Malabar Parathas (page 126)
Pineapple Achar (page 145)

Cut each eggplant into three even pieces. Stand the pieces on their ends, cut them in half down their centers, then lie them flat to slice them lengthwise into thin slivers about ⅛ inch thick.

Heat 2 Tbsp oil in a large frying pan or wok over medium-high heat, until oil is shimmering. (To test the heat, add a couple of mustard seeds. If they start to sizzle, the oil is ready.) Reduce heat to medium, add mustard seeds, and crackle them for just a few seconds until popping subsides, taking care they don't burn. Immediately add curry leaves, garlic, shallots, and chiles and sauté for 2–3 minutes, until shallots are soft and lightly browned.

Increase heat to high. Add eggplants and salt and stir-fry for 3 minutes, or until eggplants are wilted. Reduce heat to medium. Stir in turmeric, pepper, and chili powder and cook for 1 minute, until the smell of raw spice disappears. Remove from heat, then stir in lemon juice and garam masala. Drizzle remaining 1 Tbsp coconut oil on top.

This dish is great with parathas and a dollop of pineapple achar.

Pumpkin Ularthiyathu

SERVES 4–6

A warming vegan stir-fry we make when pumpkins are in season, even better the next day on a bed of fluffy rice. If pumpkins aren't in season, butternut squash makes a nice substitute.

2 cooking (pie) pumpkins

¼ cup coconut oil

1 tsp black mustard seeds

½ tsp cumin seeds

8 curry leaves

2 dried red chiles, snapped in half

1 large onion, chopped

4–6 cloves garlic, crushed

1 cup fresh or frozen (thawed) grated coconut

1 tsp red chili flakes

1 tsp salt

TO SERVE (OPTIONAL)

Rice of your choice

Plain yogurt

Halve pumpkins, remove seeds and stringy pulp, and leave the skins on. Chop into ½-inch cubes. (You'll have about 5 cups.) Set aside.

With measured spices and aromatics nearby, heat oil in a heavy-bottomed frying pan or wok over medium-high heat, until oil is shimmering. (To test the heat, add a couple of mustard seeds. If they start to sizzle, the oil is ready.) Reduce heat to medium, add mustard seeds, and crackle them just for a few seconds until popping subsides, taking care they don't burn. Immediately add cumin seeds and stir for a few seconds more. Stir in curry leaves and red chiles.

Add onions and garlic and sauté for 2–3 minutes, until golden brown. Add pumpkin, coconut, chili flakes, and salt and stir to coat. Cover and cook on low heat, stirring occasionally, for 7 minutes or until pumpkin is soft. Serve with rice and perhaps a dollop of yogurt.

Snake Gourd Stir-Fry

SERVES 4

The snake gourd (also called snake squash or serpent gourd) curls and creeps along our gardens back home in Thrissur and we often tie it down with weights to help it grow straight. It's a beautiful vegetable, nutritious and crunchy, and well worth getting to know! You can find snake gourd cut and sold in pieces in Asian grocery stores and in supermarkets with a good international produce selection, but you can substitute zucchini if you have no luck.

3–4 curry leaves

1 Indian or Thai green chile, finely chopped

1 small onion, finely chopped

¼ cup fresh or frozen (thawed) grated coconut

½ tsp ground turmeric

1 tsp salt

1½ Tbsp coconut oil

1 tsp black mustard seeds

1 tsp ground cumin

1 (12-inch) snake gourd, peeled, seeded, and cut into ¼-inch cubes (about 2 cups)

TO SERVE (OPTIONAL)

Basmati rice

In a small bowl, combine curry leaves, chiles, onions, coconut, turmeric, and salt.

Heat oil in a heavy-bottomed frying pan or wok over medium-high heat, until oil is shimmering. (To test the heat, add a couple of mustard seeds. If they start to sizzle, the oil is ready.) Reduce heat to medium, add mustard seeds, and crackle them for just a few seconds until popping subsides, taking care they don't burn. Immediately add cumin and sauté for a few seconds more. Toss in snake gourd and the coconut mixture and mix well. Cover and cook for 1 minute, until snake gourd is wilted but still firm.

Serve hot with basmati rice, if you like.

Mushroom and Pea Masala

SERVES 4

A north Indian dish, flavored with ginger-garlic paste and the north Indian spice blend chaat masala. You can use any mushroom for this recipe, or a combination of varieties. Peas are added at the end to preserve their color and crunch. Start with whipping up the spice blend if you don't have any on hand. It takes very little time and will make a world of difference to the flavor of this masala.

3 Tbsp canola oil

1 Tbsp cumin seeds

1 Tbsp Ginger-Garlic Paste (page 23)

3 onions, finely chopped

3 tomatoes, puréed

1 Tbsp House Spice Blend (page 20)

1 lb button mushrooms, quartered

½ cup fresh or frozen peas

1 tsp salt

1 tsp Chaat Masala (page 21)

1 tsp ground cumin

Chopped cilantro, for garnish

TO SERVE (OPTIONAL)
Jeera Rice (page 29)

Chapati (page 124)

Heat oil in a heavy-bottomed frying pan or wok over medium-high heat, until oil is shimmering. Add cumin seeds and crackle them for a few seconds to release their aroma. Immediately add ginger-garlic paste and sauté for 1 minute. Toss in onions and sauté for 2–3 minutes, until onions are soft and translucent. Add puréed tomatoes and cook for another 2 minutes.

Stir in the spice blend and cook for 2–3 minutes, until the oils from the spices rise to the surface. Add mushrooms, peas, and salt and cook for about 4 minutes, until mushrooms are softened.

Finish with chaat masala and cumin, and garnish with cilantro.

Serve with jeera rice and chapati (if using).

Asparagus Thoran

SERVES 4

When fresh asparagus floods our markets in May, we make this simple, delicious stir-fry, applying Kerala flavors and techniques to a favorite Ontario vegetable. Coconut is critical in this dish, so I recommend you use freshly grated. Serve this as a seasonal side at a spring dinner, or as an hors d'oeuvre topping crunchy pappadum.

1 bunch of asparagus (about 16 stalks)

10 curry leaves, coarsely chopped

3 Thai red chiles, finely chopped

2 Indian or small shallots, finely chopped

1 cup fresh grated coconut

1 tsp ground turmeric

1 tsp salt

1 Tbsp coconut oil

2 Tbsp white lentils (*urad dal*)

1 tsp fennel seeds

2 dried red chiles, snapped in half

TO SERVE (OPTIONAL)
Lemon Rice (page 29)

Meat dish of your choice

Prepare asparagus by removing woody ends and running a vegetable peeler down their lengths, from just below their tips to their bottoms. Thinly slice the spears into small rounds, reserving a few whole tips for garnish.

In a bowl, combine asparagus rounds, curry leaves, chiles, shallots, coconut, turmeric, and salt.

Heat oil in a large frying pan or wok over medium-high heat, until oil is shimmering. Reduce heat to medium, toss in lentils and fennel seeds, and roast for 2 minutes until golden brown. Add dried chiles and sauté for a few seconds. Increase heat to high, add asparagus mixture, and sauté for 2–3 minutes, until asparagus pieces are al dente.

Serve with lemon rice and a meat dish, if you wish.

Scrambled Eggs with Coconut

SERVES 4

Not really a vegetable dish unless you count the shallots, but where else to put my dad's marvelous weekend breakfast eggs, piqued lightly with chiles, fragrant with curry leaves, and moistened with freshly grated coconut? Here is as good as anywhere.

6 large eggs

3 Indian or Thai green chiles, finely chopped

2 Indian or small shallots, finely chopped

15 curry leaves, coarsely chopped

½ cup fresh grated coconut

1 tsp salt

1 Tbsp coconut oil

Black pepper, to taste

Crushed pappadum, for garnish (optional)

TO SERVE (OPTIONAL)
Chapati (page 124)

Sweet Mango Kalan (page 156)

Crack eggs into a bowl and beat with a fork. Add chiles, shallots, curry leaves, coconut, and salt and beat again.

Heat oil in a frying pan or wok over medium heat. Pour in the egg mixture and cook, stirring continuously, until eggs are softly clumped (or as you like). Add pepper and garnish with pappadum (if using).

Serve steaming hot with, if you like, chapati and a side of sweet mango kalan.

Root Vegetable Curry

SERVES 4

One of those dishes where you toss everything into the pot and let the vegetables do their thing, and then add the magic of coconut milk and a pan-load of sizzling spices. Feel free to use any root vegetables you like, though the beets do add such a pretty color.

7 Indian or Thai green chiles, halved lengthwise

2 Yukon Gold (or other yellow-fleshed) potatoes, cut into ½-inch cubes

1 carrot, cut into ½-inch cubes

1 small beet, cut into ½-inch cubes

1 onion, finely chopped

1 (½-inch) piece ginger, peeled and finely chopped

1 tsp salt

½ tsp ground turmeric

1 cup coconut milk

1 Tbsp white vinegar

1½ Tbsp coconut oil

1 tsp black mustard seeds

1 tsp cumin seeds

10 curry leaves, plus extra for garnish

3 Indian or small shallots, thinly sliced

TO SERVE (OPTIONAL)
Fried Onions (page 24)

Malabar Parathas (page 126)

Sweet Mango Kalan (page 156)

In a large frying pan or wok, combine chiles, potatoes, carrots, beets, onions, ginger, salt, and turmeric. Pour in enough cold water to cover and bring to a boil. Stir, cover, and reduce heat to medium-low. Simmer for 10 minutes, or until vegetables are softened.

Stir in coconut milk and bring back to a boil. Remove from heat, then add vinegar. Transfer mixture to a serving dish and set aside.

With measured spices and aromatics nearby, heat oil in a small frying pan (or tadka pan) over medium-high heat, until oil is shimmering. (To test the heat, add a couple of mustard seeds. If they start to sizzle, the oil is ready.) Reduce heat to medium, add mustard seeds, and crackle them for just a few seconds until popping subsides, taking care they don't burn. Immediately add cumin seeds, curry leaves, and shallots and sauté for 1 minute, until lightly browned. Pour the tempered spices over the vegetable curry.

Garnish with fried onions (if using) and a sprinkle of fresh curry leaves. Serve with warm parathas and maybe a dollop of sweet mango kalan.

Sambar

MAKES 8 CUPS

This king of stews is served at every south Indian wedding, every funeral, every festival. In fact, if there is no sambar, there is no feast. They say you can mess up a pickle, a thoran, even a biryani, but if you mess up sambar, tongues will wag.

In Kerala, we usually eat this stew as a companion to dosa (page 127). Elsewhere, sambar is drunk as a thin soup to whet the appetite.

1 cup split pigeon peas (*toor dal*), rinsed thoroughly

½ tsp ground turmeric

½ tsp red chili powder

2 tsp salt (divided)

2 carrots, cut into large chunks

1 large potato, peeled and cut into large chunks

1 moringa drumstick, peeled and cut into 1-inch sticks

10 green beans, cut into 1-inch pieces

6 okra pods, halved crosswise

3 Indian or small shallots, halved

2 ripe tomatoes, chopped

¾ cup Tamarind Paste (page 22)

½ tsp asafetida powder

2 Tbsp coconut oil

1 tsp black mustard seeds

3 dried red chiles, snapped in half

15 curry leaves

¼ cup Sambar Powder (page 21)

Chopped cilantro, for garnish

TO SERVE (OPTIONAL)
Matta Rice (page 33)

Pacha Moru (page 177)

In a large saucepan, combine pigeon peas, turmeric, chili powder, and 1 tsp salt. Add water to cover by an inch and bring to a boil. Reduce heat to medium-low, cover, and simmer for 40–45 minutes, until pigeon peas begin to soften. If needed, add more water to keep dal covered.

Add carrots, potatoes, moringa drumsticks, green beans, okra, shallots, and tomatoes. Pour in 1 cup water and increase heat to medium-high. Cook for 30 minutes, or until pigeon peas and vegetables are soft.

Add tamarind paste and bring to a boil. Season with the remaining 1 tsp salt and asafetida. Transfer stew to a serving dish and set aside.

With measured spices nearby, heat oil in a small frying pan over medium-high heat, until oil is shimmering. (To test the heat, add a couple of mustard seeds. If they start to sizzle, the oil is ready.) Lower heat to medium, add mustard seeds, and crackle them for just a few seconds until popping subsides, taking care they don't burn. Immediately add chiles and curry leaves, stirring to wilt the leaves. Stir in sambar powder and cook for 1–2 minutes, until the smell of raw spice disappears.

Pour the tempered spices over the stew and mix well. Garnish with cilantro and serve with matta rice and pacha moru, if you like.

MAKE IT QUICK! The pressure cooker is an essential piece of kitchen equipment in every Indian household and can save you significant prep and cook time. If you'd like to cook the sambar in a pressure cooker, follow the manufacturer's instructions for suggested cooking time.

Kasuri Paneer

SERVES 4

A mildly spiced and sweetly scented north Indian dish, kasuri paneer is an essential protein in a vegetarian thali. You're going to want lots of rice or bread to sop up its gorgeous creamy gravy.

1½ Tbsp Ghee (page 22)

4 cloves

3 green cardamom pods

1 star anise

½ cinnamon stick

1 cup Basic Gravy (page 23)

1½ tsp Kashmiri chili powder

2 Tbsp ketchup

1 Tbsp dried fenugreek leaves (*kasuri methi*)

1½ cups whipping cream (35%)

1 lb paneer, cut into ½-inch cubes

1 tsp ground cumin

½ tsp salt

Coarsely chopped cilantro, for garnish

TO SERVE (OPTIONAL)

Rice Pulao (page 30) or Chapati (page 124)

Pomegranate-Mint Raita (page 154)

Heat ghee in a heavy-bottomed frying pan or wok over medium heat, until hot. Add cloves, cardamom, star anise, and cinnamon. Sauté for 1–2 minutes, until fragrant. Add gravy and simmer for 2 minutes, stirring occasionally. Stir in chili powder and cook for 1 minute.

Add ketchup and fenugreek leaves and cook for 1–2 minutes. Pour in cream, then bring to a boil, stirring occasionally. Add paneer, reduce heat to medium-low, and simmer for 2 minutes, until warmed through. Stir in cumin and salt.

Transfer to a serving platter, then garnish with cilantro. Serve with rice pulao or chapati and perhaps a spoonful of raita.

Squash Rasam

SERVES 6

Traditionally made in clay pots, a rasam is probably the closest dish in our south Indian cuisine to a French consommé. It's light, restorative, and meant as an opening move in a meal to awaken the taste buds. However, poured over rice and served with bread and pickles, rasam becomes a complete meal. There are dozens of versions, and in this one the butternut squash adds sweetness to balance the sour and tangy notes.

1 small butternut squash, peeled and cut into 1-inch cubes (about 4 cups)

2 ripe tomatoes, cut into 1-inch cubes

1½ Tbsp salt

1 tsp red chili powder

¼ tsp ground turmeric

1 cup Tamarind Paste (page 22)

20 curry leaves

6 cloves garlic

2 dried red chiles, snapped in half

1 Tbsp cumin seeds

1½ tsp coriander seeds

1 tsp black peppercorns

2½ Tbsp coconut oil

1 tsp black mustard seeds

1 tsp asafetida powder

Chopped cilantro, for garnish

TO SERVE (OPTIONAL)
Pappadum

Vada (page 134)

In a large saucepan, combine squash, tomatoes, salt, chili powder, and turmeric. Cover with 8 cups water and bring to a boil. Reduce heat to medium-low, cover, and simmer for 15 minutes, or until squash is tender. Add tamarind paste and bring to a boil. Remove rasam from the heat and set aside.

Using a mortar and pestle or a food processor, pound or process curry leaves, garlic, chiles, cumin seeds, coriander seeds, and peppercorns into a coarse grind.

With measured spices and ground spices nearby, heat oil in a small frying pan over medium-high heat, until shimmering. (To test the heat, add a couple of mustard seeds. If they start to sizzle, the oil is ready.) Reduce heat to medium, add mustard seeds, and crackle them for just a few seconds until popping subsides, taking care they don't burn. Immediately add ground spices and asafetida and stir well. Add a small cupful of the rasam to the frying pan (to deglaze the pan and collect every scrap of flavor from the spices), then dump the mixture back into the soup. Bring to a gentle simmer and cook for 4–5 minutes to warm through.

Garnish with cilantro. Serve with pappadum and vada (if using).

Tawa Vegetables

SERVES 4

This north Indian vegetable dish is coated in a fragrant masala and finished with shallots, cumin, lemon, and cilantro. The vegetables are cook's choice, so go ahead and raid your fridge (or garden) or use up cooked or roasted leftovers. You'll want to make the basic gravy and house spice blend in advance.

1 large potato, cut into 1-inch cubes

1 carrot, cut into 1-inch cubes

½ cauliflower, cut into florets

¾ Tbsp salt

½ tsp ground turmeric

10 green beans, cut into 1-inch lengths

1 zucchini, cut into 1-inch chunks

½ cup fresh or frozen peas

¼ cup vegetable oil

1½ tsp cumin seeds

4 Indian or small shallots, halved

1 green bell pepper, seeded, deveined, and cut into wedges

1 cup Basic Gravy (page 23)

1 Tbsp House Spice Blend (page 20)

Juice of 1 lemon

Few sprigs of cilantro, coarsely chopped, for garnish

TO SERVE (OPTIONAL)
Malabar Parathas (page 126)

Nadan Cucumber Pulissery (page 158)

Combine potatoes, carrots, cauliflower, salt, and turmeric in a saucepan. Add enough water to cover and bring to a boil. Cook for 8–10 minutes, until al dente. Add beans, zucchini, and peas and cook for another 2–3 minutes. Drain vegetables and set aside.

Heat oil in a large frying pan or wok over medium heat, until oil is shimmering. Add cumin seeds and roast for 1–2 minutes or so, until fragrant. Add shallots and green pepper and sauté for 2 minutes to crisp up. Stir in gravy and mix well. Add the spice blend and sauté for another minute, or until the oils from the spices rise to the surface.

Add the reserved vegetables and stir to coat. Remove from heat, then add lemon juice and garnish with cilantro.

This dish goes well with parathas and nadan cucumber pulissery.

Palak Paneer

Palak is the Hindi word for spinach and paneer is a fresh Indian cheese. Here, those two stars are combined with tomatoes, onion, ginger, and chiles in a highly aromatic sauce finished with cream, a squeeze of lemon, and a pinch of fenugreek leaves.

3 handfuls baby spinach

1 ripe tomato, coarsely chopped

6 cloves garlic (divided)

2 Indian or Thai green chiles

1 (1-inch) piece ginger, peeled

2 Tbsp Ghee (page 22)

1 tsp cumin seeds

1 onion, finely chopped

½ tsp ground coriander

½ tsp ground turmeric

½ tsp red chili powder

1 cup whipping cream (35%)

1½ tsp salt

½ tsp sugar

1 lb paneer, cut into ½-inch cubes

1 tsp dried fenugreek leaves (*kasuri methi*)

Juice of ¼ lemon

TO SERVE (OPTIONAL)

Lemon Rice (page 29)

Cranberry Pickle (page 149)

Bring a saucepan of salted water to a boil. Add spinach and blanch for 1–2 minutes. Using tongs, transfer spinach to a colander set in an ice bath to shock the leaves and preserve their color.

In a food processor, combine drained spinach, tomatoes, 3 cloves garlic, chiles, and ginger. Process to a smooth paste. If needed, add up to ¼ cup water to create the right texture.

Finely chop the remaining 3 cloves of garlic. Heat ghee in a frying pan over medium heat, until shimmering. Add cumin seeds and roast for 30 seconds. Add onions and chopped garlic and sauté until lightly browned. Stir in coriander, turmeric, and chili powder and cook for 1 minute, until the smell of raw spice disappears. Add the spinach paste and ½ cup water. Cook for 5 minutes, stirring often. Pour in cream, add salt and sugar, and simmer for 3 minutes.

Stir in paneer and bring back to a gentle boil to warm through. Sprinkle fenugreek leaves on top, drizzle with lemon juice, and serve hot.

Palak paneer goes well with lemon rice and cranberry pickle.

Onam Sadya

The annual Hindu celebration of the harvest culminates in an elaborate feast, a thali of twenty-six (or more) dishes, condiments, snacks, drinks, and sweets, showcasing seasonal vegetables and fruits, all placed on a banana leaf in a particular spot and ordered with rice as the anchor. Many of the recipes in this book are essential Onam feast dishes. In a book on south Indian thalis, I'd be remiss not to highlight the Onam sadya. It is, after all, the ultimate thali. However, I went into some detail on this wonderful cultural festival in my first book, *Coconut Lagoon*— if you are curious you can learn more about it there.

My Thalis

FEASTS ON A PLATE

You may use this section as a guide only, for there are no hard rules to building a thali. The recipes in this book can come together on one plate in any manner you choose! But care is generally paid, in the creation of a thali, to what's in season, what's available locally, what's affordable, and, of course, what's the occasion. Other things to keep in mind:

CONTRASTING TEXTURES

A thali should contain crunchy things and soft things, saucy things and dry things.

VISUAL APPEAL

There should be a pleasing contrast of color on the plate or leaf.

HARMONY IN SPICING

The overall chile-heat will vary from region to region, but there should be elements in a thali that add fire and elements that soothe.

NUTRITION

A complete meal should also mean a meal that provides complete nourishment.

DELICIOUSNESS

Yes, indeed, it should taste wonderful!

Plant-Based for the Planet

An Adventurous Feast

Winter Warmers

Game Night

① Pala Goat Curry
(page 119)

② Pumpkin Ularthiyathu
(page 67)

③ Nadan Cucumber
Pulissery (page 158)

④ Meatball Curry
(page 116)

⑤ Kappa Mash
(page 64)

⑥ Roasted Beet Salad
(page 59)

⑦ Dal Makhani
(page 46)

⑧ Potato Bonda
(page 130)

⑨ mor milagai (page 15)

⑩ Basmati Rice
(page 31)

TO DRINK: Old Monk
Lassi (page 174)
or cold beer

Off To Sea

Sunday Brunch

Summer Picnic

Fish

With hundreds of miles of coastline, a network of rivers, lagoons, and lakes, and a strong fishing industry, Kerala is blessed with abundant seafood and delicious fish-based dishes. When I was a boy, my family would eat fish four or five times a week, my mother sending me or a brother on our bikes to the market for the catch of the day—maybe a bag of sardines, a small mackerel, or a few smelts. As my mum cleaned the fish the neighborhood cats were her companions, alert to dropped trimmings. The fish would then be spun out to feed six, in a saucy curry or a simple fry-up with matta rice. In the rainy seasons we'd eat salted fish. And when money was tight, dinner would be fish pickle with rice.

Fried Tilapia

SERVES 6

Don't skimp on the time the fish sits in this magnificent marinade: the longer, the better, to my mind. In our home, we often buy more fillets than we need for one meal and freeze the rest, smothered in this paste, ready for the next fry-up. The fried onions, rolled in the leftover marinade (we don't waste a bit of it!), make a fine finish.

2 lbs tilapia (about 4 fillets)

1 Indian or small shallot, finely chopped

2 Tbsp Ginger-Garlic Paste (page 23)

1 Tbsp Kashmiri chili powder

2 tsp black pepper

1½ tsp ground turmeric

1 tsp salt

1 Tbsp fresh lime juice

¼ cup coconut oil, plus extra if needed

1 large onion, sliced

1 lime, cut into wedges, for garnish

TO SERVE (OPTIONAL)
Lemon Rice (page 29)
Cranberry-Coconut Chutney (page 140)

Cut each tilapia fillet in half lengthwise. Pat dry with paper towels, then place on a baking sheet or in a shallow bowl.

Using a food processor or a mortar and pestle, combine shallots, ginger-garlic paste, chili powder, black pepper, turmeric, salt, and lime juice. Process or pound into a smooth paste. Spread the paste on both sides of the fish. Cover fish and marinate in the fridge for at least 3 hours and up to 1 day. (Fillets can also be frozen in their marinade.)

Heat oil in a large frying pan over medium-high heat, until oil is shimmering. Working in batches to avoid over-crowding, add fish to the pan (reserving excess marinade) and fry for 3 minutes on each side, adding more oil if necessary. Set cooked fillets on a serving plate and keep warm.

Meanwhile, add onions to the leftover marinade and mix well. Add a little more coconut oil to the hot frying pan if necessary, and add the onions, sautéing for 3 minutes, or until lightly browned and wilted.

Serve fish with hot onions and lime wedges. Lemon rice and cranberry-coconut chutney would pair very well.

Konkan Shrimp Curry

SERVES 4

A tangy, fiery, and delicious curry from the southern coastal state of Goa. Vegetarians can easily swap in tofu for the shrimp and you can swap out a few chiles (or remove their seeds) for a milder version.

1 cup fresh or frozen (thawed) grated coconut

2 Tbsp ground coriander

1 Tbsp cumin seeds

1 Tbsp Kashmiri chili powder

1 tsp ground turmeric

3 Tbsp coconut oil

8 Indian or small shallots, coarsely chopped

4 Indian or Thai green chiles

10 curry leaves

2 tomatoes, coarsely chopped (about 1½ cups)

⅓ cup Tamarind Paste (page 22)

1 lb medium shrimp, shelled and deveined

1½ tsp salt

TO SERVE (OPTIONAL)
Jeera Rice (page 29)

Sweet Mango Kalan (page 156)

In a small food processor, combine coconut, coriander, cumin seeds, chili powder, and turmeric. Grind into a fine powder, then add ½ cup water to create a smooth paste.

Heat oil in a heavy-bottomed frying pan over medium heat. Add shallots and sauté for 3 minutes. Add chiles, curry leaves, and the coconut paste and stir for another 3 minutes. Add tomatoes and ½ cup water and bring to a boil. Stir in tamarind paste.

Add shrimp and cook for 2–3 minutes, just until they turn pink. Season with salt and serve with jeera rice and sweet mango kalan (if using).

Salmon Vattichathu

SERVES 4

Vattichathu is a favorite of Kerala toddy shops, simple taverns where food tends to be fiery-hot to encourage the imbibing of the fermented coconut "toddy" drink. We cube fresh salmon for this recipe, though traditionally the fish used would be kingfish (or other mackerel), mahi mahi, or sardines, and the stew would be served in clay pots called *chatti*. If you find the chile count a little high for your liking, scale back!

2 Tbsp Kashmiri chili powder

1 Tbsp ground coriander

½ tsp black pepper

½ tsp ground turmeric

5 Malabar tamarinds (*kudampuli*)

1 Tbsp coconut oil

½ tsp black mustard seeds

¼ tsp fenugreek seeds

2 Indian or small shallots, coarsely chopped

6 cloves garlic, cut into matchsticks

2 Indian or Thai green chiles, halved lengthwise

1 (2-inch piece) ginger, peeled and cut into matchsticks

1 tsp salt

1 lb skinless salmon, cut into 2-inch cubes

30 curry leaves

TO SERVE (OPTIONAL)
Kappa Mash (page 64)
Desi Salad (page 159)

In a small bowl, combine chili powder, coriander, pepper, and turmeric. Add a bit of water to make a thick paste. Set aside.

In another small bowl, combine Malabar tamarinds with enough hot water to cover and set aside for 15–20 minutes to soften.

With measured spices and aromatics nearby, heat oil in a heavy-bottomed saucepan or a clay pot over medium-high heat, until oil is shimmering. (To test the heat, add a couple of mustard seeds. If they start to sizzle, the oil is ready.) Reduce heat to medium, add mustard seeds, and crackle them for just a few seconds until popping subsides, taking care they don't burn. Add fenugreek seeds and roast for another few seconds, then add shallots and sauté for 2–3 minutes, until golden brown. Add garlic, chiles, and ginger and sauté for another 2–3 minutes.

Add the reserved spice paste and sauté for 2–3 minutes, or until oil from the spices rises to the top. Add salt, the soaked Malabar tamarinds with their soaking water, and 2 cups water. Mix well and bring to a boil.

Add salmon and curry leaves. Cover, reduce heat to medium-low, and simmer for 5–7 minutes, until fish is just cooked.

This dish goes well with kappa mash and a fresh desi salad.

Lobster Moilee

SERVES 4

A *moilee* (pronounced "molly") is a simple, luxurious sauce for seafood, thought to be Goan in origin and traditionally served with rice, chapati, or lacy-thin rice crepes called *appam*. For this moilee, we've packed the sauce with juicy chunks of lobster. If you'd rather not bother with cooking, shelling, and cleaning whole lobster, you could use pre-shelled raw lobster meat, fresh or thawed from frozen, and cook the meat in the bubbling coconut sauce.

2 (1½-lb) lobsters

2 Tbsp coconut oil

4 green cardamom pods, cracked using a mortar and pestle

10 curry leaves

6 cloves garlic, cut into thin matchsticks

2 Indian or Thai green chiles, halved lengthwise

1 (1-inch) piece ginger, peeled and cut into thin matchsticks

6 Indian or small shallots, halved

1½ tsp all-purpose flour

½ tsp ground turmeric

½ tsp salt

2 cups coconut milk

Juice of ½ lime

8 cherry or grape tomatoes

1 green bell pepper, seeded, deveined, and cut into wedges

TO SERVE (OPTIONAL)
Basmati rice

Heirloom Tomato Chutney (page 143)

Bring a large pot of salted water to a boil. Add lobsters and cook for 10–12 minutes.

Fill a large bowl with ice water. Carefully transfer lobsters into the ice water and set aside until cool enough to handle. Crack the shells, clean, and remove the tomalley (the green guts). Remove the meat and rinse it. Cut the body meat into medallions, keeping the meat from the claws intact if you wish.

Heat oil in a heavy-bottomed frying pan or wok over medium-high heat, until oil is shimmering. Add cardamom, reduce heat to medium, and sauté for 1–2 minutes, until fragrant. Add curry leaves, garlic, chiles, and ginger and sauté for 1 minute. Toss in shallots and cook for 2–3 minutes, stirring occasionally, until slightly wilted.

Stir in flour, turmeric, and salt and cook for 1 minute. Pour in coconut milk and bring to a boil. Reduce heat to medium-low and simmer, stirring constantly, for 4–5 minutes, until sauce has thickened. Add lobster meat, lime juice, tomatoes, and peppers and cook for another 2 minutes (or a few minutes more if using raw lobster meat) until warmed through.

Transfer to a serving dish. Serve with basmati rice and a side of heirloom tomato chutney, if desired.

Mangalore-Style Fish Curry

SERVES 4

This curry has roots in the port city of Mangalore, in the southern state of Karnataka. Traditionally, we'd use kingfish or fresh sardines, but you could use black cod, halibut, or grouper instead. Full credit to chef Easo Johnson of the Taj Bekal Resort & Spa, who gave me this recipe.

5 small Malabar tamarinds (*kudampuli*)

6 Tbsp coconut oil (divided)

½ tsp black mustard seeds

1½ tsp coriander seeds

½ tsp cumin seeds

2 tsp red chili powder

2 tsp Kashmiri chili powder

1 tsp ground turmeric

1 cup fresh or frozen (thawed) grated coconut

5 curry leaves

1 Indian or Thai green chile

1 onion, finely chopped

1 (2-inch) piece ginger, cut into matchsticks

6 cloves garlic, cut into matchsticks

½ cup chopped tomato

1 tsp salt

2 lbs firm white fish fillets, cut into 1-inch pieces

Chopped cilantro, for garnish (optional)

TO SERVE (OPTIONAL)

Neer Dosa (page 127)

Ginger Pickle (page 151)

In a small bowl, combine Malabar tamarinds with enough hot water to cover and set aside for 15–20 minutes to soften.

With measured spices nearby, heat 2 Tbsp oil in a medium frying pan over medium-high heat, until oil is shimmering. (To test the heat, add a couple of mustard seeds. If they start to sizzle, the oil is ready.) Reduce heat to medium, add mustard seeds, and crackle them for just a few seconds until popping subsides, taking care they don't burn. Immediately add coriander and cumin seeds and roast for a few seconds. Add chili powders and turmeric and cook for 30 seconds, until the smell of raw spice disappears. Add coconut and stir for 2 minutes, until warm and lightly colored.

Add contents of the pan to a food processor. Blend, adding enough water (about ¼–½ cup) to make a smooth, thick paste. Set aside.

Heat the remaining ¼ cup oil in a large frying pan or wok over medium-high heat. Add curry leaves, green chile, onions, ginger, and garlic and sauté for 1 minute. Stir in the reserved coconut paste, then add tomatoes and soaked Malabar tamarinds with the soaking liquid. Boil for 5 minutes, then season with salt.

Add fish, reduce heat to medium-low, and simmer for 5 minutes, or until just cooked. Transfer to a serving platter. Garnish with cilantro and serve, if you wish, with neer dosa and ginger pickle.

Shrimp Coconut Masala

SERVES 4

A dish that almost makes itself, it's so effortless. Soft shrimp and crunchy slivers of coconut in a thick masala can take center stage at a dinner party, with the added benefit that the sauce tastes even better when made ahead. Then all you need to do is toss in the shrimp, steam some rice, and bingo. I also love serving this as an appetizer: a single shrimp wrapped in coconut masala and perched on a pappadum. People go nuts.

Heat oil in a frying pan or wok over medium heat. Add curry leaves, garlic, onions, chiles, ginger, and coconut and sauté for 3 minutes, until onions are translucent. Stir in coriander, turmeric, chili powder, and black pepper and cook for 2 minutes, until the smell of raw spice disappears.

Increase heat to high, then add shrimp and salt and sauté for 2 minutes, or until shrimp turn pink. Remove from heat and stir in lime juice.

Transfer to a serving platter and garnish with cilantro and crushed pappadum (if using). Serve with pappadum or chapati and perhaps a lively green chutney.

1½ Tbsp coconut oil

15 curry leaves

6 cloves garlic, cut into matchsticks

2 onions, thinly sliced

2 Indian or Thai green chiles, halved lengthwise

1 (1-inch) piece ginger, cut into matchsticks

½ cup fresh or frozen (thawed) coconut shavings

1 Tbsp ground coriander

1 tsp ground turmeric

1½ tsp red chili powder

¼ tsp black pepper

1 lb medium shrimp, shelled and deveined

1½ tsp salt

Juice of 1 lime

Coarsely chopped cilantro, for garnish

Crushed pappadum, for garnish (optional)

TO SERVE (OPTIONAL)
Pappadum or Chapati (page 124)

Green Chutney (page 146)

Steamed Clams in a Mango Coconut Sauce

SERVES 4

My mum knew how to stretch an expensive ingredient like the small clams (*kakka*) that flourished in the waterways of Kerala: she'd simply bulk up the dish with more affordable stuff—like mangoes! Here, green (young, unripe) mango gives the sauce a sour note, while coconut milk smooths and sweetens. You could easily substitute shrimp, mussels, or scallops for the littlenecks.

1 cup diced green, unripe mango

Juice of ½ lime

1 Tbsp coconut oil

½ tsp black mustard seeds

10–12 curry leaves, plus extra for garnish

3 cloves garlic, chopped

2 Indian or Thai green chiles, finely chopped

2 Indian or small shallots, coarsely chopped

2 Tbsp peeled and coarsely chopped ginger

2 cups coconut milk

1 tsp ground turmeric

1 tsp salt

4 lbs fresh littleneck clams, scrubbed

TO SERVE (OPTIONAL)
Basmati rice

Combine mango and lime juice in a small bowl and set aside to marinate.

With measured spices and aromatics nearby, heat oil in a medium frying pan over medium-high heat, until oil is shimmering. (To test the heat, add a couple of mustard seeds. If they start to sizzle, the oil is ready.) Reduce heat to medium, add mustard seeds, and crackle them for just a few seconds until popping subsides, taking care they don't burn. Immediately add curry leaves, garlic, chiles, shallots, and ginger and sauté for 2 minutes.

Transfer this mixture to a large saucepan with a lid. Add a splash of coconut milk to the frying pan and deglaze, getting every scrap of oil and spice. Pour into the saucepan, then add the remaining coconut milk, turmeric, salt, and the marinated mango and bring to a boil. Add clams, cover, and reduce heat to medium-low. Simmer for 5–8 minutes, until clams have steamed open. (Discard any that don't open.)

Serve on basmati rice, if you wish, with a garnish of curry leaves.

Poultry and Meat

Meat traditionally plays a starring role in a celebration feast but is very much the understudy in an everyday rice meal. It was on Sundays, in my Indian home, when we'd indulge in meat. The butcher would set up a makeshift stand in the market, selling cuts of goat, beef, lamb, or pork. I remember watching with fascination as he'd butcher a whole beast, weighing out cuts and haggling over price. Then there might be a pork masala for Sunday lunch, stretched out with a pile of potatoes. Or meatballs, my dad's favorite, paddling in an aromatic gravy with pillowy parathas for sopping up the sauce. Sunday supper would be a meaty dish with enough leftovers for my banana-leaf-wrapped school lunch on Mondays. One-pot dishes, like the kid-friendly Kadai Chicken (page 110) bathed in spiced yogurt, were common pleasures, while long-simmered lamb or goat were true luxuries. And, of course, these recipes all take on a greater depth of flavor when eaten the next day.

Chicken Masala

SERVES 4

A masala is a spice blend and a masala is also a dish made *with* a spice blend. This chicken masala is not a difficult recipe, far from it, but you'll want to start with making the basic gravy and mixing up the house spice blend in advance. If you do so the day before, it will all come together quickly and brilliantly. I've suggested using boneless chicken thighs, but you could easily use a whole sectioned chicken or boneless breasts, adjusting the cooking time to suit the cut.

Heat oil in a large frying pan or wok over medium heat, until oil is shimmering. Add cardamom, cloves, and star anise and stir for 30 seconds, or until fragrant. Stir in curry leaves and chiles, then add gravy and mix well.

Add the spice blend and cook for 2–3 minutes, until the oils from the spices rise to the surface. Add tomatoes, chicken, and salt and stir to coat. Pour in 2 cups water and simmer for 10–15 minutes, until the chicken is cooked.

Transfer to a serving dish and serve with parathas, if you like. Rice pulao and a green chutney would also be brilliant with this dish.

2 Tbsp coconut oil

5 green cardamom pods

3 cloves

1 star anise

15 curry leaves

2 dried red chiles, snapped in half

2 cups Basic Gravy (page 23)

3 Tbsp House Spice Blend (page 20)

1 large ripe tomato, coarsely chopped

1 lb boneless, skinless chicken thighs, cut into bite-sized cubes

1 tsp salt

TO SERVE (OPTIONAL)

Malabar Parathas (page 126)

Rice Pulao (page 30)

Green Chutney (page 146)

Kadai Chicken

SERVES 4

A one-pot chicken recipe named for the traditional vessel in which it is cooked, made rich and tangy with yogurt and aromatic with garam masala. We use boneless thighs for this dish, but if you prefer bone-in thighs, or are using a whole, sectioned chicken, cook the stew a little longer. As with most dishes of amazing flavor, marination is key.

1 lb boneless, skinless chicken thighs, cut into bite-sized pieces

½ cup Greek yogurt, plus extra if needed

2 Tbsp Ginger-Garlic Paste (page 23)

1 Tbsp salt (divided)

2 tsp coconut oil

½ tsp cumin seeds

4 large onions, finely chopped

4 large tomatoes, puréed (roughly 2 cups)

2–3 tsp House Spice Blend (page 20)

½ tsp Garam Masala (page 20)

Cilantro, for garnish

TO SERVE (OPTIONAL)

Malabar Parathas (page 126)

Cucumber-Tomato Raita (page 155)

In a bowl, combine chicken, yogurt, ginger-garlic paste, and 1½ tsp salt. Mix well, cover, and marinate in the fridge for at least 6 hours.

Heat oil in a large frying pan or wok (or a kadai if you have one!) over medium-high heat, until oil is shimmering. Reduce heat to medium, add cumin seeds, and roast for 20–30 seconds until fragrant. Add onions and sauté for 3–4 minutes, until golden brown. Add puréed tomatoes and the spice blend and cook for 4–5 minutes, until the liquid has slightly reduced and thickened and the oils from the spices rise to the surface.

Add chicken (no need to wipe off the yogurt) and remaining 1½ tsp salt. Mix well, cover, and reduce heat to low. Simmer for 20–30 minutes, until chicken is cooked through. (If you find it getting dry, add more yogurt and water.) Stir in garam masala and garnish with cilantro.

This goes well with parathas and cucumber-tomato raita.

Kozhi Mappas

SERVES 4

A rock star Kerala chicken dish with amazing aromas. Once you've organized your mise en place (measuring out all the ingredients), this recipe comes together easily and appeals to everyone. We've used a whole chicken, chopped into smaller pieces to give lots of surface area for the spices and aromatics to dig in, but bone-in chicken thighs would work as well.

3 Tbsp coconut oil (divided)

5–6 cloves garlic, thinly sliced

4 Indian or Thai green chiles, halved lengthwise

1 (2-inch) piece ginger, peeled and cut into matchsticks

3 onions, sliced

20 curry leaves (divided)

2 tsp ground coriander

1 tsp Garam Masala (page 20)

½ tsp ground turmeric

½ tsp ground fennel

1 (2-lb) whole chicken, cut into 10–12 pieces, rinsed well, and patted dry

1 Tbsp salt

1 cup coconut milk

1 tsp black mustard seeds

2–3 Indian or small shallots, finely chopped

TO SERVE (OPTIONAL)
Matta Rice (page 33)

Ginger Pickle (page 151)

Heat 2 Tbsp oil in a large frying pan or wok over medium heat, until oil is shimmering. Add garlic, chiles, ginger, onions, and 10 curry leaves and sauté for 2–3 minutes, until onions have browned. Stir in coriander, garam masala, turmeric, and fennel and cook for 1 minute, until the smell of raw spice disappears. Add chicken, salt, and ½ cup water, then cover and simmer for 40 minutes, or until chicken is cooked through. Stir in coconut milk and heat through.

To finish, have measured seeds and aromatics nearby. Heat the remaining 1 Tbsp oil in a small frying pan (or tadka) over medium-high heat, until oil is shimmering. (To test the heat, add a couple of mustard seeds. If they start to sizzle, the oil is ready.) Reduce heat to medium, add mustard seeds, and crackle them for just a few seconds until popping subsides, taking care they don't burn. Immediately add shallots and the remaining 10 curry leaves and sauté for 3–4 minutes, until shallots turn golden brown. Pour this mixture over the stew and stir to combine.

You could serve this with matta rice and ginger pickle.

Pork and Potato Masala

SERVES 4

If ever there was the perfect dish with a good beer, this is it. In true Indian fashion (and for anyone budget-aware!), we stretch the pork to feed a crowd with a pile of potatoes. Make sure you use a fatty cut of pork for this recipe. The potatoes will love it.

SEASONING BLEND
3 Tbsp ground coriander
2 tsp ground turmeric
1½ tsp red chili powder
1 Tbsp salt

PORK AND POTATO MASALA
1½ lbs pork butt or shoulder, cut into 1-inch cubes
3 Tbsp coconut oil (divided)
2 onions, sliced (divided)
2 Tbsp Ginger-Garlic Paste (page 23, divided)
1 qty Seasoning Blend (see here, divided)
½ tsp black pepper

2 tsp Garam Masala (page 20)
15 curry leaves
2–3 Yukon Gold (or other yellow-fleshed) potatoes, peeled and cut into 1-inch cubes (about 2 cups)

TO SERVE (OPTIONAL)
Jeera Rice (page 29) and/or Malabar Parathas (page 126)
Pineapple Achar (page 145)

SEASONING BLEND Combine spices in a small bowl and mix well.

PORK AND POTATO MASALA Pat pork dry with a paper towel.

Heat 1 Tbsp oil in a large heavy-bottomed frying pan over medium-high heat, until oil is shimmering. Add pork, half the onions, 1 Tbsp ginger-garlic paste, half the seasoning blend, pepper, and garam masala. Stir well and cook for 4–5 minutes, until pork is seared on all sides.

Add curry leaves and enough water to cover the meat. Bring to a boil, then reduce heat to medium-low. Cover and simmer 30 minutes, until meat is cooked through. Add potatoes and cook for another 20–30 minutes, until meat and potatoes are tender. Set aside.

Heat the remaining 2 Tbsp coconut oil in another large frying pan or wok over medium-high heat. Add the remaining onions and sauté for 2–3 minutes, until onions have wilted. Add the remaining 1 Tbsp ginger-garlic paste and sauté for another minute. Reduce heat to medium-low, then add remaining seasoning blend and cook for 1–2 minutes, stirring, until the smell of raw spice disappears. Add the pork and potatoes. Stir gently for 5 minutes, until liquid evaporates and meat and potatoes begin to brown and get a little crispy.

This dish is best served with jeera rice, parathas, and a side of pineapple achar.

Pork Vindaloo

SERVES 6

This was the very first stew I cooked for my family after returning home from culinary school. I wanted to show off a new dish I had mastered, one that was totally unfamiliar to my parents. The introduction was a hit, and this remains a family favorite. Vindaloo hails from the state of Goa and has a reputation for scorching heat, though really it's the tang of the vinegar and earthy sweet flavors of the cloves and cinnamon that are the stars.

2 lbs pork shoulder or butt, cut into 1-inch cubes

5 dried red chiles, snapped in half

2 onions, chopped

15 cloves

2 cinnamon sticks

1 Tbsp cumin seeds

1 Tbsp black mustard seeds

1 Tbsp Kashmiri chili powder

1½ tsp black peppercorns

½ tsp ground turmeric

½ cup white wine vinegar (divided)

3 Tbsp Ginger-Garlic Paste (page 23)

1½ Tbsp salt (divided)

3 Tbsp coconut oil

6–8 Indian or small shallots, chopped

20 curry leaves

6 large tomatoes, chopped (about 3 cups)

1 tsp Garam Masala (page 20)

2 Tbsp chopped cilantro

TO SERVE (OPTIONAL)
Jeera Rice (page 29)

Charred Corn Raita (page 155)

Pat pork dry with a paper towel and place in a large bowl. Set aside.

In a food processor, combine chiles, onions, cloves, cinnamon, cumin seeds, mustard seeds, chili powder, peppercorns, turmeric, ¼ cup vinegar, and ginger-garlic paste. Process to a smooth paste.

Add the paste and 1½ tsp salt to the bowl of pork and stir to coat. Cover and marinate in the fridge for at least 6 hours and up to 24 hours.

Heat oil in a large heavy-bottomed pan over medium-high heat. Add shallots and curry leaves and sauté for 2 minutes. Add pork with its marinade and stir for 10 minutes, until meat has browned. Reduce heat to low, then add tomatoes. Cover and cook for another 45 minutes, or until meat is tender. Stir in the remaining ¼ cup vinegar, the remaining 1 Tbsp salt (or to taste), garam masala, and cilantro.

Serve with jeera rice and a side of charred corn raita, if you wish.

MAKE IT QUICK! The pressure cooker is an essential piece of kitchen equipment in every Indian household and can save you significant prep and cook time. Once the pork has marinated, proceed using a pressure cooker and follow the manufacturer's instructions for suggested cooking time.

Meatball Curry

SERVES 4

My dad's favorite curry—a weekday comfort dish, kid-friendly, easy to make ahead, and excellent with parathas or chapatis. Temper the chile-fire to suit your diners.

1 lb lean ground beef

6 Indian or small shallots, finely chopped

3 Indian or Thai green chiles, finely chopped

1 (½-inch) piece ginger, finely chopped

1½ tsp Garam Masala (page 20, divided)

1½ tsp salt, plus extra to taste

1 tsp black pepper, plus extra to taste

2 tsp coconut oil

1 cup Basic Gravy (page 23)

15 curry leaves

½ tsp ground turmeric

1 cup coconut milk

Juice of ½ lemon

TO SERVE (OPTIONAL)
Malabar Parathas (page 126)

Baby Arugula and Sunflower Seed Raita (page 155)

In a bowl, combine beef, shallots, chiles, ginger, 1 tsp garam masala, salt, and pepper and mix well. Divide into 12 balls.

Heat oil in a large heavy-bottomed frying pan over medium heat, until oil is shimmering. Add meatballs and sear for 6 minutes, using tongs to rotate them to ensure even browning. Using a slotted spoon, transfer browned meatballs to a plate and set aside.

Drain oil from pan and return to the heat. Add gravy and stir well. Stir in curry leaves and turmeric and cook for 1–2 minutes. Pour in coconut milk and 1 cup water and bring to a boil. Add meatballs to the sauce and simmer for 8 minutes, or until cooked through. Sprinkle the remaining ½ tsp garam masala on top. Add lemon juice and season to taste with salt and pepper.

This dish goes well with parathas and a side of arugula and sunflower seed raita.

Pala Goat Curry

SERVES 4

Whenever our chef de cuisine Rajesh Gopi has time, he makes this rich dark curry for the staff meal, scented with his secret spice blend (page 20) and doused with a pan full of tempered aromatics. Named in honor of his hometown of Pala, on the banks of the Meenachil River, there's a touch of Goan flavor in the vinegar tang and noticeable heat from the fresh green chiles. If you don't have access to goat meat, you could use lamb.

2 lbs bone-in stewing goat meat

20 curry leaves

12 Indian or small shallots, finely chopped

12 cloves garlic, crushed

6 Indian or Thai green chiles, halved lengthwise

1 (2-inch) piece ginger, crushed

3 Tbsp House Spice Blend (page 20)

1 Tbsp salt, plus extra to taste

½ cup plain yogurt

1 Tbsp coconut oil

1 tsp black mustard seeds

½ tsp ground turmeric

2 tomatoes, coarsely chopped

2 Tbsp white wine vinegar or coconut vinegar

TO SERVE (OPTIONAL)

Malabar Parathas (page 126)

Ginger Pachadi (page 144)

In a large bowl, combine goat meat, curry leaves, shallots, garlic, chiles, ginger, spice blend, salt, and yogurt. Mix well. Cover and leave to marinate in the fridge for at least 6 hours and up to 1 day.

Add goat with its marinade to a heavy-bottomed saucepan. Add enough water to cover and bring to a boil. Reduce heat to medium-low and cover. Gently simmer for 1 hour, stirring occasionally, until meat is falling off the bone tender. (Add more water if it dries out.)

With measured spices nearby, heat oil in a medium frying pan over medium-high heat, until oil is shimmering. (To test the heat, add a couple of mustard seeds. If they start to sizzle, the oil is ready.) Reduce heat to medium, add mustard seeds, and crackle them for just a few seconds until popping subsides, taking care they don't burn. Remove pan from heat and stir in turmeric. Add tomatoes with a pinch of salt. Return pan to medium heat and cook for 2 minutes. Pour this mixture over the goat curry, then add vinegar and mix well.

Serve with parathas and ginger pachadi (if using).

MAKE IT QUICK! The pressure cooker is an essential piece of kitchen equipment in every Indian household and can save you significant prep and cook time. If you'd like to cook the curry in a pressure cooker, follow the manufacturer's instructions for suggested cooking time.

Kashmiri Lamb

SERVES 6

This gently spiced, deeply fragrant lamb stew hails from India's most northerly region of Kashmir. The marination of the meat in the spiced yogurt is key to the flavor and tenderness, so budget time for that step.

2 lbs boneless lamb leg or shoulder, cut into ½-inch cubes

3 Tbsp Ginger-Garlic Paste (page 23, divided)

1½ cups plain yogurt (divided)

Juice of 1 lime

1 Tbsp salt

3 Tbsp Ghee (page 22)

4 Indian bay leaves

2 cinnamon sticks

10 green cardamom pods

1 black cardamom pod

3 onions, chopped

3 Tbsp ground coriander

2 Tbsp Kashmiri chili powder

1 Tbsp Garam Masala (page 20)

1 Tbsp ground fennel

1–2 tomatoes, puréed (about 1 cup)

Cilantro leaves, for garnish

TO SERVE (OPTIONAL)
Rice Pulao (page 30)

Pomegranate-Mint Raita (page 154)

In a large bowl, combine lamb, 2 Tbsp ginger-garlic paste, ½ cup yogurt, lime juice, and salt. Mix well, then cover and marinate in the fridge for at least 6 hours and up to 1 day.

With measured spices and aromatics nearby, heat ghee in a wok or heavy-bottomed frying pan over medium heat, until sizzling. Add whole spices and roast for 1 minute, or until fragrant. Add onions and cook for 3–4 minutes, until golden brown.

Add the remaining 1 Tbsp ginger-garlic paste and cook for 2 minutes. Stir in the lamb with its marinade. Increase heat to high and sear the meat, stirring constantly, for about 10 minutes.

Add coriander, chili powder, garam masala, and fennel and cook for 1 minute, or until the smell of raw spice disappears. Pour in tomatoes and 1 cup water. Mix well, cover, and simmer for 30–40 minutes, until lamb is tender. Add more water if stew becomes dry. Fold in the remaining 1 cup yogurt and bring back to a gentle boil.

Remove from heat, transfer to a serving dish, and garnish with cilantro. Serve with rice pulao and pomegranate-mint raita, if you like.

Breads, Snacks, and Crunchy Things

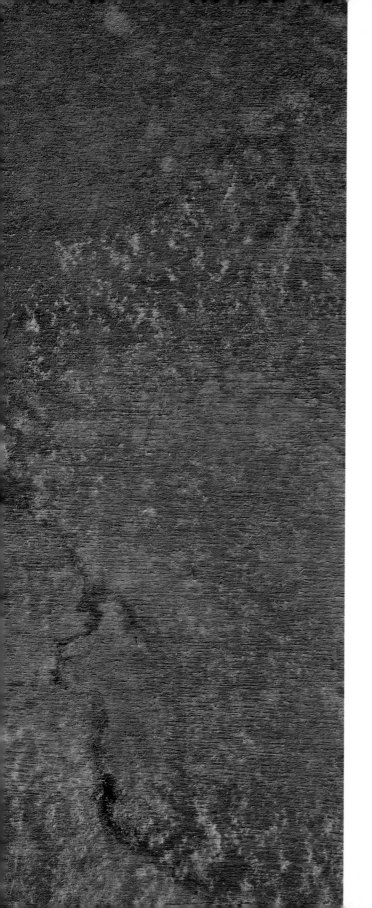

There are about thirty different breads in India—flat and puffy, soft and crisp, pan-fried, deep-fried, or baked—each unique to its region. As we eat with our fingers in most of India, bread is more than a staple: it is the scoop used to ferry food to the mouth, the mop for the last dribs of a luscious sauce. This chapter also contains recipes for some of our favorite crunchy things—teatime treats, late-night snacks, or dinner-party starters served with an array of chutneys.

Chapati

SERVES 8

Other than the kneading, which is critical for a soft result, chapati dough is a cinch to make. Traditionally, a *chakla* and *belan* (page 11) are used to make chapatis, but a cutting board and any rolling pin will do fine.

1 cup atta or finely milled wheat flour, plus extra for dusting

Pinch of salt

Combine flour and salt in a bowl. Gradually add in ¼ cup lukewarm water, mixing and kneading by hand for at least 2 minutes or until dough is soft but not sticky. (If sticky, add a bit more flour.) Cover the dough with a dish towel and set aside for at least 10 minutes to rest.

Divide dough into 8 pieces. On a lightly floured work surface, roll each piece of dough into a thin circle, about ⅛ inch thick.

Heat a flat griddle or nonstick frying pan over medium heat. Add a round of dough to the pan. Cook for 30 seconds, using the back of a spoon to press the surface in different spots, until bubbles form. Flip and cook for another 30 seconds, pressing the spoon down from spot to spot. Flip again and cook until the chapati bubbles are pale brown. Transfer to a plate and repeat with remaining rounds of dough.

Malabar Parathas

MAKES 16

Coiled, layered, and buttery—crunchy and caramelized on the outside, soft and flaky on the inside—the paratha is our star bread in south India, perfectly suited to scooping up coconut-rich curries, loose dal stews, or hearty meat masalas.

4½ cups all-purpose flour, plus extra for dusting

2 large eggs

½ cup milk

1 Tbsp sugar

2 tsp salt

¼ cup warm Ghee (page 22), plus extra for rubbing and brushing

Place flour in a large bowl and make a well in the center.

In a second bowl, whisk together eggs, milk, sugar, and salt. Pour the mixture into the well and combine using your hands. Pour in ghee and 1 cup room-temperature water and mix until a slightly sticky dough forms.

Transfer dough to a lightly floured work surface and knead for 5 minutes, until soft, smooth, and no longer sticky. (Add more flour if needed.) Divide dough into 16 balls. Oil your hands with a little ghee and rub over each ball. Cover balls with a clean dish towel and allow to rest for 30 minutes.

Using a rolling pin, roll out a dough ball to a ⅛-inch thickness. Using a knife or pastry cutter, cut ½-inch-wide strips of dough. Wind one strip around your finger. Add a second strip over top and continue winding, until all the strips for one dough ball are used. Place on parchment-lined baking sheet. Repeat cutting and coiling with remaining dough balls until you have made 16 coiled raw parathas. Cover with plastic wrap or a damp towel and set aside at room temperature for 2–3 hours.

Using a rolling pin, gently roll out the raw parathas to about ¼ inch thick. Heat a large frying pan over medium heat. Add a raw paratha and cook for 1 minute. Brush a little ghee on the surface, flip and fry for another minute, until the paratha forms light brown spots on the surface (from the caramelization of the sugar). Flip again a few more times, until both sides are golden brown and crispy. Transfer to a plate.

While the bread is still hot, use a clapping motion with your hands to gently scrunch its edges toward the center. (This releases the layers, making the paratha lighter and flakier.) Wrap the paratha in aluminum foil to keep warm, then repeat with the remaining raw parathas.

Place parathas in a basket and serve immediately.

Neer Dosa

MAKES 10

Here's an easy recipe for the great south Indian dosa, without the long fermentation of the traditional recipe. This one is lighter and goes beautifully with any rich coconut milk curry. A great bread for those who want gluten-free.

1 cup patna rice, washed
½ cup fresh grated coconut
½ tsp salt
2 Tbsp Ghee (page 22)

In a large bowl, combine patna rice and enough water to cover and soak for 4–5 hours. Drain, then rinse under cold running water.

Transfer rice to a small food processor, add ½ cup water and grind into a thick paste. Add coconut and salt and process again to a very fine consistency.

Transfer to a bowl and gradually whisk in 2 cups water, until you have a thin batter about the consistency of buttermilk. If necessary, add up to ¼ cup more water to achieve this consistency.

Have ghee next to the stove and a tray ready to receive the dosa. Heat a nonstick pan over medium heat. Using a paper towel, carefully spread a drizzle of ghee all over the pan. Give the batter a good stir and, using a small ladle, pour it in a thin stream to just cover the pan, tilting pan as you pour to fill in all the holes. Drizzle a little ghee around the edges with a spoon. Once the edges of the dosa start to come up off the pan, fold the dosa in half and in half again and remove to the tray. Repeat with remaining batter, stirring the batter every time, and lightly oiling the pan between each dosa. Take care not to stack the dosa, as they may stick together.

Aloo Tikki

MAKES 8–10

These bring back memories of being stuck in Delhi in March of 2020 while hosting a food tour of India. We were an anxious bunch as the world shut down, desperate for ways to get home. To cheer us all up, our hostess fried up these crunchy treats and served them with tamarind chutney. It worked. For a while, anyway. Aloo tikki are popular street food fare in north India: crispy potato patties stuffed with a spiced pea filling and delicious dunked in pretty much any chutney.

4 Yukon Gold (or other yellow-fleshed) potatoes, peeled and quartered

½ tsp grated ginger

¾ tsp salt

2 Tbsp breadcrumbs, plus extra if needed

½ cup frozen peas

½ tsp Chaat Masala (page 21)

½ tsp Garam Masala (page 20)

¼ tsp red chili powder

Small handful of cilantro, chopped

Vegetable oil, for frying

TO SERVE (OPTIONAL)
Date-Tamarind Chutney (page 148), Green Chutney (page 146), and/or Heirloom Tomato Chutney (page 143)

Add potatoes to a large saucepan of salted water. Bring to a boil and cook for 15–20 minutes, until tender. Drain, then return to pan and allow to dry out from their residual heat before mashing.

Transfer potatoes to a bowl, add ginger and salt, and mash. Add breadcrumbs and mix until potatoes no longer feel sticky in your hands. Add more breadcrumbs if needed.

Divide potato mixture into 8–10 balls.

Bring a saucepan of water to a boil. Add peas and cook for 30 seconds. Drain, then transfer to a bowl and mash with a fork. Add chaat masala, garam masala, chili powder, and cilantro and mix well.

Make a small depression with your finger in a potato ball, fill with a teaspoon of spiced peas, and close the potato back over the depression, pinching to seal the filling inside. Gently flatten into a patty, about ½ inch thick. Repeat with remaining potato balls and filling.

Add oil to a depth of 1 inch to a deep frying pan over medium-high heat. Heat until it reaches a temperature of 350°F. Working in batches to avoid overcrowding, gently lower patties into the pan and shallow-fry for 2–3 minutes on each side, until crisp. Using a slotted spoon, transfer aloo tikki to a plate lined with paper towels and drain excess oil.

Serve with your choice of date-tamarind chutney, green chutney, or heirloom tomato chutney—or all three!

Potato Bonda

SERVES 4

A famously addictive south Indian tea-time snack, bonda are spiced potato balls wrapped in a crunchy casing of gram flour. Walking home from school, the mom-and-pop food stalls in my village would be frying up bonda, and their glorious aroma would set a young boy's stomach grumbling.

POTATO BALLS

4 Yukon Gold (or other yellow-fleshed) potatoes, peeled and quartered

2 Tbsp vegetable oil

½ tsp black mustard seeds

10 curry leaves

2 Indian or Thai green chiles, finely chopped

1 onion, finely chopped

1 tsp salt

1 tsp ground turmeric

½ tsp red chili powder

1 Tbsp chopped cilantro

GRAM BATTER

1 cup gram flour

½ tsp salt

½ tsp ground turmeric

½ tsp red chili powder

¼ tsp asafetida powder

ASSEMBLY

Neutral oil such as vegetable or peanut oil, for deep-frying

TO SERVE (OPTIONAL)

Date-Tamarind Chutney (page 148) or Cranberry-Coconut Chutney (page 140)

MAKE IT AHEAD! These are popular at parties and can be made in advance so you're spending less prep time in the kitchen and more time with your guests. Make the mashed potato balls and batter ahead of time and refrigerate. As your guests arrive, simply dunk the potatoes in the batter and fry to order.

POTATO BALLS Put potatoes in a saucepan of salted water and bring to a boil. Cook for 15–20 minutes, or until tender. Drain, then return to pan and allow to dry out from their residual heat before mashing.

With measured spices and aromatics nearby, heat oil in a small frying pan (or tadka pan) over medium-high heat, until oil is shimmering. (To test the heat, add a couple of mustard seeds. If they start to sizzle, the oil is ready.) Reduce heat to medium, add mustard seeds, and crackle them for just a few seconds until popping subsides, taking care they don't burn. Immediately add curry leaves, chiles, and onions and sauté for 5 minutes. Stir in salt, turmeric, and chili powder.

Add this mixture to the mashed potatoes and combine well. Stir in cilantro and set aside.

GRAM BATTER In a bowl, combine all ingredients. Add ¾ cup water and mix until smooth. The consistency should be between a pancake and a crepe batter and thick enough to coat a spoon. If needed, add more water to get the right consistency.

ASSEMBLY Heat enough oil in a deep fryer or deep, heavy-bottomed saucepan to cover the potato balls (about 4 inches deep) until it reaches a temperature of 350°F.

Meanwhile, make lime-sized balls of the potato mixture. Dip each ball into the gram batter to coat. Using a slotted spoon, carefully lower the balls into the hot oil. If necessary, work in batches to avoid overcrowding. Deep-fry for 2–3 minutes, turning once, until light golden brown. Using the same slotted spoon, transfer the balls to a plate lined with paper towels and drain excess oil.

Serve with your choice of chutney.

Pappadum

When I was a kid, my mother would give me or a brother 50 paises to buy fresh pappadum for the day's meals. Every neighborhood would have a house where these legume flour flatbreads were made, rolled paper thin and dried to a crisp in the hot sun. I remember the big blue tarp in the front yard of the pappadum vendor and the old woman crouched down, turning the raw round wafers over and over in the sunshine, rushing to cover them whenever the rains came. We'd buy a stack, they'd get wrapped in newspaper, and my mum would fry them up right away.

Many brands of pappadum are now factory made, but a women's cooperative called Lijjat has been making them by hand since 1959. Launched in Mumbai by seven women looking for a way to make a living from their cooking skills, today Lijjat employs tens of thousands and has eighty-two branches in seventeen states. It operates, still, as a women-run cooperative.

Pappadum play many roles in our meals and vary from region to region in their spelling, their ingredients, their spicing, their size, and the way they are cooked. In the south, they are typically smaller disks and unseasoned, while in the north they are often bigger and spicier.

At Thali, we typically use three types of pappadum: Madras pappadum are made simply of white lentils (*urad dal*), rice flour, and salt. These are best deep-fried for a few seconds in hot vegetable oil until they puff up. Punjabi masala pappadum (or papad as they are called in the north) are made of white lentils with asafetida powder and coarsely cracked pepper, while jeera pappadum/papad, also made of white lentils, are studded with cumin seeds and chiles. These pappadum are better grilled on the flat-top or barbecue or cooked right over a gas flame. They take a little longer to cook than the Madras pappadum, about 15 seconds on each side, depending on the heat level, and can also be shaped while still hot: rolled like a cigar or a cone, or made into a bowl to hold a salad or rice dish.

It takes a little practice but frying up raw pappadum at home is a skill worth learning. Whip up a few chutneys (pages 140–148) to go with them.

Vada

MAKES 8

Nothing beats the crunch of a great vada dunked in yogurt or a spicy tomato chutney. These are highly aromatic, lightly fiery south Indian fritters, studded with vegetables, chiles, and curry leaves and fried to a golden brown. Typically, vadas are devoured for breakfast or enjoyed with an afternoon chai, though many thalis begin with a vada as a sort of *amuse bouche*.

Here, we've shaped them into little donuts, but you could make them more disk-shaped if you'd rather not fuss with the hole.

1 cup white lentils (*urad dal*)

1 tsp salt

6 curry leaves, finely chopped

2 Indian or Thai green chiles, finely chopped

1 small onion, finely chopped

1 small carrot, finely chopped

1 (½-inch) knob ginger, peeled and finely chopped

1 Tbsp finely chopped cilantro

Neutral oil such as vegetable oil or peanut oil, for deep-frying

TO SERVE (OPTIONAL)

Plain yogurt or Basic Raita (page 154)

Heirloom Tomato Chutney (page 143)

Soak the lentils in cold water for at least 4 hours. Drain, then rinse under cold running water.

Place lentils and salt in a small food processor and grind into a fine paste, scraping down the sides of the bowl with a spatula as required.

In a large bowl, combine curry leaves, chiles, onions, carrots, ginger and cilantro. Add lentil paste and mix well. Cover, then set aside at room temperature for 2 hours.

Heat enough oil in a deep fryer or deep, heavy-bottomed saucepan to cover the patties (about 3 inches deep). Oil should reach a temperature of 350°F.

With damp hands, scoop a golf-ball-sized dollop of the dough and make a donut shape. Insert a thumb in the center to make a small hole. Gently lower each donut as you shape it into the hot oil, a few at a time, and fry for 2–3 minutes, until golden brown on each side. If necessary, work in batches to avoid overcrowding. Using a slotted spoon, transfer vada to a plate lined with paper towels and drain excess oil. Repeat until all the vada are made.

Serve immediately, perhaps with plain yogurt (or basic raita) and a bowl of heirloom tomato chutney.

Kerala Plantain Chips

MAKES ABOUT 2 CUPS (OR ENOUGH FOR 4 TO SNACK ON)

Whenever my Kerala family comes for a visit, they bring plantain chips. I know it sounds a bit crazy, but just the sound of people crunching down on these golden chips fills my heart! An essential part of any feast, any celebration in Kerala, it's a sound that's joyful and festive. During the harvest season, when plantain is abundant, there are stalls everywhere with bubbling vats of coconut oil, frying up freshly sliced plantain and sweetly perfuming the air.

1 Tbsp ground turmeric
1 Tbsp salt, plus extra to taste
3 firm green plantains
3 cups coconut oil

In a large bowl, combine turmeric, salt, and 4 cups water. Mix well and set aside.

To peel the plantain, trim off the ends and score the skin from top to bottom along its ridges, taking care not to cut into the flesh. Using a mandolin, thinly slice the plantains. Add plaintains to the turmeric water and set aside for at least 10 minutes to color and flavor the chips. Drain, then pat dry with paper towels.

Heat oil in a deep fryer or large, heavy-bottomed saucepan until it reaches a temperature of 350°F. (To test if oil has reached correct temperature, add a single slice of plantain: if oil starts bubbling and plantain immediately pops to the surface, the oil is ready.)

Line a baking sheet with paper towels. Working in batches to avoid overcrowding, carefully lower plantains into the hot oil. Deep-fry for 3–4 minutes, stirring often with a slotted spoon to prevent slices from sticking together, until the chips are crisp and light brown. Using the slotted spoon, transfer chips to the prepared baking sheet and drain excess oil. Repeat with remaining plantain. Season with more salt if you wish.

Once cool, chips can be stored in an airtight container for 2 weeks.

Condiments

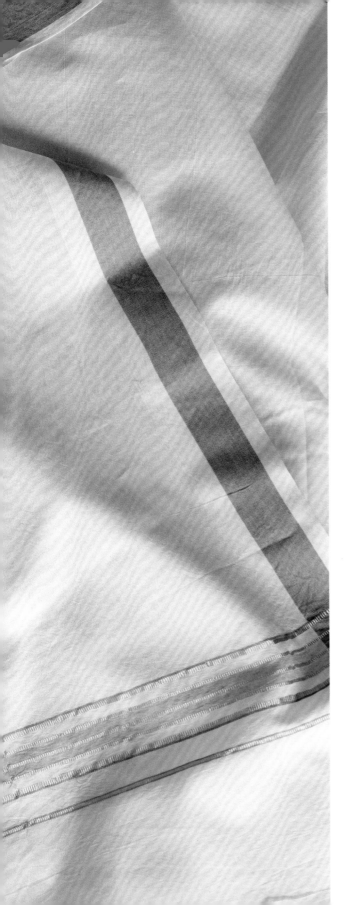

Chutneys, raitas, and pickles are essential sides to add flavor, serving to fire up a dish or chill it out. These are the extra dishes that take food from tasty to magnificent, adding layers of flavor, texture, and variety to a simple meal. And yes, usually adding chile heat too! We never had commercially made pickles when I was a child. In any fruit's season—mango, pineapple, lime, gooseberry, lovi-lovi—every household would preserve the bounty in a pickle. Even today, if there's a jar of homemade pickle in the house and I'm too weary to cook, I make a batch of rice, add a dollop of pickle, and *boom!* I have a meal.

Cranberry-Coconut Chutney

MAKES ABOUT 3 CUPS

When I discovered cranberries after moving to Canada, they reminded me of the much-loved lovi-lovi trees growing in our backyard in Thrissur. My mum would make a spicy chutney with their scarlet berries and, from time to time, a wine. This is her recipe for lovi-lovi (batoko plum) chutney, using our beautiful Ontario cranberries. We've made this recipe with fresh gooseberries as well.

7 Indian or Thai green chiles (adjust to your spice level)

4 Indian or small shallots, coarsely chopped

1 (1-inch) piece ginger, peeled and coarsely chopped

2 cups fresh or frozen (thawed) grated coconut

1½ cups fresh or frozen cranberries

1½ tsp salt

Add all ingredients to a food processor and process until smooth. Store chutney in an airtight container in the fridge for maximum of 2 days.

Heirloom Tomato Chutney

MAKES ABOUT 2–3 CUPS

We are incredibly lucky in Ottawa to be rich in farmers' markets. And in the fall, when their heirloom tomatoes are in abundance, we buy crates of them and make masses of this chutney.

3 large heirloom tomatoes, coarsely chopped

2 cloves garlic

2 Indian or Thai green chiles

1 onion, coarsely chopped

1 (¼-inch) piece ginger, peeled

1½ tsp coconut oil

1 tsp black mustard seeds

12 curry leaves

2 dried red chiles, snapped in half

1½ tsp white lentils (*urad dal*)

1 tsp Kashmiri chili powder

1 tsp salt

½ tsp sugar

In a food processor, combine tomatoes, garlic, green chiles, onions, and ginger and purée until smooth. Transfer to a saucepan and set aside.

With measured spices nearby, heat oil in a small frying pan (or tadka pan) over medium-high heat, until oil is shimmering. (To test the heat, add a couple of mustard seeds. If they start to sizzle, the oil is ready.) Reduce heat to medium, add mustard seeds, and crackle them for just a few seconds until popping subsides, taking care they don't burn. Immediately add curry leaves, red chiles, and lentils and roast until dal has taken on some color and the curry leaves are curled and crisp. Add chili powder and cook for 30 seconds, until the smell of raw spice disappears.

Pour this mixture into the saucepan of tomatoes and bring to a boil. Reduce heat to medium-low, add salt and sugar, and simmer for 3–4 minutes. Set aside to cool.

This chutney will sit happily in the fridge for about a week.

Ginger Pachadi

MAKES ABOUT 2½ CUPS

A traditional south Indian condiment, bright and fresh tasting, and a good companion for heavier dishes to help with digestion. Serve pachadi chilled with dishes such as Kadai Chicken (page 110), Pala Goat Curry (page 119), and Duck Biryani (page 38).

1½ Tbsp coconut oil

1 tsp black mustard seeds

15 curry leaves

2 dried red chiles, snapped in half

1 (3-inch) piece ginger, peeled and coarsely chopped

1 large onion, coarsely chopped

1 tsp salt (divided)

2 cups plain yogurt

With measured spices and aromatics nearby, heat oil in a small frying pan (or tadka pan) over medium-high heat, until oil is shimmering. (To test the heat, add a couple of mustard seeds. If they start to sizzle, the oil is ready.) Reduce heat to medium, add mustard seeds, and crackle them for just a few seconds until popping subsides, taking care they don't burn. Immediately add curry leaves and chiles and sauté for a few seconds. Add ginger and sauté for 2–3 minutes, until crisp. Add onions and ½ tsp salt and cook for 1 minute. Set aside to cool.

In a bowl, stir the remaining ½ tsp salt into the yogurt. Add the ginger-onion mixture. Remove from heat, then transfer to an airtight container and refrigerate until chilled. Store pachadi in the fridge for up to 3 days.

Pineapple Achar

MAKES ABOUT 3 CUPS

The toughest part of this recipe may be finding a sweet, ripe pineapple. Once you've accomplished this, the rest is easy. My wife Suma grew up in the pineapple-rich foothills of Munnar, and this sweet and sour (and spicy!) pickle is a treasured family recipe. Stained pink with Kashmiri red chiles and speckled with black mustard seeds, pineapple achar pairs beautifully with fish dishes.

½ cup vegetable oil

1 Tbsp + 1 tsp black mustard seeds (divided)

Handful of curry leaves

4 dried red chiles, snapped in half

2 tsp Kashmiri chili powder

1 tsp ground turmeric

1 ripe pineapple, peeled, cored, and cut into ¼-inch dice

½ cup white wine vinegar

1 Tbsp sugar

1½ tsp salt

1 tsp fenugreek seeds

1 tsp asafetida powder

With measured spices nearby, heat oil in a heavy-bottomed saucepan or wok over medium-high heat, until oil is shimmering. (To test the heat, add a couple of mustard seeds. If they start to sizzle, the oil is ready.) Reduce heat to medium, add 1 Tbsp mustard seeds, and crackle them for just a few seconds until popping subsides, taking care they don't burn. Immediately add the curry leaves, chiles, chili powder, and turmeric and cook for 1 minute, until the smell of raw spice disappears. Add pineapple, vinegar, sugar, and salt. Reduce heat to low and cook for 20 minutes, or until the pineapple has turned into a chunky purée.

Using a mortar and pestle, coarsely grind the remaining 1 tsp mustard seeds, fenugreek seeds, and asafetida. Add to the pan and stir.

You could serve this right away, but the flavors deepen if the pineapple achar is allowed to sit for at least a day. It can be kept for up to a week in the fridge.

Green Chutney

MAKES 1½ CUPS

Wonderfully versatile as a condiment for pappadum or Potato Bonda (page 130) or as a marinade for grilled paneer, chicken, or fried fish. Mixed with yogurt, it becomes a green sauce for a lamb curry or for dunking Aloo Tikki (page 129). Combined with grated coconut, it makes a lovely side for Dal Makhani (page 46) or Neer Dosa (page 127). Add a few more chiles if you like it fiery and sub in one bunch of fresh mint for one of cilantro if you wish.

2 bunches cilantro, leaves and tender stems only

2 Indian or Thai green chiles

2 cloves garlic

1 onion, chopped

1 ripe tomato, chopped

Juice of ½ lime

1 tsp salt

½ tsp sugar

In a food processor, combine cilantro, chiles, garlic, onions, and tomatoes. Add ¼ cup water and process into a fine paste. Transfer to a bowl, then stir in lime juice, salt, and sugar and mix well. Green chutney will last 3–4 days in the fridge.

FROM TOP: GREEN
CHUTNEY WITH
YOGURT, COCONUT
GREEN CHUTNEY,
VEGAN GREEN CHUTNEY

Date-Tamarind Chutney

MAKES ABOUT 1 CUP

Medjool dates give this all-purpose chutney body, the jaggery adds a caramel sweetness, and the tamarind brings sour notes for balance. Served with all manner of north Indian snacks, you can store this chutney in the fridge for up to two weeks.

½ cup seedless tamarind pulp

½ cup pitted Medjool dates

¼ cup grated jaggery or brown sugar

½ tsp ground cumin

½ tsp ground coriander

½ tsp red chili powder

½ tsp Chaat Masala (page 21)

½ tsp salt

In a saucepan, combine tamarind, dates, and 3 cups water. Bring to a boil, then reduce heat to low and simmer for 10 minutes, until tamarind and dates are softened. Add jaggery (or brown sugar) and stir to dissolve. Simmer for 2 minutes.

Add cumin, coriander, chili powder, chaat masala, and salt. Simmer for another 2 minutes. Remove from heat.

Add ½ cup water and use a food processor to blend the mixture into a smooth paste. The chutney should be thick enough to coat the back of a spoon. Strain through a sieve to remove any tamarind fibers, cool, and serve right away—or, for deeper flavor, the next day. You can store this chutney in the fridge for up to 2 weeks.

Cranberry Pickle

MAKES ABOUT 2 CUPS

Suma makes a big batch of this pickle at Christmas and gives it away as gifts. If you have coconut vinegar (available in most supermarkets), it adds a rounded sweetness and gentle tartness to the pickle.

3 cups fresh or frozen cranberries

1½ Tbsp salt (divided)

¼ cup vegetable oil

1 tsp black mustard seeds

2 Tbsp Ginger-Garlic Paste (page 23)

1 Tbsp Kashmiri chili powder

1 Tbsp red chili powder

1 tsp asafetida powder

¼ cup coconut vinegar or white vinegar

Combine cranberries, 1½ tsp salt, and 5 cups water in a large saucepan and bring to a boil. Reduce heat to medium-low and simmer for 10 minutes, until berries have softened. Drain, then set aside.

With measured spices and aromatics nearby, heat oil in a frying pan or wok over medium-high heat, until oil is shimmering. (To test the heat, add a couple of mustard seeds. If they start to sizzle, the oil is ready.) Reduce heat to medium, add mustard seeds, and crackle them for just a few seconds until popping subsides, taking care they don't burn. Immediately add ginger-garlic paste and sauté for 2 minutes, until golden brown. Add chili powders and cook for 1 minute, until the smell of raw spice disappears. Stir in asafetida.

In a bowl, combine vinegar and 1 cup water, then pour into the pan of spices. Bring to a boil, then add cranberries and the remaining 1 Tbsp salt. Reduce heat to medium-low and simmer for 3 minutes, stirring occasionally.

You can serve immediately or, time permitting, let the pickle rest a few days in the fridge to let the flavors meld. Can be stored in a jar with tight-fitting lid in the fridge for up to 4 weeks.

Ginger Pickle

MAKES 1 CUP

This pickle, also known as *inji curry*, takes a bit of work. But given it can keep for a long time, and given its sheer heavenly flavor, the results are worth the effort.

¼ cup tamarind pulp

1 cup coconut oil

1½ cups peeled and thinly sliced ginger

2 Indian or Thai green chiles, finely chopped

1 Tbsp red chili powder

1 tsp salt

½ tsp fenugreek seeds

1 tsp grated jaggery or brown sugar

1 tsp black mustard seeds

20 curry leaves

2 dried red chiles, snapped in half

Combine tamarind pulp and 1 cup hot water in a bowl and set aside for 15–20 minutes to soften. Loosen the pulp with your fingers to help break it up further, then press the pulp through a fine-mesh strainer into a bowl. Discard the contents of the strainer and retain the tamarind paste in the bowl.

Heat oil in a small saucepan over medium-high heat, until oil is shimmering. Reduce heat to medium, then add ginger and sauté for 5 minutes, or until dark brown and crispy. Adjust heat, if necessary, to prevent ginger from burning. Using a slotted spoon, transfer ginger to a baking sheet lined with paper towels and drain excess oil. Blot dry with more paper towels, then set aside to cool. Using a spice mill or mortar and pestle, grind into a fine powder. Reserve oil.

In another small saucepan, combine tamarind paste, green chiles, chili powder, salt, and fenugreek seeds and boil for 2 minutes. Add ginger powder and simmer for 3 minutes. Remove from heat, then stir in jaggery (or brown sugar).

With measured spices nearby, heat 2 Tbsp of reserved oil in a small frying pan (or tadka pan) over medium-high heat, until oil is shimmering. (To test the heat, add a couple of mustard seeds. If they start to sizzle, the oil is ready.) Reduce heat to medium, add mustard seeds, and crackle them for just a few seconds until popping subsides, taking care they don't burn. Immediately add curry leaves and red chiles and sauté for 1 minute. Pour tempered spices on top of ginger pickle and stir gently to combine.

Serve immediately or cool down and store in sterilized jars with tight-fitting lids in the fridge for up to 4 weeks.

Raitas

A soothing foil to the heat of a curry, a creamy contrast to the sour notes of a pickle, a chilled raita is an essential condiment for brightening any Indian meal. It's also terrific on its own, with a bowl of rice or a hunk of bread. There are hundreds of regional, seasonal variations, of course, but here we offer our basic raita recipe along with some of our favorite variations. Simply add key ingredients to our foundational recipe to create an entire realm of new flavor profiles.

① Charred Corn Raita
(page 155)

② Cucumber-Tomato
Raita (page 155)

③ Basic Raita
(page 154)

④ Pomegranate-Mint
Raita (page 154)

⑤ Boondi Mango Raita
(page 155)

⑥ Baby Arugula and
Sunflower Seed Raita
(page 155)

Basic Raita

MAKES 2 CUPS

2 cups plain yogurt

¼ tsp salt

½ tsp cumin seeds, coarsely crushed using a mortar and pestle

¼ tsp Chaat Masala (page 21)

Chopped cilantro, for garnish

Combine yogurt, salt, crushed cumin seeds, and chaat masala in a bowl. Stir well. Garnish with chopped cilantro.

This basic raita can be stored for up to a week in the fridge.

Variations

Pomegranate-Mint Raita

MAKES 3 CUPS

1 qty Basic Raita (see here) minus the chaat masala and cilantro

Seeds from 2 pomegranates (see Note)

Chopped mint leaves, for garnish

Pomegranate juice, for garnish

HOW TO SEED A POMEGRANATE Cut fruit in half. Holding the halves upside down over a strainer-lined bowl, use a wooden spoon or the back of a chef's knife to repeatedly tap the skin and dislodge the seeds. Remove any of the white pith that falls into the strainer. Drizzle a bit of pomegranate juice from the bowl on top of the raita just before serving. (Or simply drink it up!)

Boondi Mango Raita

MAKES 3 CUPS

Boondi is a crunchy besan-flour snack, available at Indian groceries.

1 qty Basic Raita
(see here)

1 ripe mango, cut into
¼-inch cubes

¼ cup boondi, for garnish

Cucumber-Tomato Raita

MAKES 4 CUPS

1 qty Basic Raita (see here)

2 ripe tomatoes, finely chopped

1 cucumber, finely chopped

1 red onion, finely chopped

1 tsp cumin seeds, coarsely ground using a mortar and pestle

Chaat Masala (page 21), for garnish

Charred Corn Raita

MAKES 3 CUPS

1 charred cob of corn, kernels removed (see Note)

1 qty Basic Raita (see here)

Juice of ½ lime

CHARRED CORN Season corn cob with a pinch of salt. Using a cast-iron frying pan, grill, or barbecue, roast the cob over medium-high heat, turning often, until kernels are evenly charred. Set aside to cool, then use a knife to shave off kernels.

Baby Arugula and Sunflower Seed Raita

MAKES 3 CUPS

1 qty Basic Raita (see here)

2 cups chopped baby arugula

Handful of roasted sunflower seeds, for garnish

Sweet Mango Kalan

MAKES ABOUT 3 CUPS

Sweet mangoes, plucked from our own trees in Kerala, are high up on the list of things I miss about life in India. This recipe brings me back to my garden and to my mother's kitchen. It takes ripe mangoes and simmers them in a spiced yogurt and coconut gravy, lightly sweetened with jaggery. It's unthinkable to celebrate the Onam sadya harvest festival without mango kalan included in the banana leaf feast.

1 cup fresh or frozen (thawed) grated coconut

¼ tsp cumin seeds

5 Indian or Thai green chiles

2 cups peeled and diced ripe mangoes

1 Tbsp grated jaggery or brown sugar

1 Tbsp salt (divided)

1½ tsp ground turmeric

1 tsp black pepper

2 cups plain yogurt

½ tsp ground fenugreek

2 Tbsp coconut oil

½ tsp black mustard seeds

15 curry leaves

2 dried red chiles, snapped in half

½ tsp fenugreek seeds

Using a mortar and pestle or a food processor, combine coconut, cumin seeds, and about ½ cup water. Pound or process to a smooth paste. Set aside.

Combine green chiles, mangoes, jaggery (or brown sugar), 1½ tsp salt, turmeric, and pepper in a heavy saucepan. Add 1½ cups water and simmer over medium heat, stirring occasionally, for 20 minutes, or until mango is very soft. Add the coconut-cumin paste, yogurt, and ground fenugreek. Reduce heat to medium-low and simmer for 5 minutes, or until the mixture is thick and creamy. Stir in the remaining 1½ tsp salt. Remove from heat and transfer to a serving bowl.

With measured spices and aromatics nearby, heat oil in a small frying pan (or tadka pan) over medium-high heat, until oil is shimmering. (To test the heat, add a couple of mustard seeds. If they start to sizzle, the oil is ready.) Reduce heat to medium, add mustard seeds, and crackle them for a few seconds just until popping subsides, taking care they don't burn. Immediately add curry leaves, red chiles, and fenugreek seeds and sauté, shaking the pan, for no more than 30 seconds. Pour the tempered spices on top of the mango-yogurt mixture. Mango kalan will keep for 3–4 days in the fridge.

Nadan Cucumber Pulissery

MAKES ABOUT 4 CUPS

A Kerala version of a north Indian raita, where the yogurt is combined with grated coconut and green chiles and finished with sizzling mustard seeds and crisped curry leaves. Served with rice and bread, pulissery is the ultimate comfort condiment!

6 cloves garlic, finely chopped

2 Indian or Thai green chiles, halved lengthwise

2 field cucumbers, peeled, seeded, and cut into 1-inch chunks

1 (1-inch) piece ginger, peeled and finely chopped

¼ tsp red chili powder

2 Indian or small shallots, chopped (divided)

2 tsp salt (divided)

1 tsp ground turmeric (divided)

¾ cup fresh or frozen (thawed) grated coconut

1 Tbsp cumin seeds

2 cups plain yogurt

2 tsp coconut oil

½ tsp black mustard seeds

10 curry leaves

1 dried red chiles, snapped in half

½ tsp fenugreek seeds

TO SERVE (OPTIONAL)
Matta Rice (page 33)

In a large saucepan, combine garlic, chiles, cucumbers, ginger, chili powder, half the shallots, 1 tsp salt, ½ tsp turmeric, and 1 cup water. Mix well. Bring to a boil, then reduce heat to medium-low and cover. Simmer for 5 minutes, or until cucumber has softened. Set aside.

Using a mortar and pestle or a food processor, combine remaining half of the shallots, coconut, cumin seeds, and ½ cup water. Pound or process to a fine paste. Add coconut mixture to the saucepan with the cucumber mixture and stir to combine.

In a bowl, whisk together yogurt, remaining 1 tsp salt, and remaining ½ tsp turmeric. Add to the saucepan and bring to a boil, stirring frequently. Transfer to a serving bowl.

With measured spices nearby, heat oil in a small frying pan (or tadka pan) over medium-high heat, until oil is shimmering. (To test the heat, add a couple of mustard seeds. If they start to sizzle, the oil is ready.) Reduce heat to medium, add mustard seeds, and crackle them for just a few seconds until popping subsides, taking care they don't burn. Immediately add curry leaves, red chiles, and fenugreek seeds. Sauté for a few seconds, or until curry leaves begin to curl. Pour these tempered spices over the yogurt curry and cover for 5 minutes to allow the flavors to infuse. Serve at room temperature or chilled over hot matta rice.

This cooling yogurt curry will keep in the fridge for a week.

Desi Salad

SERVES 4

This simple, traditional Indian salad is wonderful for adding a fresh tang to a meat curry or a fish fry.

2 red onions, thinly sliced	2 Tbsp vegetable oil
Salt	1 Tbsp white vinegar
2 Indian or Thai green chiles, finely chopped	1 tsp sugar
1 tomato, quartered, seeded, and thinly sliced	½ tsp Chaat Masala (page 21)
½ carrot, cut into matchsticks	Handful of cilantro, coarsely chopped

Combine onions and a pinch of salt in a fine-mesh strainer over a bowl. Set aside for 10 minutes, then gently press out the liquid.

In a serving bowl, combine chiles, onions, tomatoes, and carrots. Add oil, vinegar, sugar, chaat masala, ½ tsp salt, and cilantro. Mix well and serve chilled.

Sweets
and
Drinks

There is no doubt we like our sweets in India. At almost every meal, there is a fruity pudding or a scented custard, and a myriad of little pick-me-up treats can be enjoyed at a market stall, a street-side vendor, or at home with an afternoon chai. My mother and aunties lovingly competed for the prize of best dessert at any family gathering. And there's no doubt keeping hydrated is key in India, so drinks, savory and sweet, play a pleasurable part in cooling off in the tropical heat.

Jackfruit Payasam

SERVES 4

When the jack trees in our Thrissur garden were heavy with ripe fruit, my mother would make this dessert: a chilled pudding fragrant with cardamom and cumin, crowned with roasted cashews and raisins. I'm suggesting using either canned or frozen jackfruit, as sourcing the ripe, fresh fruit can be challenging and preparing it is a bit of a sticky job. If you want to make this dessert vegan, use vegetable ghee.

2 Tbsp Ghee (page 22) or vegetable ghee

Handful of raw cashew nuts

Handful of raisins

1½ cups canned or frozen jackfruit, drained and chopped

7 green cardamom pods, cracked open using a mortar and pestle

¼ cup grated jaggery or brown sugar

2 cups coconut milk

1 tsp ground ginger

1 tsp ground cumin

Heat ghee in a frying pan over medium-low heat. Add cashews and roast for 5 minutes, stirring, or until golden brown. Add raisins and sauté for 1 minute, or until plump. Using a slotted spoon, transfer nuts and raisins to a plate, keeping as much ghee in the pan as possible.

Add jackfruit to the same pan with the residual ghee and sauté over low heat for 3 minutes, until fruit has softened. Add cardamom and jaggery (or brown sugar). Sauté for another 2–3 minutes, until jaggery has dissolved and jackfruit is caramelized.

Pour in coconut milk and ½ cup water and bring to a boil. Reduce heat to medium-low and simmer for 5 minutes, stirring occasionally, until sauce has thickened to a pudding consistency. Stir in ginger and cumin. Set aside to cool, then transfer to a serving bowl and refrigerate until chilled.

Serve payasam with roasted cashews and raisins, and add a crumble of jaggery if you wish it sweeter.

Pineapple Kesari

SERVES 8

When my grandfather would return from playing cards at his club, he'd take a little nap. When he woke up, he'd make everyone coffee and this kesari—and whatever little trouble the day might have delivered, all was put to right. *Kesari* translates to yellow or saffron-colored, and *rava* is the south Indian word for semolina. We roast the semolina ahead of cooking it, as that results in a more flavorful and less gloopy dessert.

4 Tbsp Ghee
(page 22, divided)

Handful of raw cashew nuts

Handful of raisins

1 ripe pineapple, cut into ½-inch cubes

1 cup coarse semolina (*rava* or *sooji*)

1 cup sugar

¼ tsp pineapple extract (optional)

3 drops yellow food coloring (optional)

Heat 2 Tbsp ghee in a frying pan over medium heat. Add cashews and roast, stirring, for 5 minutes, or until golden brown. Add raisins and sauté for 1 minute, or until plump. Using a slotted spoon, transfer nuts and raisins to a plate, keeping as much ghee in the pan as possible. Add pineapple to the pan and sauté over medium-high heat for 2–3 minutes until lightly caramelized. Transfer to a bowl and set aside.

Warm the remaining 2 Tbsp ghee in a medium saucepan over medium heat. Add semolina and roast, stirring, for 2 minutes. Pour in 2½ cups water in a slow, steady stream, stirring constantly, and bring to a boil. Add sugar, then reduce heat to medium-low. Simmer gently for 5 minutes, stirring frequently, until semolina is cooked and pudding has thickened. Add pineapple extract and food coloring (if using) and pour into a serving bowl or square pan.

Set aside to cool, then cut the kesari into squares, and serve with pineapple, cashews, and raisins.

Rice Phirni

SERVES 6

North India's most treasured dessert, *phirni* is regal rice pudding, slow-cooked in milk, delicately scented with rosewater and saffron, and topped with nuts and rose petals. A dessert that traces its roots to the Middle East, phirni would traditionally be served in small earthen bowls called *sakoras*. For those who think they hate rice pudding, this will change your mind.

½ cup basmati rice, washed, rinsed, and soaked for 1 hour (page 31)

4 cups milk

Pinch of saffron (roughly 15 threads)

1 cup sugar

4–5 green cardamom pods, crushed using a mortar and pestle

2 tsp rosewater

2 Tbsp Ghee (page 22)

Handful of almonds and pistachios, coarsely chopped, for garnish

Rose petals, for garnish (optional)

Drain rice. Transfer to a food processor and pulse once or twice, until rice is slightly broken down.

Bring milk to a simmer in a heavy-bottomed saucepan. In a small bowl, combine saffron and 2 Tbsp of the warm (not yet hot) milk. Soak 10–15 minutes.

Meanwhile, add the processed rice and sugar to the pan. Cover and cook on low heat for 20 minutes, stirring occasionally, until the mixture coats the back of a spoon.

Add saffron (with its soaking liquid) and crushed cardamom. Stirring continuously, simmer for another 8 minutes. Remove from heat, then stir in rosewater. Transfer to a serving bowl and refrigerate for 2 hours, or until chilled and set.

Heat ghee in a small saucepan over medium heat. Add almonds and pistachios and sauté for 2–3 minutes, until golden brown.

Garnish rice phirni with roasted nuts and perhaps some rose petals and serve.

Cashew Balls

MAKES 20 BALLS

In honor of my late aunty Mary, a treasured recipe for an addictive sweet. These nutty confections, known also as *unda*, are an anytime pleasure and lovely served with an afternoon chai or a coffee. The crunchiness comes from roasted rice, the flavor from the cashews, and the sweetness from jaggery. (Demerara or your everyday brown sugar is an acceptable substitute.)

½ cup parboiled or matta rice, washed well and soaked for an hour

1 cup fresh grated coconut

3 cups roasted unsalted cashew nuts

1 cup grated jaggery or brown sugar

Drain rice and pat dry with a paper towel, removing as much moisture as possible. Add rice to a heavy-bottomed frying pan and dry-roast over medium heat for 8 minutes, stirring frequently, until lightly browned, slightly crisp, and puffed up. Remove from pan and set aside to cool.

Add coconut to the pan and stir until lightly toasted. Remove from pan and set aside to cool.

Using a food processor, grind rice into a coarse powder, then transfer to a large bowl. Add cashews to the food processor and pulse to a fine powder. Transfer to the rice bowl. Add jaggery (or brown sugar) and coconut to the food processor and process into a fine powder. Add it to the bowl. Using your hands, mix well.

Shape the mixture into firm, lime-sized balls. Serve immediately. (Cashew balls can be refrigerated for up to 3 days. To freeze them, place balls on a baking sheet to freeze individually, then store in an airtight container in freezer for up to a month.)

Mango-Coconut Mousse

SERVES 6

When we were looking for a showstopper dessert for New Year's Eve, we turned to Marie Wang, a baker and one of our most devoted volunteer cooks at Food for Thought. She came up with this spectacular layered mousse and we served it in brandy glasses. It's a bit of an effort—a perfect recipe when you have the time and want to wow.

COOKIE CRUMBLE
¼ cup (½ stick) butter, room temperature
⅓ cup brown sugar
½ cup all-purpose flour
½ cup almond flour
Pinch of salt

MANGO MOUSSE
1½ tsp unflavored gelatin powder
2 cups fresh (or canned) mango purée (divided)
¼ cup sugar
1 cup whipping cream (35%)

COCONUT MOUSSE
2½ tsp unflavored gelatin powder
2 cups pure coconut cream (divided)
1 egg white, room temperature
⅓ cup sugar
1 cup whipping cream (35%)
1 ripe mango, diced, for garnish
Lime zest, for garnish

COOKIE CRUMBLE Preheat oven to 350°F. Line a baking sheet with parchment paper.

Using a stand mixer, combine butter and brown sugar. Add both flours and salt and mix well. Scatter small clumps of the dough over the prepared sheet.

Bake for 15–25 minutes, until golden. (You might want to check the cookie crumble at the 10-minute mark to ensure even browning. Rotate the pan if necessary.) Set aside to cool. While it cools, make the mango mousse.

MANGO MOUSSE In a small bowl, stir gelatin into 2 Tbsp warm water. Set aside.

In a saucepan, combine ¼ cup mango purée and sugar, simmer over medium heat, and stir to dissolve sugar. Add the gelatin mixture and stir for 1 minute, or until gelatin is fully incorporated.

Remove from heat and stir in the remaining 1¾ cups mango purée. Set aside for 20 minutes to cool down.

Whip cream in a bowl until soft peaks form. Fold into mango mixture.

Divide cookie crumble between 6 brandy glasses. Carefully spoon the mango mousse over the cookie crumble, cover, and put glasses in fridge for at least 15 minutes. While the mango mousse is setting, make the coconut mousse layer.

COCONUT MOUSSE In a small bowl, combine gelatin and 2 Tbsp warm water and mix well. Set aside.

Whisk coconut cream to make sure it's smooth. Heat ¼ cup coconut cream in a small saucepan over medium heat, until warmed through. Add the gelatin mixture and stir for 1 minute, or until gelatin is fully incorporated.

Remove from heat.

Stir in the remaining 1¾ cups coconut cream and set aside for 15 minutes to completely cool but not set.

Meanwhile, beat egg white in a medium bowl. Add sugar slowly and beat until soft peaks form. Whip whipping cream in a separate bowl until soft peaks form.

Fold the whipped egg white into the coconut cream. Fold in the whipped cream.

Remove brandy glasses from fridge and carefully spoon a layer of coconut mousse over the set mango mousse. Cover and set back in the fridge for 2 hours before serving.

Spiced Tea

SERVES 4

We'd typically brew this tea during the rainy seasons in Kerala and sweeten it with jaggery. Feel free to use any sweetener (white or brown sugar or honey) or, if you prefer, none. This tea does wonders for a head cold.

1 tsp ground fennel

½ tsp ground ginger

½ tsp ground cardamom

½ tsp ground cloves

½ tsp ground cinnamon

4 Tbsp grated jaggery (optional)

Combine all ingredients in a saucepan. Add 4 cups water and bring to a boil, then reduce heat to medium-low. Simmer for 3 minutes.

Strain through a fine-mesh strainer to remove any jaggery remnants (if using) and serve hot.

Mint-Lime Rickey

SERVES 4

A refreshing drink on a hot summer day.

2 limes

Small bunch of mint leaves

1 (1-inch) piece ginger,
peeled and coarsely
chopped

½ cup sugar

Ice cubes, to serve

Using a sharp knife, remove skin and white pith from the limes.

Combine limes, mint leaves, ginger, sugar, and 2 cups water in a blender and blend until smooth. Pour into chilled glasses over ice.

Old Monk Lassi

The famous dark and stormy rum from Uttar Pradesh is slipped elegantly into a mango-ginger lassi.

1 (1-inch) piece ginger, peeled

2 cups plain yogurt

¼ cup sugar

1½ cup fresh (or canned) mango purée

6 oz Old Monk or any dark rum of your choice (divided)

Juice of 1 lime (divided)

Ice cubes

Rose petals, for garnish (optional)

Grate ginger into a small bowl and squeeze gratings with your fingers to make ginger juice.

Combine ginger juice, yogurt, sugar, and mango purée in a blender and process until smooth.

In a cocktail shaker, combine 3 oz of the mango-ginger smoothie with 1½ oz rum, 1½ tsp lime juice, and ice. Cover and shake well. Transfer to a glass and garnish with rose petals (if using) and perhaps a lime slice. Repeat with remaining lassi mixture.

Golden Milk

SERVES 2

Turmeric has been used medicinally and ceremonially for centuries, and in my house, it's a daily drink in flu season. It's also used as a dye, so wear gloves if you don't want your fingers stained yellow! Using freshly grated turmeric is absolutely the way to go (it will make a big difference).

2 cups milk or non-dairy milk of your choice

1 (2-inch) piece turmeric root, peeled and rinsed

Sweetener such as jaggery, brown sugar, honey, or maple syrup (optional)

Heat milk in a saucepan over medium heat.

Using a mortar and pestle, crush turmeric. (Alternatively, grate it.) Add turmeric to the pan and bring to a boil. Reduce heat to medium-low and simmer for 2 minutes. Set aside for 3 minutes to cool slightly and to further infuse the milk with the turmeric.

Strain the warm golden milk into glasses. Sweeten with the sweetener of your choice (if using).

TRY TRADITIONAL! When straining the milk into glasses, try pouring the milk from a height. This helps to lighten and froth the milk a little.

Pacha Moru

SERVES 4

Always on our lunch table at home, and served at every feast, pacha moru can be a rejuvenating drink or a flavorful, fiery, and tangy sauce served over rice.

1 cup plain yogurt

8 curry leaves, chopped

¾ tsp salt

6 Indian or small shallots

1 Indian or Thai green chile

1 (½-inch) piece ginger, peeled

Combine yogurt, curry leaves, salt, and 1¼ cups water in a bowl.

Using a mortar and pestle or small food processor, combine shallots, chile, and ginger. Pound or process into a coarse paste. Add paste to yogurt, stir well, and transfer to glasses. Serve chilled.

Peach Cardamom Lassi

SERVES 3

At the height of peach season in Ontario, we drink this dreamy lassi pretty much every day.

4 peaches, peeled and pitted, or 2 cups frozen

½ cup sugar

2 cups plain yogurt

4 green cardamom pods

Ice cubes, to serve

Mint, for garnish (optional)

Add peaches, sugar, yogurt, and cardamom to a blender or food processor and blend into a smooth drink. Pour into chilled glasses over ice and garnish with mint if you like.

Saudi Shake

I was one of many who left India in the 1990s looking for better-paying kitchen work in Saudi Arabia. There, I met and fell in love with dates, and I was hardly alone: the many who returned opened bakeries and drinks stands and the Saudi Shake (in many iterations) became hugely popular. This is my family's no-fuss recipe for the rich and creamy date-nut milkshake, perfect to whizz up for my kids after summer soccer practice.

12 Medjool dates, pitted
3 cups milk
½ cup roasted, unsalted nuts
(we use a mix of cashews and pistachios)
Ice cubes, to serve

Combine dates and milk in a bowl. Set aside to soak in the fridge for at least 2 hours.

Transfer date mixture to a blender or food processor and blend until smooth. Add nuts and whizz one more time. Pour into chilled glasses over ice.

Resources

If you're looking for fresh Indian spices and/or pantry essentials but are unable to find the quality you want locally, I recommend the following online resources.

CANADA

APNIROOTS
apniroots.com

SINGAL'S
singals.ca

THE SPICE TRADER
thespicetrader.ca

SPICE TREKKERS
spicetrekkers.com

UNITED STATES

GROCERY BABU
grocerybabu.com

ISHOPINDIAN
ishopindian.com

KHANA PAKANA
shop.khanapakana.com

UNITED KINGDOM

THE ASIAN COOKSHOP
theasiancookshop.co.uk

SPICES OF INDIA
spicesofindia.co.uk

AUSTRALIA

HINDUSTAN IMPORTS
hindustan.com.au

INDIA AT HOME
indiaathome.com.au

MY INDIAN BAZAAR
myindianbazaar.com.au

Acknowledgments

My thanks . . .

To my restaurants team, many on board since the very beginning, and without whom there would be no Coconut Lagoon, no Thali, and no cookbooks: to my brother Thomas and brother-in-law Sudeep and to their wonderful families; to my right-hand man and chef de cuisine Rajesh Gopi, and my front-of-house superstar Malkit Singh; and to Makori, Myint ta, Renju, Yassein, Saji, Yoga, Tenzin, Chirag, Binu, Anu, and Ray.

To our restaurant guests, so generous and constant, even during the pandemic closures when your takeaway orders were a much-appreciated lifeline. Thank you as well for your caring support when Coconut Lagoon suffered its terrible fire. By the time you are cooking from this book, it will be back in full swing!

To Anne DesBrisay, my coauthor on this and on my first cookbook, *Coconut Lagoon*, for your enthusiasm and wisdom, and for transforming my thoughts and ideas into beautiful words.

To the team at Figure 1 Publishing, it's been a pleasure to partner with you on both cookbook projects. Special thanks to Naomi MacDougall for her design brilliance and to Michelle Meade for her editing prowess, and thanks to Pam Robertson and Renate Preuss for their fine-tuning.

To Christian Lalonde, I am so fortunate that you were available yet again to lend your astonishing talent to the photography in *My Thali*, and to food stylist Sylvie Benoit and prop stylist Irene Garavelli who worked so hard on every tiny, beautiful detail.

To the fabulous team at our food security charity Food for Thought for stepping up when the need was great. And to the chefs from all over the city who showed up at our shuttered Thali restaurant to help feed the hungry during the darkest days of the pandemic: too many to name, but I thank you all.

To Jim Watson, mayor of Ottawa 2010–2022; the culinary team at Algonquin College; the Shepherds of Good Hope; and the Ottawa Mission—Food for Thought would not be where it is today without your support.

And finally, to my family—my wife Suma and our children Marieann, Mathew, and Michael—*dhanyavaad*. You are everything.

SNAKE GOURD

Metric Conversion Chart

VOLUME

Imperial or U.S.		Metric
⅛ tsp	→	0.5 mL
¼ tsp	→	1 mL
½ tsp	→	2.5 mL
¾ tsp	→	4 mL
1 tsp	→	5 mL
1½ tsp	→	8 mL
1 Tbsp	→	15 mL
1½ Tbsp	→	23 mL
2 Tbsp	→	30 mL
¼ cup	→	60 mL
⅓ cup	→	80 mL
½ cup	→	125 mL
⅔ cup	→	165 mL
¾ cup	→	185 mL
1 cup	→	250 mL
1¼ cups	→	310 mL
1⅓ cups	→	330 mL
1½ cups	→	375 mL
1⅔ cups	→	415 mL
1¾ cups	→	435 mL
2 cups	→	500 mL
2¼ cups	→	560 mL
2⅓ cups	→	580 mL
2½ cups	→	625 mL
2¾ cups	→	690 mL
3 cups	→	750 mL
4 cups	→	1 L
5 cups	→	1.25 L
6 cups	→	1.5 L
7 cups	→	1.75 L
8 cups	→	2 L

WEIGHT

Imperial or U.S.		Metric
½ oz	→	15 g
1 oz	→	30 g
2 oz	→	60 g
3 oz	→	85 g
4 oz (¼ lb)	→	115 g
5 oz	→	140 g
6 oz	→	170 g
7 oz	→	200 g
8 oz (½ lb)	→	225 g
9 oz	→	255 g
10 oz	→	285 g
11 oz	→	310 g
12 oz (¾ lb)	→	340 g
13 oz	→	370 g
14 oz	→	400 g
15 oz	→	425 g
16 oz (1 lb)	→	450 g
1¼ lbs	→	570 g
1½ lbs	→	670 g
2 lbs	→	900 g
3 lbs	→	1.4 kg
4 lbs	→	1.8 kg
5 lbs	→	2.3 kg
6 lbs	→	2.7 kg

LIQUID MEASURES (FOR ALCOHOL)

Imperial or U.S.		Metric
½ fl oz	→	15 mL
1 fl oz	→	30 mL
1½ fl oz	→	45 mL
2 fl oz	→	60 mL
3 fl oz	→	90 mL
4 fl oz	→	120 mL
6 fl oz	→	180 mL

CANS AND JARS

Imperial or U.S.		Metric
14 oz	→	398 ml
28 oz	→	796 ml

LINEAR

Imperial or U.S.		Metric
⅛ inch	→	3 mm
¼ inch	→	6 mm
½ inch	→	12 mm
¾ inch	→	2 cm
1 inch	→	2.5 cm
1¼ inches	→	3 cm
1½ inches	→	3.5 cm
1¾ inches	→	4.5 cm
2 inches	→	5 cm
2½ inches	→	6.5 cm
3 inches	→	7.5 cm
4 inches	→	10 cm
5 inches	→	12.5 cm
6 inches	→	15 cm
7 inches	→	18 cm
8 inches	→	20 cm
9 inches	→	23 cm
10 inches	→	25 cm
11 inches	→	28 cm
12 inches (1 foot)	→	30 cm
13 inches	→	33 cm
16 inches	→	41 cm
18 inches	→	46 cm

TEMPERATURE

(for oven temperatures, see list at right)

Imperial or U.S.		Metric
90°F	→	32°C
120°F	→	49°C
125°F	→	52°C
130°F	→	54°C
135°F	→	57°C
140°F	→	60°C
145°F	→	63°C
150°F	→	66°C
155°F	→	68°C
160°F	→	71°C
165°F	→	74°C
170°F	→	77°C
175°F	→	80°C
180°F	→	82°C
185°F	→	85°C
190°F	→	88°C
195°F	→	91°C
200°F	→	93°C
225°F	→	107°C
250°F	→	121°C
275°F	→	135°C
300°F	→	149°C
325°F	→	163°C
350°F	→	177°C
360°F	→	182°C
375°F	→	191°C

OVEN TEMPERATURE

Imperial or U.S.		Metric
200°F	→	95°C
250°F	→	120°C
275°F	→	135°C
300°F	→	150°C
325°F	→	160°C
350°F	→	180°C
375°F	→	190°C
400°F	→	200°C
425°F	→	220°C
450°F	→	230°C

Index

Page numbers in italics refer to photos. Page numbers in bold refer to discussions of an ingredient.

in Konkan shrimp curry, 96
in kozhi mappas, 112
in lemon rice, 29
in lobster moilee, 99
in Mangalore-style fish curry, 101
in meatball curry, 116
in nadan cucumber pulissery, 158
in Pala goat curry, 119
in palak paneer, 80
in pineapple achar, 145
in pork and potato masala, 113
in pork vindaloo, 115
in potato bonda, 130
in root vegetable curry, 75
in salmon vattichathu, 98
in sambar, 76
in sambar powder, 21
in shrimp coconut masala, 102
in snake gourd stir-fry, 69
in squash rasam, 78
in steamed clams in a mango
    coconut sauce, 105
in sweet mango kalan, 156
in tawa vegetables, 79
in vegetable uppumavu, 36

ULARTHIYATHU
    kale, 60
    pumpkin, 67
uppumavu, vegetable, 36, 37
urad dal. *See* white lentils (urad dal)

vada, 134, *135*
vattichathu, salmon, 98
VEGETABLE(S). *See also*
        specific vegetables
    root, curry, *74*, 75
    tawa, 79
    uppumavu, 36, *37*
vindaloo, pork, *114*, 115

wax beans, in vegetable uppumavu, 36
WHITE LENTILS (URAD DAL)
    in asparagus thoran, 71
    in chickpea sundal, 44
    in curd rice, 32
    in heirloom tomato chutney, 143
    in kale ularthiyathu, 60
    in lemon rice, 29

in sambar powder, 21
in vada, 134

yam, yellow, in Kerala-style aviyal, 61
yellow bell pepper, in Food for
    Thought curry, 49
YOGURT
    in basic raita, 154
    in curd rice, 32
    in duck masala, 38
    in ginger pachadi, 144
    in kadai chicken, 110
    in Kashmiri lamb, 120
    in Kerala-style aviyal, 61
    in nadan cucumber pulissery, 158
    in Old Monk lassi, 174
    in pacha moru, 177
    in Pala goat curry, 119
    in peach cardamom lassi, 179
    in sweet mango kalan, 156

ZUCCHINI
    in Food for Thought curry, 49
    in tawa vegetables, 79

About the Authors

JOE THOTTUNGAL is a Canadian chef, the owner of the restaurants Coconut Lagoon and Thali in Ottawa, and the culinary director of the charity Food for Thought. Joe was born in Kerala, trained in Mumbai, and worked in the Middle East before moving to Canada, where he earned his Certified Chef de Cuisine (CCC) designation. In 2008, Joe was named Ottawa Chef of the Year by the Canadian Culinary Federation. In 2016, he and his Coconut Lagoon team won the top prize at Ottawa's Gold Medal Plates competition, followed by a silver medal at the 2017 Canadian Culinary Championships. Joe has an honorary degree from Algonquin College, where he occasionally teaches classes. In 2020, he received the Outstanding Individual Philanthropist Award by the Association of Fundraising Professionals, and a United Way East Ontario Community Builder Award for his work on addressing food insecurity in the city. In 2022, Joe was honored with the Order of Ottawa.

His first cookbook, *Coconut Lagoon*, celebrates regional south Indian cuisine and won gold at the Taste Canada Awards in 2020. Joe lives in Ottawa with his wife and three children.

ANNE DESBRISAY is an award-winning food writer and editor, the author of *Ottawa Cooks* and three editions of the restaurant guidebook *Capital Dining*. She is the coauthor of *Atelier* by Marc Lepine; *Coconut Lagoon* by Joe Thottungal; and *Great Scoops!* by the Merry Dairy team. As a journalist, Anne was a restaurant critic for thirty years, first for *enRoute Magazine*, then the *Ottawa Citizen* and *Ottawa Magazine*. She is a senior editor and feature writer for *Taste & Travel* magazine and a well-fed judge for the Canadian Culinary Championships and for Canada's Great Kitchen Party. For "outstanding contributions to the hospitality, culinary, and food industry within the printed world and the electronic media," Anne received a Gold Award from the Ontario Hostelry Institute. She lives in Ottawa.